AF361884

Demand Response in

Demand Response in Smart Grids

Editors

Pedro Faria
Zita Vale

MDPI • Basel • Beijing • Wuhan • Barcelona • Belgrade • Manchester • Tokyo • Cluj • Tianjin

Editors
Pedro Faria
GECAD—Research Group on Intelligent
Engineering and Computing for Advanced
Innovation and Development
Portugal

Zita Vale
GECAD—Research Group on Intelligent
Engineering and Computing for Advanced
Innovation and Development, Polytechnic of
Porto (P.PORTO)
Portugal

Editorial Office
MDPI
St. Alban-Anlage 66
4052 Basel, Switzerland

This is a reprint of articles from the Special Issue published online in the open access journal *Energies* (ISSN 1996-1073) (available at: https://www.mdpi.com/journal/energies/special_issues/DR_SG).

For citation purposes, cite each article independently as indicated on the article page online and as indicated below:

LastName, A.A.; LastName, B.B.; LastName, C.C. Article Title. *Journal Name* **Year**, *Volume Number*, Page Range.

ISBN 978-3-0365-5055-8 (Hbk)
ISBN 978-3-0365-5056-5 (PDF)

Contents

About the Editors . . . . **vii**

Pedro Faria and Zita Vale
Demand Response in Smart Grids
Reprinted from: *Energies* **2023**, *16*, 863, doi:10.3390/en16020863 . . . . **1**

Hari Prasad Devarapalli, V. S. S. Siva Sarma Dhanikonda and Sitarama Brahmam Gunturi
Non-Intrusive Identification of Load Patterns in Smart Homes Using Percentage Total
Harmonic Distortion
Reprinted from: *Energies* **2020**, *13*, 4628, doi:10.3390/en13184628 . . . . **5**

Huichao Ji, Junyou Yang, Haixin Wang, Kun Tian, Martin Onyeka Okoye and Jiawei Feng
Electricity Consumption Prediction of Solid Electric Thermal Storage with a
Cyber–Physical Approach
Reprinted from: *Energies* **2019**, *12*, 4744, doi:10.3390/en12244744 . . . . **21**

Eunjung Lee, Jinho Kim and Dongsik Jang
Load Profile Segmentation for Effective Residential Demand Response Program: Method and
Evidence from Korean Pilot Study
Reprinted from: *Energies* **2020**, *13*, 1348, doi:10.3390/en13061348 . . . . **39**

Xiaofeng Liu, Qi Wang and Wenting Wang
Evolutionary Analysis for Residential Consumer Participating in Demand Response
Considering Irrational Behavior
Reprinted from: *Energies* **2019**, *12*, 3727, doi:10.3390/en12193727 . . . . **57**

Omid Abrishambaf, Pedro Faria and Zita Vale
Ramping of Demand Response Event with Deploying Distinct Programs by an Aggregator
Reprinted from: *Energies* **2020**, *13*, 1389, doi:10.3390/en13061389 . . . . **77**

Anders Clausen, Aisha Umair, Yves Demazeau and Bo Nørregaard Jørgensen
Impact of Social Welfare Metrics on Energy Allocation in Multi-Objective Optimization
Reprinted from: *Energies* **2020**, *13*, 2961, doi:10.3390/en13112961 . . . . **95**

Michel Zade, Zhengjie You, Babu Kumaran Nalini, Peter Tzscheutschler and Ulrich Wagner
Quantifying the Flexibility of Electric Vehicles in Germany and California—A Case Study
Reprinted from: *Energies* **2020**, *13*, 5617, doi:10.3390/en13215617 . . . . **115**

Alejandro Tristán, Flurina Heuberger and Alexander Sauer
A Methodology to Systematically Identify and Characterize Energy Flexibility Measures in
Industrial Systems
Reprinted from: *Energies* **2020**, *13*, 5887, doi:10.3390/en13225887 . . . . **137**

**Abdellatif Elmouatamid, Radouane Ouladsine, Mohamed Bakhouya, Najib El Kamoun,
Mohammed Khaidar and Khalid Zine-Dine**
Review of Control and Energy Management Approaches in Micro-Grid Systems
Reprinted from: *Energies* **2021**, *14*, 168, doi:10.3390/en14010168 . . . . **173**

**Ildar Daminov, Rémy Rigo-Mariani, Raphael Caire, Anton Prokhorov
and Marie-Cécile Alvarez-Hérault**
Demand Response Coupled with Dynamic Thermal Rating for Increased Transformer Reserve
and Lifetime
Reprinted from: *Energies* **2021**, *14*, 1378, doi:10.3390/en14051378 . . . . **203**

About the Editors

Pedro Faria

Dr. Pedro Faria works in the field of power systems with focus on energy markets, smart grids, and demand response. Current work includes renewable-based distributed generation, energy storage, and electric vehicles. In these fields, optimization, clustering, and classification methods have been applied to real and simulated environment problems. Those include methods based on artificial intelligence, namely meta-heuristics and data mining. He has been developing business models for the modeling, aggregation, and remuneration of consumers participating in electricity markets and in demand response programs. He has also worked in real-time simulation of power and energy systems, namely using the OPAL-RT platform and hardware in the loop (HIL) techniques. Pedro Faria has participated in a significant number of national and international research projects, contributing with models and their implementation, testing, demonstration, and piloting. In those projects, he had scientific management and coordination responsibilities, being namely the leader of work packages and tasks in international projects, the GECAD leader of one national project, and the principal investigator of one national project. Pedro Faria has participated in several panels in relevant international conferences, as invited panelist. He has supervised 1 PhD student and 10 Master students toward conclusion. Pedro Faria is author of 1 patent and of more than 250 scientific papers, 75 of them in indexed journals, where he has edited 3 special issues.

Zita Vale

Prof. Zita Vale is Principal Coordinator Professor at the School of Engineering (ISEP) of the Polytechnic of Porto (IPP) and Director of GECAD Research Group on Intelligent Engineering and Computing for Advanced Innovation and Developments. She is also member of LASI—Associate Laboratory of Intelligent Systems, where she chairs the line on "Smart cities, mobility and energy", one of the five LASI research lines. Her main interests regard the application of artificial intelligence techniques to power systems, including knowledge-based systems, multi-agent systems, neural networks, meta-heuristics, optimization, machine learning, and knowledge discovery techniques. She has been involved in more than 60 R&D projects, of which she coordinated more than 30 projects. The main application fields of these projects are smart grids, with intensive use of renewable energy sources, distributed energy resources and distributed generation; addressing the management of energy resources; the negotiation of DER in electricity markets, demand response, and electric vehicles; electricity markets, addressing prices and tariffs; and decision support for market participants, ancillary services, derivatives market, pricing, and market simulation. She has published over 900 works, including more than 180 papers in international scientific journals. She has been contributing to renowned international conferences as a program committee member, program chair, reviewer, and by organizing special and panel sessions. She has also been a keynote speaker at several conferences and guest editor and/or editorial board member of scientific international journals. She also contributes to several scientific and technical committees and working groups. Currently, she chairs the Institute of Electrical and Electronic Engineers (IEEE) Power and Energy Society (PES) Working Group on Data Analysis and Mining and Task Force on Open Data Sets.

Editorial

Demand Response in Smart Grids

Pedro Faria and Zita Vale *

Research Group on Intelligent Engineering and Computing for Advanced Innovation and Development (GECAD), Intelligent Systems Associated Laboratory (LASI), Polytechnic of Porto, 4200-072 Porto, Portugal
* Correspondence: zav@isep.ipp.pt

1. Introduction

The Special Issue "Demand Response in Smart Grids" includes 10 papers [1–10].

Demand Response (DR) and Smart Grids (SG) are two rather broad key topics in the operation of power and energy systems. Although new DR approaches are introduced daily, more work is needed to reach its full potential, offering advantages for all involved. Successful implementation of SG requires widespread use of DR, taking advantage of the flexibility of large- and medium-size consumers as well as targeting small-size consumers. Effective approaches are needed to implement adequate strategies and methods to design and manage DR. As part of the power and energy ecosystem, DR is a very valuable resource which, when coordinated with the increasing penetration of renewable energy and market-driven business models, can significantly increase the system efficiency while keeping energy costs reasonable. The Special Issue "Demand Response in Smart Grids" addresses aspects related to demand flexibility, DR, and their importance for efficient SG operation.

The topics of this Special Issue include:

- Demand response.
- Electricity markets.
- Electric vehicles.
- Energy storage.
- Real-time simulation.
- Smart grids.

Nine research papers and one review have been published in this Special Issue, with the following statistics:

- Submissions: 18; published: 10; rejected: 8.

2. Published Papers Highlights

This paper provides a summary of the *Energies* Special Issue, covering the published articles [1–10], which address several topics related to demand response and smart grids. Table 1 identifies the most relevant topics in each publication. These topics have been selected by the Editors as the most relevant.

As seen in Table 1, most of the publications focus on demand response and load patterns and profiling, while two papers include electric vehicles. Most of the research is dedicated to the end-user aspects and their comfort, and operation and control.

In [1], Devarapalli et al. explore non-intrusive identification of load patterns. The percentage Total Harmonic Distortion (THD) is used for DR management from a Power Quality perspective. The results demonstrate that percentage THD identifies a different combination of loads, as well as alternate load combinations. Recommendations are provided to the consumer to reduce harmonic pollution in the distribution grid.

Citation: Faria, P.; Vale, Z. Demand Response in Smart Grids. *Energies* **2023**, *16*, 863. https://doi.org/10.3390/en16020863

Received: 28 September 2022
Accepted: 26 October 2022
Published: 12 January 2023

Table 1. Topics covered in each publication.

Topic	Publication									
	[1]	[2]	[3]	[4]	[5]	[6]	[7]	[8]	[9]	[10]
Demand response	x		x	x	x	x	x	x		x
Distributed generation					x			x	x	
Electricity markets			x	x	x		x			
Electric vehicles							x		x	
End-user and their comfort	x	x		x	x		x	x		
Energy storage		x					x	x	x	
Load patterns and profiling	x	x	x		x	x	x			x
Operation and control	x			x	x	x	x		x	x
Total	4	3	3	4	6	3	7	4	4	3

In [2], the authors propose a cyber–physical approach to the prediction of electricity consumption of a solid electric thermal storage (SETS) system. The prediction error of the physical model is used as an input of the cyber model to calibrate the prediction. K-means and support vector machine are used. The proposed solution receives the temperature as input, where the case study used a 1 MW SETS system, obtaining a reduction in the maximum relative errors (MRE) to 4.8%.

A pilot study is presented in [3], which proposes a new clustering method for consumer segmentation. The focus is on residential consumers participating in DR. The approach is a two-stage k-means procedure including consumption features and load patterns. Segmentation results are used to identify groups for participation in DR. In a pilot study implemented in Korea, the proposed method shows demand reduction increased by 33.44% with the proposed methodology, DR programs issuers, such as retail utilities or independent system operators, can select targeted customers, improving the economic efficiency.

In [4], two stages are implemented in a method that considers the bounded rationality of residential users, utilizing a dynamic model to decide whether to participate in the DR programs. The evolutionary process of consumers participating in DR is analyzed. DR participants compete to maximize profit, for which a non-cooperative game model is proposed. A distributed algorithm is used to achieve the Nash equilibrium.

The methodology presented in [5] addresses several groups of consumers being sequentially activated to reach the desired consumption reduction by an aggregator. Real-time simulation is used to obtain more accurate results in what concerns the electric grid components modeling. With the provided model, an aggregator is able to efficiently manage the available Dr resources, dispatching them by groups according to the actual response of each group.

The problem of energy allocation in an optimization model is proposed in [6], and is supported by social welfare metrics. Multi-objective optimization is used to obtain Pareto sets of solutions. Commercial greenhouse growers are considered in the case study. The solution maximizes the social welfare among the solutions in the Pareto set. In the end, the impact of each social welfare metric on the optimization outcome is investigated, which will affect energy allocation.

Quantification of the flexibility provided by electric vehicles is analyzed in [7], for a case study on Germany and California. Fixed and dynamic prices are considered for three charging power levels. The vehicle charging is optimized, and the flexibility of each vehicle is calculated every 15 min. Flexibility is mostly available during the early morning or in the evening.

Focusing on industrial systems, [8] identifies the available energy flexibility opportunities in industrial environments. The implemented methodology is flexible, as it can be used for any type of industry. Qualitative and quantitative aspects are tackled in an audit with experts working in the facility. With the obtained flexibility mapping, participation in DR programs becomes easier.

A review on energy management approaches is provided in [9]. The control aspects in microgrids are summarized where relevant. The findings include the need for more predictive approaches to improve energy management in buildings.

Finally, in [10], the optimal operation of a transformer is addressed in conjunction with DR programs. According to the proposed methodology, when a violation in the transformer operation parameters is reached, the minimum DR use is determined to ensure normal operation.

Author Contributions: Investigation, P.F. and Z.V.; Writing—original draft, P.F.; Writing—review and editing, Z.V. All authors have read and agreed to the published version of the manuscript.

Conflicts of Interest: The authors declare no conflict of interest.

References

1. Devarapalli, H.; Dhanikonda, V.; Gunturi, S. Non-Intrusive Identification of Load Patterns in Smart Homes Using Percentage Total Harmonic Distortion. *Energies* **2020**, *13*, 4628. [CrossRef]
2. Ji, H.; Yang, J.; Wang, H.; Tian, K.; Okoye, M.; Feng, J. Electricity Consumption Prediction of Solid Electric Thermal Storage with a Cyber–Physical Approach. *Energies* **2019**, *12*, 4744. [CrossRef]
3. Lee, E.; Kim, J.; Jang, D. Load Profile Segmentation for Effective Residential Demand Response Program: Method and Evidence from Korean Pilot Study. *Energies* **2020**, *13*, 1348. [CrossRef]
4. Liu, X.; Wang, Q.; Wang, W. Evolutionary Analysis for Residential Consumer Participating in Demand Response Considering Irrational Behavior. *Energies* **2019**, *12*, 3727. [CrossRef]
5. Abrishambaf, O.; Faria, P.; Vale, Z. Ramping of Demand Response Event with Deploying Distinct Programs by an Aggregator. *Energies* **2020**, *13*, 1389. [CrossRef]
6. Clausen, A.; Umair, A.; Demazeau, Y.; Jørgensen, B. Impact of Social Welfare Metrics on Energy Allocation in Multi-Objective Optimization. *Energies* **2020**, *13*, 2961. [CrossRef]
7. Zade, M.; You, Z.; Kumaran Nalini, B.; Tzscheutschler, P.; Wagner, U. Quantifying the Flexibility of Electric Vehicles in Germany and California—A Case Study. *Energies* **2020**, *13*, 5617. [CrossRef]
8. Tristán, A.; Heuberger, F.; Sauer, A. A Methodology to Systematically Identify and Characterize Energy Flexibility Measures in Industrial Systems. *Energies* **2020**, *13*, 5887. [CrossRef]
9. Elmouatamid, A.; Ouladsine, R.; Bakhouya, M.; El Kamoun, N.; Khaidar, M.; Zine-Dine, K. Review of Control and Energy Management Approaches in Micro-Grid Systems. *Energies* **2021**, *14*, 168. [CrossRef]
10. Daminov, I.; Rigo-Mariani, R.; Caire, R.; Prokhorov, A.; Alvarez-Hérault, M. Demand Response Coupled with Dynamic Thermal Rating for Increased Transformer Reserve and Lifetime. *Energies* **2021**, *14*, 1378. [CrossRef]

Article

Non-Intrusive Identification of Load Patterns in Smart Homes Using Percentage Total Harmonic Distortion

Hari Prasad Devarapalli [1,2,*], **V. S. S. Siva Sarma Dhanikonda** [2] **and Sitarama Brahmam Gunturi** [1]

[1] Tata Consultancy Services, Hyderabad 500081, T.S., India; sitaramabrahmam.gunturi@tcs.com
[2] Department of Electrical Engineering, National Institute of Technology Warangal, Warangal 506004, T.S., India; sivasarma@gmail.com or dvss@nitw.ac.in
* Correspondence: hdevarapalli@ieee.org; Tel.: +91-94904-37035

Received: 3 August 2020; Accepted: 3 September 2020; Published: 6 September 2020

Abstract: Demand Response (DR) plays a vital role in a smart grid, helping consumers plan their usage patterns and optimize electricity consumption and also reduce harmonic pollution in a distribution grid without compromising on their needs. The first step of DR is the disaggregation of loads and identifying them individually. The literature suggests that this is accomplished through electric features. Present-day households are using modern power electronic-based nonlinear loads such as LED (Light Emitting Diode) lamps, electronic regulators and digital controllers to reduce the electricity consumption. Furthermore, usage of SMPS (Switched-Mode Power Supply) for computing and mobile phone chargers is increasing in every home. These nonlinear loads, while reducing electricity consumption, also introduce harmonic pollution into the distribution grid. This article presents a deterministic approach to the non-intrusive identification of load patterns using percentage Total Harmonic Distortion (THD) for DR management from a Power Quality perspective. The percentage THD of various combinations of loads is estimated by enhanced dual-spectrum line interpolated FFT (Fast Fourier Transform) with a four-term minimal side-lobe window using a LabVIEW-based hardware setup in real time. The results demonstrate that percentage THD identifies a different combination of loads effectively and advocates alternate load combinations for recommending to the consumer to reduce harmonic pollution in the distribution grid.

Keywords: demand response; load disaggregation; percentage total harmonic distortion and non-intrusive identification of load pattern

---

1. Introduction

Demand Response (DR) provides an opportunity for consumers to play a significant role in the operation of the electric grid by reducing or shifting their electricity usage from peak time to off peak and/or altering their usage pattern in response to time-based tariffs or other forms of financial incentives to improve Power Quality (PQ). DR also plays an important role in a smart grid in helping consumers plan their consumption pattern and optimize electricity usage without compromising on their needs [1–3], and is made possible through (i) identification of unnecessary consumption of electricity at an individual appliance level, (ii) alerting consumers with timely information that helps to balance the load between appliances and (iii) leading to reduced bills. DR is approached through the following four steps:

- Identification of load features;
- Load disaggregation;
- Developing insights into consumption behavior;
- Actionable recommendations.

1.1. Identification of Load Features

In the past 30 years, researchers have tried several diverse electrical and non-electrical features to uniquely identify all types of home appliances with different operation modes [4]. Event-based techniques have been employed to identify turn-on and turn-off events.

The identification of electrical appliances is possible through their electrical behavior (active power (P) reactive power (Q), voltage (V), current (I), harmonics, power factor (pf), phase angle, etc.). Moreover, PQ and VI trajectories were used as described in [5–7]. These parameters can be at a steady state or transient (turned on). There is some research using non-electrical behaviors such as light emitted (lumens), heat generated (joules), vibration and sound (noise), EMF (Electromagntic Field) produced, etc. If we employ non-electrical behavior to recognize appliances, then we need respective sensors to collect data for analysis. While this method can be considered as non-intrusive from an electricity point of view, it is an expensive proposition and may be invasive from a personal privacy point of view.

A detailed literature review was presented by Antonio Ruano et al. in [4]. As per the review article, the observations of various load features for disaggregation are 1. active power and reactive power, 2. voltage, current and fundamental phase angle, 3. average displacement power factor, 4. VI trajectory and current vectors, 5. steady-state current transients, 6. PQ disturbance trajectories, 7. third and fifth harmonic amplitudes, 8. average Total Harmonic Distortion (THD), 9. the maximum magnitude of the first eight harmonics 10. The rate of change of transient signals, 11. Shannon and Reini entropies, 12. the spectral band energy of the current spectrum, 13. wavelet transform coefficients, 14. features of power spectrum density, 15. occupancy data, 16. usage patterns and 17. Electro-Magnetic Interference (EMI) signals.

It is quite evident that, initially, researchers tried the active power of the load and its steady-state active power (P) features that are readily available in an energy meter for load disaggregation, which requires a low sampling rate at the time of data acquisition. Later, it was found that, due to the additive nature of active power, it is only useful and deterministic when the power ratings of appliances are distinct, and the sum of the power ratings of various combinations of appliances in operation is also distinct. Therefore, the active power feature is suitable mostly for high-power appliances and is not helpful to discern the simultaneous operation of appliances with the same ratings. Later, researchers employed transient power features [8] (ON slope at the start and step changes) to detect appliance status at a higher sampling frequency (high to very high). Load features such as steady-state power characteristics versus transient power characteristics, capturing events like turn-on step changes, are quite cumbersome. Some researchers used multiple harmonics [9] and the harmonic phase [10,11], since they are induced by nonlinear loads, which are small in amplitude.

1.2. Load Disaggregation

Once the features are decided as discussed above, they are used to discern the start time and duration of operation of the appliances. There have been several attempts by various researchers to disaggregate loads using multiple electrical and non-electrical features [4]. Usually, a power supplier provides a single energy meter outside the premises and records the cumulative energy consumption since its installation. Periodic billing (bill cycles) is calculated by subtracting the current reading from the reading taken at the previous bill cycle and applying a tariff. This procedure does not allow the consumer (i) to realize opportunities for saving energy at an appliance level and (ii) to prevent unnecessary consumption before it happens. The most effective method for load disaggregation is to meter the power at every appliance, termed appliance-level monitoring (ALM) [12]. G.W. Hart [13] recognized that ALM is highly prohibitive due to (a) the high number of appliances in use, (b) the various types of appliances and (c) inconsistent times and durations of operation of appliances at home. He proposed, for the first time, in 1992 that there is a need for identifying individual loads without sub-metering [13,14] and termed this Non-Intrusive Appliance Load Monitoring (NIALM), which is now referred to as Non-Intrusive Load Monitoring (NILM).

There are various ways to discern appliances using digital signal processing (DSP), wavelet transform (WT) and artificial intelligence (AI). Recently, machine learning (ML) [15] and deep learning (DL) techniques have been trialed for this purpose. In particular, two types of ML, namely supervised learning (e.g., classification) and unsupervised learning (e.g., clustering), have been widely used for NILM [16–24]. Load features and the selected disaggregation technique decide the sampling rate for data acquisition for the desired accuracy.

A concise and updated review of the various features reported in the literature for NILM and a comprehensive feature selection from a benchmarked dataset are reported in [25]. A multi-objective evolutionary algorithm is proposed in [26], where five objective functions using active power, apparent power, reactive power, current waveform, and harmonics as load signatures are established to identify several electrical appliances. Antonio Ruano et al. [4] carried out an exhaustive review of disaggregation approaches, including the most recent ones, namely machine learning and deep learning, and they are 1. the Hidden Markov Model (HMM), 2. Deep Neural Networks (D-NN), 3. k-NN (K-nearest neighbors) classifiers, 4. Naïve Bayes Classifiers, 5. Tensor and Matrix Factorization, 6. NN auto encoders, 7. Graphic Signal Processing, 8. Multi-Layer Perceptron, 9. linear searches of databases, 10. Maximum a Posteriori probability, 11. fuzzy "C" means, 12. Discriminative Disaggregation Sparse Coding, 13. The Factorial Hidden Markov Model, 14. hierarchical HMM, 15. variants of HMM, 16. Viterbi Decoding, 17. density-based spatial clustering of applications with noise, 18. quadratic discriminant analysis, 19. Modified Combinatorial Optimization, 20. Long- and Short-Term Memory Recurrent Neural Networks, 21. Karhunen–Loève Spectral Decomposition, 22. iterative subsequence dynamic time warping, 23. particle filtering, 24. Maximum a Posteriori (MAP) criteria, 25. Gaussian Process Classifiers, 26. rule-based classifiers, 27. decision trees, 28. Adaboost classifiers, 29. several supervised classifiers, 30. Self-Optimizing Mapping (SOM), 31. Particle Swarm Optimization, 32. Ant Colony Optimization, 33. Siamese Artificial NN, 34. PCA (Principal component analysis), 35. location-aware energy disaggregation frameworks, and microscopic power features and pattern recognition (reported for NILM in [27]).

It is noted that all of these methods are non-deterministic and hence non-repeatable, which are useful for predictions with some certainty, but not to the extent that the inferences and insights of these methods are beneficial for their deployment in production for consumers to respond in real time and realize the benefits then and there. Furthermore, these methods draw correlations, not causations, to effectively convince consumers to change their behavior.

A comparison between traditional non-deterministic NILM methods versus proposed deterministic methods is illustrated in Table 1, and gives reasons for the limitations of machine learning and deep learning techniques over the proposed deterministic experimental approach.

Table 1. Comparison between traditional non-deterministic Non-Intrusive Load Monitoring (NILM) methods versus proposed deterministic method.

Factor	Non-Experimental, Statistical Approach	Experimental Approach
Establishes	Correlations	Causation
Deals with	Aggregates, general understanding	Individuals, hence specific
Application	Understanding of macroscopic phenomena	Real-time measurement and micro-level control
Bias	Not easy to eliminate bias	Unbiased
Data	Approximations	Real data
Results	Probability, hence not conclusive	Exact, so conclusive
Regularity and repeatability	Mostly	Always

The realization of benefits and the value of DR lie in the accurate measurement of a load feature that could be used by employing a suitable load feature that can determine, with certainty, the consumer demand and suggest alternate propositions for a better response, either from an energy conservation or PQ perspective.

1.3. Developing Insights on Consumption Behavior for Reducing Harmonic Pollution

Most often, the purpose of NILM is stated as potential savings in energy consumption [28]. All the techniques listed in the previous section are data intensive and require high computing power for application in real-time DR and are uneconomical for this purpose. Moreover, these methods are developed for energy conservation only. DR requires a simplification of the approach and should be fast enough to recommend possible opportunities for savings.

1.4. Actionable Recommendations

After disaggregation, the appliance usage pattern of consumers should be analyzed to ascertain opportunities for energy savings or PQ improvement and should recommend the same. The consumer may decide to act upon these recommendations. The consumer benefit lies in the reduction in the unnecessary usage of high-power appliances such as geysers, air conditioning and HVAC (heating, ventilation, and air conditioning). However, the power ratings of these appliances are becoming greener and smarter day by day. At the same time, energy savings on account of the usage of low-power appliances (SMPS, mobile chargers, CFL (compact fluorescent lamp), LED) are not remunerative enough to the consumer. By reviewing the previous literature [29], Kelly, J. et al. assessed that the reduction in domestic electricity consumption on average is 0.7–4.5% only. Usually, consumers are conscious about the usage of high-power appliances, whereas low-power appliances are not considered from an energy savings point of view. Zhuang, M. [30] highlights that there are other good reasons why we need NILM of appliances that employ electronic controls for load demand forecasting accuracy, to provide better criteria for utilities to decide on generation. For the grid operators, NILM additionally allows flexible resource management for DR and tackles the uncertainty derived from renewable sources, apart from energy savings.

1.5. Summary and Proposal

Interestingly, all of the research is focused on energy savings from the high-power appliances of a household, which are usually not a big concern for utilities. The simplification of the DR approach with reasonable accuracy and reliability is the main characteristic that drives its adoption. A schematic of the DR management system is shown in Figure 1, and the generic process is self-explanatory.

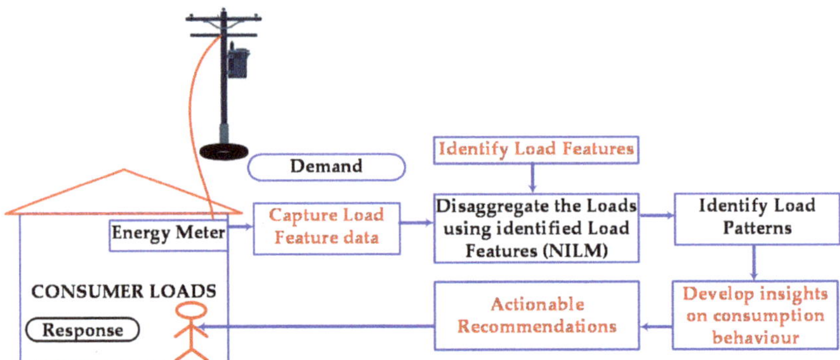

Figure 1. Schematic of Demand Response (DR) management.

The literature suggests that harmonics, harmonic phasors and percentage THD features are used for NILM. A.S. Bouhouras et al. [31] employed a simple lookup table and used the summation of the first three odd harmonic phasors. A non-intrusive discriminant analysis of loads based on PQ data was reported in [32], where classifiers based on linear and quadratic discriminant analysis were implemented. Household appliance classification using lower odd-numbered harmonics and the bagging decision tree approach was detailed in [33]. A.S. Bouhouras et.al [34] employed entropies and spectral band energy to specified frequency bands of the current spectrum to simplify the identification of appliances and accomplished good results. All of these methods adopted deep learning, machine learning and/or statistical methods. Moreover, these approaches considered DR management from an energy conservation perspective but not from a PQ perspective.

Consumers are not paying for the harmonic power induced by nonlinear appliances, which are up to 20% above their real consumption, as energy meters ignore harmonic currents. Harmonics, due to the low electronic power loads of consumers, are disregarded due to their low power consumption; however, the harmonic diction levels are very high. This issue was highlighted by a survey report in 2015 by the Government of India in the name of Swachh Power [35]. There are various ISO(International Organization for Standardization) standards and guidelines such as IEC (International Electrotechnial Commission) 61000-3-2, 61000-4-30 and IEEE (The Institute of Electrical and Electronics Engineers) 519-2014 for harmonic estimation and control [35].

The THD percentages of loads were established by O Deepu et.al [36] for three different loads—CFL, personal computer, and uninterrupted power supply—and were found to be distinct. The percentage THD of current harmonics for these loads is found to be 72, 125, and 35, respectively. Marko Dimitrijević [17] measured the first 39 harmonics for six different appliances and various combinations of these appliances in operation.

In the opinion of authors of this article, the quality of power is more critical for utilities than for power savings in smart homes. The authors would like to approach DR from a PQ point of view rather than to reduce power consumption. If any conventional methods or intrusive methods are used to identify the load patterns, they would not throw light on the harmful impact of harmonics introduced by the pervasive use of low-power nonlinear loads, and hence would not lead to improving PQ from a harmonics perspective. Two standards, namely, IEEE 519-2014 and IEC 61000-3-2 and the Central Electricity Authority guidelines recommend reducing the percentage THD from both utility and consumer perspectives. The understanding of the authors of this article is that this approach is very much needed to impress upon consumers and utilities the urgency of this issue. The industry can adopt this solution if percentage THD is measured, and we offer insights into further actions that can be taken.

Low-power non-linear load harmonic compensation is not addressed individually; therefore, the authors understand that accurate estimation of low-power harmonics and their corresponding percentage THD will identify the consumer loads effectively. The authors propose to use the percentage THD of steady-state harmonics, which is distinct for individual appliances and different combinations of appliances. In this article, the authors propose a novel solution that uses the uniqueness of percentage THD to identify which set of appliances is in operation at any point in time without any ambiguity. This solution is the first of its kind, to the best of our knowledge, that establishes the efficient disaggregation of energy consumption at the appliance level. DR management using percentage THD is depicted in Figure 2.

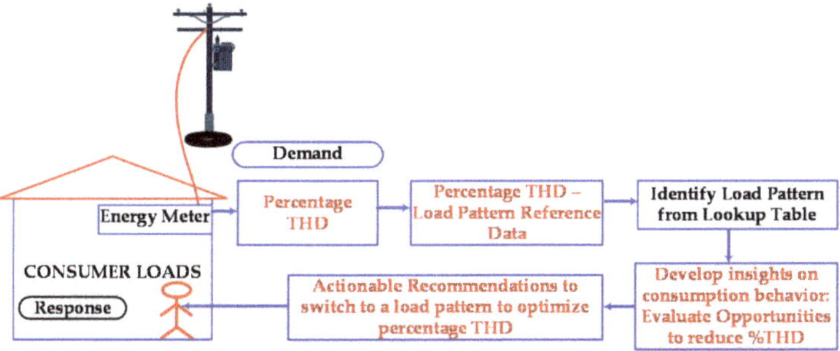

Figure 2. DR management in view of load monitoring using % Total Harmonic Distortion (THD).

NILM has been applied mostly to high-power appliances (50 W and above) from a power conservation perspective. There is insufficient research on percentage THD conservation induced by low-power nonlinear loads (less than 100 W), which is a big challenge of today in our distribution systems. For the first time, we tackle a problem that is not addressed as required by IEEE 519-2014, IEC 61000-3-2 standards and Central Electricity Authority guidelines. NILM is discussed in this article to bring the state-of-the-art research into context and to contrast with the proposed approach.

The major contributions are:

- The measurement of percentage THD using enhanced dual-spectrum line interpolated FFT (EDLIFFT) with a four-term minimal side-lobe window (4MSW) [37] for various real-world loads.
- The development of real-time load pattern identification for DR using a LabVIEW-based virtual instrumentation test bed.
- The recommendation of load patterns for DR management using a lookup table.

This paper is organized as follows: Section 2 discusses the method for the measurement of percentage THD. Section 3 describes the real-time measurement of percentage THD using EDLIFFT with a 4MSW on NI-LabVIEW for various combinations of real-world appliances. Section 4 discusses the results and establishes that percentage THD is a reliable single feature to identify the load consumption pattern; Section 5 concludes with a brief recap.

2. Measurement of Percentage THD, a Single Feature for Load Consumption Pattern Identification

PQ data are useful not only for assessing the quality of power for consumption and compensation aspects, but also for load pattern identification [32]. In particular, harmonic data can be found among PQ data. Among harmonic data, such as harmonic amplitudes, harmonic phases and percentage THD, the latter is found to be helpful to uniquely identify load patterns. Therefore, accurate measurement of percentage THD is essential for load identification. In general, the percentage THD of a current's harmonic signal is measured using FFT. If the measurements of harmonic orders are not accurate then the percentage THD value will be erroneous. FFT measurements have limitations such as spectral leakage and the picket fence effect [37,38]. Recently, EDLIFFT with a 4MSW for real-time harmonic estimation was proposed in [37], which overcame both the drawbacks of FFT. By using this algorithm, the harmonic orders are measured and the percentage THD of any given load pattern is computed as in Equation (1) [39,40]:

$$Percentage\ Current\ THD = \frac{\sqrt{\sum_{n=2}^{H} I_n}}{I_{fund}} \tag{1}$$

where

H = harmonic order

$I_n = n^{th}$ harmonic current

I_{fund} = fundamental current

Thus, we can employ percentage THD as a simple database lookup to find which combinations of appliances are in operation at any point in time. The rated power of all appliances is a known value, so we can compute the energy consumed by individual appliances for the time they spend operation. This lookup table is used to create the disaggregated load by appropriating the rated power of the appliance. The total sum of the active power at any point is compared to the power measured at the energy meter. This information, when applied with contextual data like occupancy, ambient temperature, etc., can deduce unnecessary use of appliances and can be shared with the customer suggest potential savings.

3. Real-Time Experimentation for Non-Intrusive Identification of Load Pattern (NIILP) Using Percentage THD Measurement

In this section, real-time experimentation for NIILP using percentage THD measurement is described. A National Instruments (NI) compact reconfigurable input–output system (cRIO) 9082-based virtual instrumentation experimental setup is developed for validating the proposed NIILP. It is one of the potential real-time hardware tools for percentage THD measurement, as per the requirements of international standards such as IEEE 519,1159 and IEC 61000-4-60. The NI-cRIO 9082 has a Field-Programmable Gate Array (FPGA) architecture which is equipped with an Intel Core-i7 dual-core Central Processing Unit (CPU) with a frequency of 1.33 GHz, 2 GB of DRAM (Dynamic random access memory), 32 GB of ROM (read only memory), and a Xilinx Spartan-6 LX150 FPGA. It consists of a reconfigurable embedded chassis with an integrated intelligent real-time controller and data acquisition modules for analog signal acquisition [41,42]. EDLIFFT with 4 MSW was deployed in the LabVIEW-configured host computer and interfaced to the NI LabVIEW-powered NI-cRIO 9082 through the TCP/IP interface, as illustrated in Figure 3. Most of the typical real-world loads are CFL, LED, fan and PC. Hence, the CFL, LED, exhaust fan and SMPS of the personal computer, which is connected to the single-phase 230 V, 50 Hz utility supply mains, are considered for computing the percentage THD using EDLIFFT with a 4MSW. The load current waveforms are acquired from supply mains and processed to the NI-cRIO 9082 using the NI-9227 current input module.

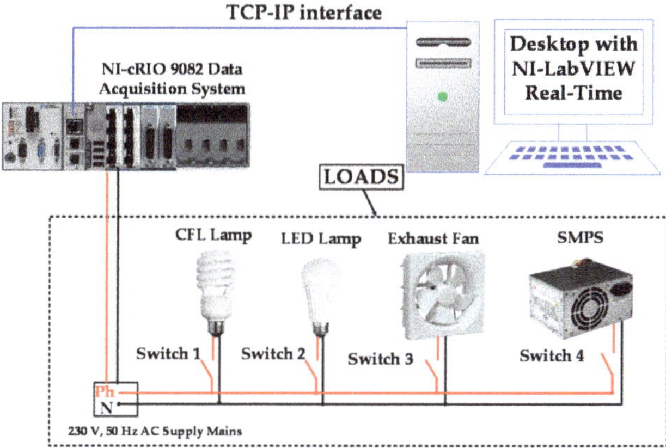

Figure 3. Hardware schematic for percentage THD computation.

A detailed flow-chart of the percentage THD measurement method is described in Figure 4. The percentage THD is measured by using the data obtained from EDLIFFT with 4 MSW. The individual switches turn the loads ON and OFF to verify the different load combinations. Thereby, the percentage THD of each combination is computed by the algorithm given in Figure 4.

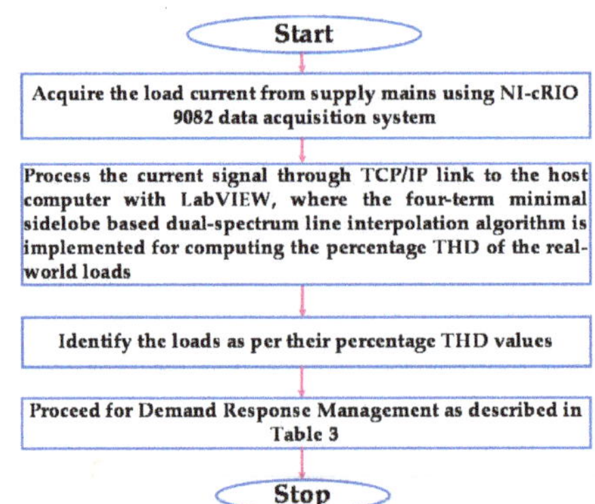

Figure 4. Flow chart for percentage THD computation.

4. Results and Discussion

The real-world load current waveforms acquired by the NI-cRIO 9082 for various load combinations are depicted in Figures 5–7. Initially, single-load operation is acquired and the individual load current waveforms of the CFL, LED, exhaust fan and SMPS of the PC are illustrated in Figure 5. From the individual load waveforms depicted in Figure 5a–d, it can be observed that no two waveforms are found to be the same shape due to the harmonic pollution. Moreover, these waveforms are highly nonlinear. Therefore, the percentage THD values of these loads are found to be unique in nature.

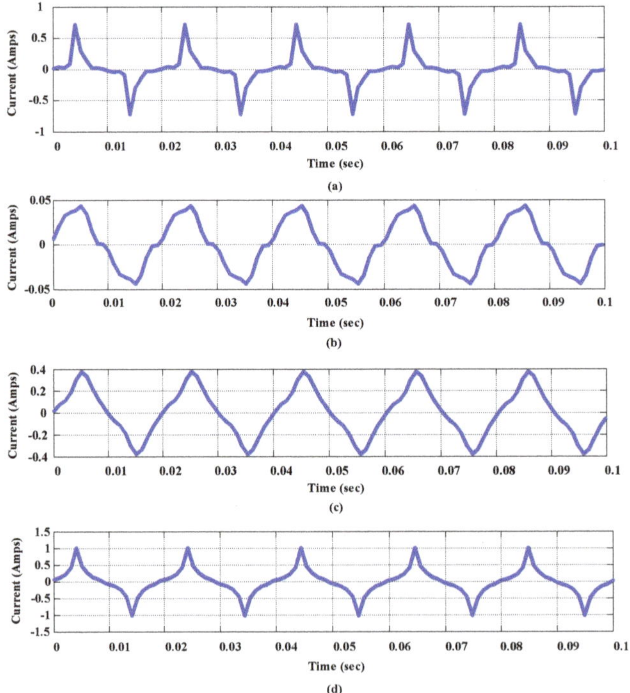

Figure 5. Individual single loads. (**a**) CFL load waveform; (**b**) LED load waveform; (**c**) exhaust fan waveform; (**d**) SMPS of the PC waveform.

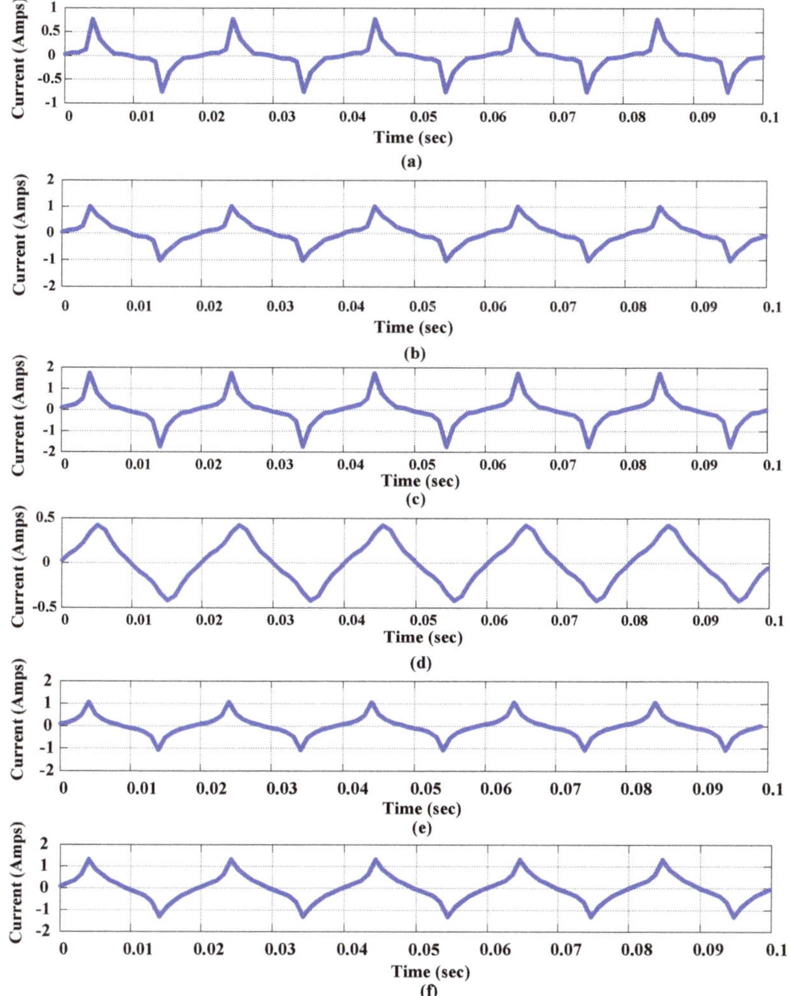

Figure 6. Any two loads. (**a**) CFL + LED load waveform; (**b**) CFL + exhaust fan load waveform; (**c**) CFL + SMPS of the PC load waveform; (**d**) LED + exhaust fan load waveform; (**e**) LED + SMPS of the PC load waveform; (**f**) exhaust fan + SMPS of the PC load waveform.

Any two real-world load combinations are monitored by turning ON their corresponding load switches. The waveforms acquired from the NI-cRIO 9082 for any two real-world load combinations are shown in Figure 6. From the figures, it is observed that no two waveforms are found to be the same shape due to the harmonic distortion. Therefore, the percentage THD values of these load patterns are found to be unique in nature.

Any three load combinations are monitored by turning ON their corresponding load switches and the three load combination waveforms of the CFL, LED, exhaust fan and SMPS of the PC are depicted in Figure 7. It is observed from Figure 7 that no two waveforms are found to be the same due to the harmonic pollution.

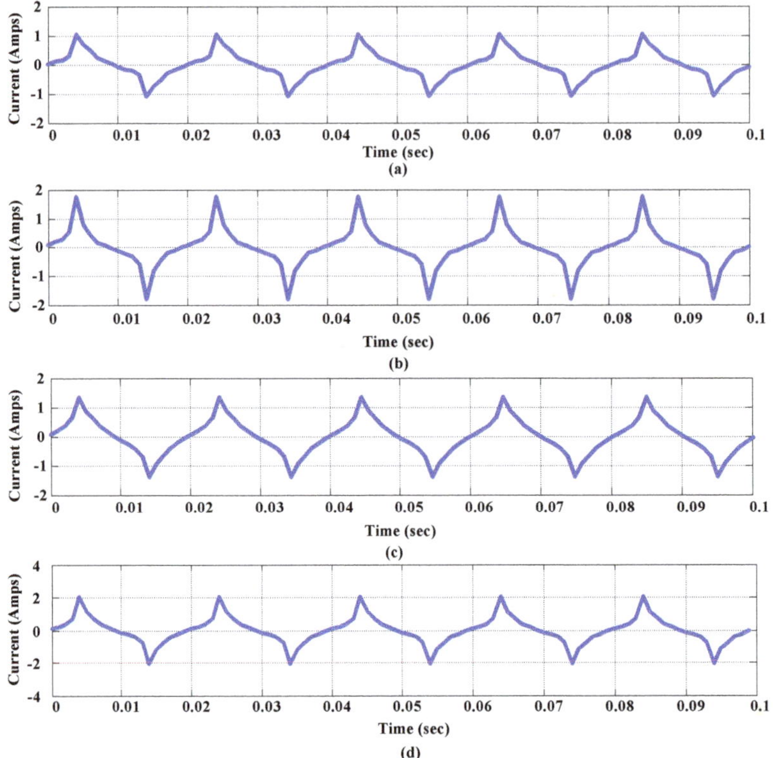

Figure 7. Any three loads. (**a**) CFL + LED + exhaust fan load waveform; (**b**) CFL + LED + SMPS of the PC load waveform; (**c**) LED + exhaust fan + SMPS of the PC load waveform; (**d**) CFL + exhaust fan + SMPS of the PC load waveform.

The load waveform at the supply mains when all the loads are active is illustrated in Figure 8. This waveform is also found to be quite different from the other load pattern waveforms.

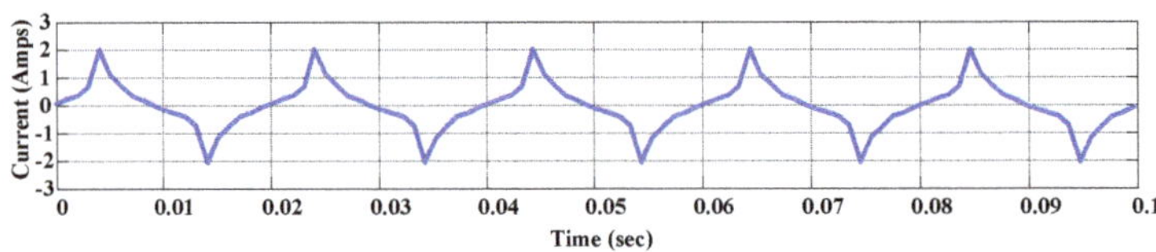

Figure 8. All four loads. CFL + LED + exhaust fan + SMPS of the PC load waveform.

From single-load operation to four-load operation, the load current waveforms are unique, indicating that the percentage THD is different for all load combinations. Therefore, percentage THD can be safely used for load identification effectively. The percentage THDs measured by EDLIFFT with a 4MSW using an NI-cRIO 9082 data acquisition system described in Figure 4 are tabulated in Table 2.

Table 2. Percentage THD for all 15 load patterns in real time.

S.No	Combinations of Different Appliances	CODE	Power (Watts)	%THD
1	CFL	1 0 0 0	85	123.395
2	LED	0 1 0 0	9	19.7279
3	Exhaust Fan	0 0 1 0	20	21.3272
4	SMPS of PC	0 0 0 1	200	92.763
5	CFL + LED	1 1 0 0	94	112.457
6	CFL + Exhaust Fan	1 0 1 0	105	72.007
7	CFL+ SMPS of the PC	1 0 0 1	285	105.402
8	LED + Exhaust Fan	0 1 1 0	29	20.3095
9	LED + SMPS of the PC	0 1 0 1	209	86.1947
10	Exhaust FAN + SMPS of the PC	0 0 1 1	220	59.6673
11	CFL+ LED + Exhaust Fan	1 1 1 0	114	68.3618
12	CFL+ LED + SMPS of PC	1 1 0 1	294	101.076
13	LED + Exhaust Fan + SMPS of PC	0 1 1 1	229	57.0561
14	CFL+ Exhaust Fan + SMPS of PC	1 0 1 1	305	80.1096
15	CFL+ LED + Exhaust Fan + SMPS of PC	1 1 1 1	314	77.6529
			Standard Deviation	**33.069**

Our experiment was conducted and established that percentage THD uniquely identifies all combinations of loads. The standard deviations of the percentage THD values indicate that they are all uniquely different. Hence, they are used in the primary key of a lookup table to discern the loads in operation based on the percentage THD. A load power versus percentage THD plot is depicted in Figure 9, where the percentage THDs are relatively different for the different load powers. The actionable recommendations are based on the possible shifting of load demand from the red region to the yellow region and from the yellow region to the green region for a better response in order to reduce harmonic pollution to the extent possible in the load consumption scenarios.

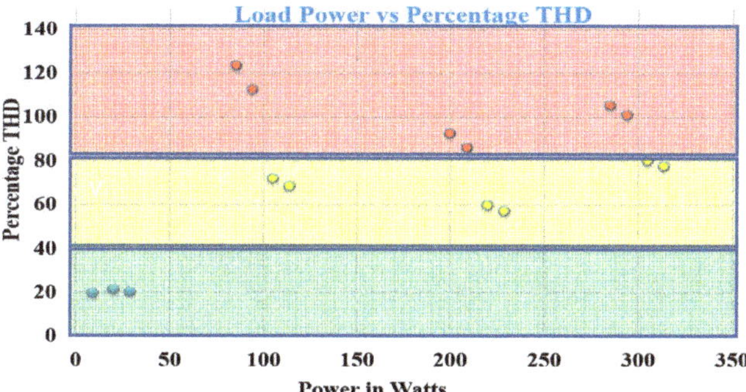

Figure 9. Load power versus percentage THD of the demand.

The recommendation chart for DR management in Table 3 establishes the opportunity to reduce percentage THD within premises. The consumer benefits from the increased life of his appliances and utilities benefit from the increased life of their equipment in the distribution system.

Table 3. Actionable insights for DR management.

Actionable Insights and Benefits										
Demand					Response				Benefits	
S.No	CODE	Power	%THD	Actionable Insights	S.No	CODE	Power	%THD	Change in %THD	Change in Power
1	1 0 0 0	85	123.395	Turn off CFL	2	0 1 0 0	9	19.7279	−84.0124	−89.4118
2	0 1 0 0	9	19.7279	NR [1]	NR	0 1 0 0	9	19.7279	0	0
3	0 0 1 0	20	21.3272	NR	NR	0 0 1 0	20	21.3272	0	0
4	0 0 0 1	200	92.763	Turn off LED for daytime	9	0 0 0 1	209	86.1947	−7.08073	4.5
5	1 1 0 0	94	112.457	Turn off CFL	2	0 1 0 0	9	19.7279	−82.4574	−90.4255
6	1 0 1 0	105	72.007	Turn off CFL	8	0 1 1 0	29	20.3095	−71.7951	−72.381
7	1 0 0 1	285	105.402	Turn off CFL for daytime	4	0 0 0 1	200	92.763	−11.9912	−29.8246
8	0 1 1 0	29	20.3095	Turn off LED for daytime	3	0 0 1 0	20	21.3272	5.010955	−31.0345
9	0 1 0 1	209	86.1947	Turn off LED for daytime	-	0 1 0 1	209	86.1947	0	0
10	0 0 1 1	220	59.6673	NR	-	0 0 1 1	220	59.6673	0	0
11	1 1 1 0	114	68.3618	Turn off CFL	8	0 1 1 0	29	20.3095	−70.2912	−74.5614
12	1 1 0 1	294	101.076	Turn off CFL	9	0 1 0 1	209	86.1947	−14.7229	−28.9116
13	0 1 1 1	229	57.0561	Turn off LED for daytime	10	0 0 1 0	220	59.6673	4.576548	−3.93013
14	1 1 0 1	305	80.1096	Turn off CFL	13	0 1 1 1	229	57.0561	−28.7774	−24.918
15	1 1 1 1	314	77.6529	Turn off CFL	13	0 1 1 1	229	57.0561	−26.5242	−27.0701

[1] NR = no recommendation.

The change in power consumption and the change in percentage THD columns show these direct benefits by responding positively to the proposed recommendations, as appropriate.

C. Nalmpantis et al. [15] proposed qualitative and quantitative metrics for NILM, and the authors of this study applied the same metrics, which are presented in Table 4 below. The quantitative metrics clearly demonstrate the effectiveness of the proposed experimental approach over non-deterministic NILM methods.

Table 4. Quantitative metrics for NILI–percentage THD.

Quantitative Metric Category	Quantitative Metrics	%THD	Other NILM Methods
Feature selected	THD Sampling rate	Medium	High
Accuracy	Disaggregation percentage(D)	100	<100
	Disaggregation Error (DE)	0	>0
	Precision(P)–TP [1]/(TP + FP [2])	1	<1
	Recall (R)-P/TP+FN	1	<1
	Accuracy (Acc) = (TP + TN [3])/(TP+TN+FP+FN [4])	1	<1
	F-measure (f1) 2 *P *R/(P + R)	1	<1
No training	User interaction	Low	
Real-time capabilities	Depends on algorithm's computational complexity (computational cost)	Low	Low
Scalability	Algorithm computational complexity (simple algorithm scales better)	High	High
Identification factor	Standard deviation (FATσ) of %THD	33.069	NA
Generalization	Generalization over unseen houses	High	Medium

[1] TP = true positive; [2] FP = false positive; [3] TN = true; [4] FN = false negative.

5. Conclusions

This paper presents a deterministic approach, using percentage THD to identify load consumption patterns through EDLIFFT with a 4MSW in the NI-LabVIEW program, for various combinations of loads in real time, and demonstrates that the percentage THD value effectively identifies various load combinations. The proposed method for the non-intrusive identification of the load pattern is essential for responsible electricity consumption, as it contributes to raising awareness about the quality of electricity, encourages countries/companies to use harmonic-free devices and calls for policy changes to promote harmonic-free appliances (compensation at the source) so the grid can become free of harmonics. Harmonics contribute 20% more than the real consumption billed to the consumer, which can be reduced to help with the cost of energy for utilities, leading to better margins, which are otherwise considered as losses. Utilities and consumers also benefit from the increased lifetime of their equipment and appliances, respectively. Our DR chart highlights the change in power and the change in percentage THD, then the customer acts on the DR management system's recommendations. Standard deviation, a measure of the dispersion of values, indicates the differentiation of load patterns without ambiguity. Since both percentage THD against all possible combinations of loads are measured and put into a lookup table and recommendations are also chosen in advance, the performance of DR is fairly good for real-time DR management from a PQ perspective.

Author Contributions: Conceptualization, H.P.D.; formal analysis, H.P.D. and S.B.G.; investigation, H.P.D. and S.B.G.; methodology, H.P.D.; project administration, H.P.D.; resources, V.S.S.S.S.D.; supervision, V.S.S.S.S.D. and S.B.G.; validation, H.P.D. and V.S.S.S.S.D.; writing—original draft, H.P.D.; writing—review and editing, V.S.S.S.S.D. and S.B.G. All authors have read and agreed to the published version of the manuscript.

Funding: This research received no external funding.

Acknowledgments: The authors gratefully acknowledge the Department of Electrical Engineering, National Institute of Technology, Warangal, for their support.

Conflicts of Interest: The authors declare no conflict of interest.

References

1. Siano, P. Demand response and smart grids—A survey. *Renew. Sustain. Energy Rev.* **2014**, *30*, 461–478. [CrossRef]
2. Du, P.; Lu, N.; Zhong, H. *Demand Response in Smart Grids*; Springer Nature: Cham, Switzerland, 2019.

3.   D'Hoop, G.; Deblecker, O.; Thomas, D. Power quality improvement of a microgrid with a demand-side-based energy management system. In *Micro-Grids—Applications, Operation, Control and Protection*; Intech Open: London, UK, 2019.

4.   Ruano, A.; Hernandez, A.; Ureña, J.; Ruano, M.; Garcia, J. NILM techniques for intelligent home energy management and ambient assisted living: A review. *Energies* **2019**, *12*, 2203. [CrossRef]

5.   Iksan, N.; Sembiring, J.; Haryanto, N.; Supangkat, S.H. Appliances identification method of non-intrusive load monitoring based on load signature of VI trajectory. In Proceedings of the 2015 International Conference on Information Technology Systems and Innovation (ICITSI), Bandung, Indonesia, 16–19 November 2015; IEEE: New Jersey, NJ, USA, 2015; pp. 1–6.

6.   Baptista, D.; Mostafa, S.S.; Pereira, L.; Sousa, L.; Morgado-Dias, F. Implementation strategy of convolution neural networks on field programmable gate arrays for appliance classification using the voltage and current (VI) trajectory. *Energies* **2018**, *11*, 2460. [CrossRef]

7.   Bouhouras, A.S.; Gkaidatzis, P.A.; Chatzisavvas, K.C.; Panagiotou, E.; Poulakis, N.; Christoforidis, G.C. Load signature formulation for non-intrusive load monitoring based on current measurements. *Energies* **2017**, *10*, 538. [CrossRef]

8.   Le, T.T.H.; Kim, H. Non-intrusive load monitoring based on novel transient signal in household appliances with low sampling rate. *Energies* **2018**, *11*, 3409. [CrossRef]

9.   Djordjevic, S.; Simic, M. Nonintrusive identification of residential appliances using harmonic analysis. *Turk. J. Electr. Eng. Comput. Sci.* **2018**, *26*, 780–791. [CrossRef]

10.   Đorđević, S.D.; Dimitrijević, M.; Litovski, V. A non-intrusive identification of home appliances using active power and harmonic current. *Facta Univ. Ser. Electron. Energetics* **2016**, *30*, 199–208.

11.   Shi, X.; Ming, H.; Shakkottai, S.; Xie, L.; Yao, J. Nonintrusive load monitoring in residential households with low-resolution data. *Appl. Energy* **2019**, *252*, 113283. [CrossRef]

12.   Haq, A.U.; Jacobsen, H.A. Prospects of appliance-level load monitoring in off-the-shelf energy monitors: A technical review. *Energies* **2018**, *11*, 189. [CrossRef]

13.   Hart, G.W. Non-intrusive load monitoring. *Proc. IEEE* **1992**, *80*, 1870–1891. [CrossRef]

14.   Zoha, A.; Gluhak, A.; Imran, M.A.; Rajasegarar, S. Non-intrusive load monitoring approaches for disaggregated energy sensing: A survey. *Sensors* **2012**, *12*, 16838–16866. [CrossRef]

15.   Nalmpantis, C.; Vrakas, D. Machine learning approaches for non-intrusive load monitoring: From qualitative to quantitative comparison. *Artif. Intell. Rev.* **2019**, *52*, 217–243. [CrossRef]

16.   Chang, H.H. Non-intrusive demand monitoring and load identification for energy management systems based on transient feature analyses. *Energies* **2012**, *5*, 4569–4589. [CrossRef]

17.   Dimitrijević, M.; Stošović, M.A.; Stevanović, D. Classification of Nonlinear Loads using Current Spectrum. *Power* **2019**, *1*, 2.

18.   Zheng, Z.; Chen, H.; Luo, X. A supervised event-based non-intrusive load monitoring for non-linear appliances. *Sustainability* **2018**, *10*, 1001. [CrossRef]

19.   Çavdar, İ.H.; Faryad, V. New design of a supervised energy disaggregation model based on the deep neural network for a smart grid. *Energies* **2019**, *12*, 1217. [CrossRef]

20.   Fagiani, M.; Bonfigli, R.; Principi, E.; Squartini, S.; Mandolini, L. A non-intrusive load monitoring algorithm based on non-uniform sampling of power data and deep neural networks. *Energies* **2019**, *12*, 1371. [CrossRef]

21.   Wu, Q.; Wang, F. Concatenate convolutional neural networks for non-intrusive load monitoring across complex background. *Energies* **2019**, *12*, 1572. [CrossRef]

22.   Kim, J.G.; Lee, B. Appliance classification by power signal analysis based on multi-feature combination multi-layer LSTM. *Energies* **2019**, *12*, 2804. [CrossRef]

23.   Rafiq, H.; Shi, X.; Zhang, H.; Li, H.; Ochani, M.K. A Deep Recurrent Neural Network for Non-Intrusive Load Monitoring Based on Multi-Feature Input Space and Post-Processing. *Energies* **2020**, *13*, 2195. [CrossRef]

24.   Puente, C.; Palacios, R.; González-Arechavala, Y.; Sánchez-Úbeda, E.F. Non-Intrusive Load Monitoring (NILM) for Energy Disaggregation Using Soft Computing Techniques. *Energies* **2020**, *13*, 3117. [CrossRef]

25.   Sadeghianpourhamami, N.; Ruyssinck, J.; Deschrijver, D.; Dhaene, T.; Develder, C. Comprehensive feature selection for appliance classification in NILM. *Energy Build.* **2017**, *151*, 98–106. [CrossRef]

26.   Li, L.; Yang, L.; Chen, H.; Li, M.; Zhang, C. Multi-objective evolutionary algorithms applied to non-intrusive load monitoring. *Electr. Power Syst. Res.* **2019**, *177*, 105961. [CrossRef]

27. de Souza, W.A.; Garcia, F.D.; Marafão, F.P.; Da Silva, L.C.P.; Simões, M.G. Load disaggregation using microscopic power features and pattern recognition. *Energies* **2019**, *12*, 2641. [CrossRef]

28. Chui, K.T.; Lytras, M.D.; Visvizi, A. Energy sustainability in smart cities: Artificial intelligence, smart monitoring, and optimization of energy consumption. *Energies* **2018**, *11*, 2869. [CrossRef]

29. Kelly, J.; Knottenbelt, W. Does disaggregated electricity feedback reduce domestic electricity consumption. A systematic review of the literature. *arXiv* **2016**, arXiv:1605.00962.

30. Zhuang, M.; Shahidehpour, M.; Li, Z. An overview of non-intrusive load monitoring: Approaches, business applications, and challenges. In Proceedings of the 2018 International Conference on Power System Technology (POWERCON), Guangzhou, China, 6–8 November 2018; IEEE: New Jersey, NJ, USA, 2018; pp. 4291–4299.

31. Bouhouras, A.S.; Gkaidatzis, P.A.; Panagiotou, E.; Poulakis, N.; Christoforidis, G.C. A NILM algorithm with enhanced disaggregation scheme under harmonic current vectors. *Energy Build.* **2019**, *183*, 392–407. [CrossRef]

32. Jimenez, Y.; Cortes, J.; Duarte, C.; Petit, J.; Carrillo, G. Non-intrusive discriminant analysis of loads based on power quality data. In Proceedings of the 2019 IEEE Workshop on Power Electronics and Power Quality Applications (PEPQA), Manizales, Colombia, 30–31 May 2019; pp. 1–5.

33. Kang, H.; Kim, H. Household Appliance Classification using Lower Odd-Numbered Harmonics and the Bagging Decision Tree. *IEEE Access* **2020**, *8*, 55937–55952.

34. Bouhouras, A.S.; Milioudis, A.N.; Labridis, D.P. Development of distinct load signatures for higher efficiency of NILM algorithms. *Electric Power Syst. Res.* **2014**, *117*, 163–171. [CrossRef]

35. Available online: https://www.powergridindia.com/sites/default/files/footer/climate_change/Swachh_Power.pdf (accessed on 16 June 2019).

36. Deepu, O.; Sindhu, T.K. Modeling of Nonlinear Loads and Analysis of Harmonics in a Small Scale IT Park. *Resonance* **2013**, *100*, 7.

37. Varaprasad, O.V.S.R.; Sarma, D.V.S.S.S.; Paredes, H.K.M.; Simões, M.G. Enhanced Dual-Spectrum Line Interpolated FFT with Four-Term Minimal Sidelobe Cosine Window for Real-Time Harmonic Estimation in Synchrophasor Smart-Grid Technology. *Electronics* **2019**, *8*, 191.

38. Varaprasad, O.V.S.R.; Sarma, D.V.S.S.S.; Panda, R.K. Advanced windowed interpolated FFT algorithms for harmonic analysis of electrical power system. In Proceedings of the 2014 Eighteenth National Power Systems Conference (NPSC), Guwahati, India, 18–20 December 2014; IEEE: New Jersey, NJ, USA, 2018; pp. 1–6.

39. Khokhlov, V.; Meyer, J.; Grevener, A.; Busatto, T.; Rönnberg, S. Comparison of Measurement Methods for the Frequency Range 2–150 kHz (Supraharmonics) Based on the Present Standards Framework. *IEEE Access* **2020**, *8*, 77618–77630. [CrossRef]

40. IEEE. Recommended practices and requirements for harmonic control in electric power systems. In *IEEE Std 519-2014 (Revision of IEEE Std 519-1992)*; IEEE: New York, NY, USA, 2014.

41. Available online: http://www.ni.com/pdf/manuals/376904a_03.pdf (accessed on 9 August 2018).

42. Kelly, J.; Aldaiturriaga, E.; Ruiz-Minguela, P. Applying international power quality standards for current harmonic distortion to wave energy converters and verified device emulators. *Energies* **2019**, *12*, 3654. [CrossRef]

Article

Electricity Consumption Prediction of Solid Electric Thermal Storage with a Cyber–Physical Approach

Huichao Ji [1,2], Junyou Yang [1,*], Haixin Wang [1], Kun Tian [1], Martin Onyeka Okoye [1] and Jiawei Feng [1]

[1] School of Electrical Engineering, Shenyang University of Technology, Shenyang 110870, China; huichaoji@neepu.edu.cn (H.J.); haixinwang@sut.edu.cn (H.W.); shiguangliushi1@126.com (K.T.); martinokoye@yahoo.com (M.O.O.); jiaweifeng_sut@163.com (J.F.)

[2] School of Automation Engineering, Northeast Electric Power University, Jilin 132012, China

* Correspondence: junyouyang@sut.edu.cn; Tel.: +86-139-4008-8191

Received: 11 November 2019; Accepted: 9 December 2019; Published: 12 December 2019

Abstract: This paper proposes a cyber–physical approach to enhance the prediction accuracy of electricity consumption of solid electric thermal storage (SETS) system, which integrates a physical model and a data-based cyber model. In the cyber–physical model, the prediction error of the physical model is used as an input of the cyber model to further calibrate the prediction error. Firstly, customers' behavior characteristics are extracted by the integration of K-means and one-versus-one support vector machine. Secondly, based on the behavior characteristics and ambient temperature, the physical model is developed to predict daily electricity consumption. Finally, the error levels of physical model are classified, together with the temperature and prediction values of the physical model, are selected as the inputs of the cyber model using the back propagation (BP) neural network to calibrate the results of the physical model. The effectiveness of the proposed cyber–physical model (CPM) is verified by a 1 MW SETS system. The simulation results show that, compared with the physical model (PM) and cyber model (CM), the maximum relative errors (MRE) with the CPM are reduced to 25.4% and 4.8%, respectively.

Keywords: solid electric thermal storage; cyber–physical model; K-means; support vector machine; neural network

1. Introduction

Thermal energy storage is considered as one of the advanced energy technologies [1]. Electric energy can be stored in the form of heat during off-peak demand periods and used for heating of rooms during peak demand periods. The improvement of thermal storage is useful to reasonably arrange the electricity consumption of thermal storage loads and promote the thermal storage peak shaving incentive mechanism realization. Therefore, solid electric thermal storage (SETS) has become one of the most promising solutions as a flexible demand response (DR) in demand-side management (DSM) [2]. Consequently, SETS prediction is an important precondition for peak shaving in DSM. Due to the electric–thermal time shift characteristics of SETS, the electricity prediction can provide multiple options for peak shaving and dispatching of the power system. However, there are few studies on the SETS prediction, and the prediction accuracy needs to be improved. Therefore, if the SETS prediction is applied to the actual operation of the power system, it is necessary to enhance the prediction accuracy of SETS based on advanced algorithms.

There are few studies on the electricity prediction of those devices. A physical model (PM) of SETS is enterated into an energy management system for isolated microgrids [3]. SETS is used to accommodate wind power to supply heat load in isolated microgrids [4]. A PM for residential forced-air electric furnaces is built to predict the thermal energy storage, which is an early application of SETS [5]. This work does not take into account the effect of continuously changing the ambient temperature on thermal energy storage. A PM is integrated into the TRNSYS calculation tool to evaluate the optimal thermal energy storage of forced-air electric furnaces with changing ambient temperature [6]. Therefore, it is necessary to improve the basis of the existing methods, and enhance the prediction accuracy. However, the customers' behavior characteristics of SETS are not considered in the above-mentioned PMs, which is very important to improve the accuracy of prediction.

In [7], a sparse continuous conditional random fields method was proposed to predict electric load with the identification of behavior. The data from advanced metering infrastructure is used to understand the power consumption patterns to improve the load forecasting accuracy in [8]. The prediction accuracy would be significantly enhanced with the consideration of behavior. However, the working mode of SETS are completely different from those of conventional electric loads [9]. SETS is charged by the off-peak electricity, and its thermal energy is released all-day. During the off-peak hours, usually from 21:00 to 6:00, the heating elements quickly heat the dense bricks to a high temperature owning to its cheaper electricity prices. During the peak period from 6:00 to 21:00, the heating elements are switched off, and SETS continues to release its thermal energy to warm the rooms. Many behavior characteristics of SETS directly affect the heat load demand, such as all-day continuous work (e.g., convenience store), and holiday and non-holiday period (e.g., star hotel), which need to be considered. The conventional models of predicting electric load are not adequate for SETS. Due to the wide geographical distribution of SETS installation, the PM cannot consider all the situations comprehensively. The prediction accuracy of the PM still needs to be improved. Under this motivation, this paper considers the behavior characteristics into the PM to enhance the accuracy of SETS.

Another prediction approach of electric load is based on cyber models (CMs), such as auto-regression algorithm [10,11], fuzzy algorithm [12], support vector machine [13], extreme learning machine [14], stochastic methods [15], and multi-stages estimators of nonlinear additive models [16]. Among the existing methods, machine learning (neural network) is commonly used in heat load prediction. A data-driven approach with machine learning is presented to predict the heat load in the rooms [17], and a bi-directional long short-term memory recurrent neural network is proposed to combine the correlation between past information and future information to predict the thermal storage time in [18]. A linear regression model with the ambient temperature is proposed to predict heat load in [19]. These prediction methods are based on historical data combined with multiple machine learning methods. However, it is difficult to predict a sudden increase and decrease of heat load for the above CMs.

With the comprehensive consideration, the PM is sensitive to the sudden increase and decrease of heat load by adding the restriction conditions of electric behaviour, but it cannot consider all the influencing factors. On the other hand, the CM can make good use of historical data to reflect the influencing factors on the SETS, which lacks the guidance of the PM. Modern smart grids have applied cyber–physical systems (CPS) to energy systems including modeling energy systems, energy efficiency, energy resource management, and energy control [20]. Therefore, this paper used a method that combines model-based and data-driven [21], that integrates the PM of SETS and the CM.

The difficulty in using the cyber–physical model (CPM) is how to integrate the cyber components (including influencing factors) and physical components (including behavior characteristics and average power consumption) of SETS. In this paper, we develop a PM of SETS to predict daily average power consumption based on the ambient temperature. The integration of K-means and one-versus-one support vector machine is applied to extract the behavior of SETS. The error levels of the PM and influencing factors are considered as the input of the Back Propagation (BP) neural network-based CM to further enhance the prediction accuracy of electricity consumption. Our contributions are summarized as follows:

- To the best of the authors' knowledge, this is the first work to use the cyber–physical approach to predict the SETS' load change. The physical and cyber components of SETS are integrated.

- Using the existing knowledge of thermodynamics, a SETS PM is developed by considering the customers' behavior characteristics.
- The load data of 1MW SETS is used to validate the CPM, and the results show that, compared with the PM and the CM, the maximum relative errors (MRE) with the CPM are reduced to 25.4% and 4.8%, respectively.

The rest of this paper is organized as follows: The PM of SETS, including its structure, principle, formula derivation, the customers' behavior characteristics of SETS, and influencing factors are presented in Section 2. The cyber–physical approach combining the PM and CM of SETS is proposed in Section 3. Simulations are conducted to validate the effectiveness of the cyber–physical approach in Section 4. The conclusion is drawn in Section 5.

2. PM of SETS

In the PM of SETS, the structure of the temperature-based energy flow of the SETS is proposed in Section 2.2. (Principle). The heat conduction of SETS is described in Section 2.3. (Thermal Energy Storage). The heat transfer formulas of the heat exchanger are given in Section 2.4.1 (Heat Transfer). The Formula (14) is developed by the authors in Section 2.4.2 (Customers Heating). The behavior extraction of SETS is firstly considered in the model.

2.1. SETS Structure

Figure 1 shows a schematic diagram of SETS. It contains electric heating wires, magnesia bricks, the temperature sensor of bricks, two layers of thermal isolation (including perlite and ceramic fiber) wrapped in steel plates, heat exchanger, and fan. When the SETS operates, the resistive heating wires generate thermal energy, which will be stored in the bricks for discharging. The cold air is blown into the interior by the fan, and transferred into hot air by the heating wires for changing the entrance temperature of the heat exchanger. The cold water is circulated into the heat exchanger through the return water pipe and heated by the hot air to improve the customers' room interior temperature. SETS becomes a convenient DSM tool for utility companies.

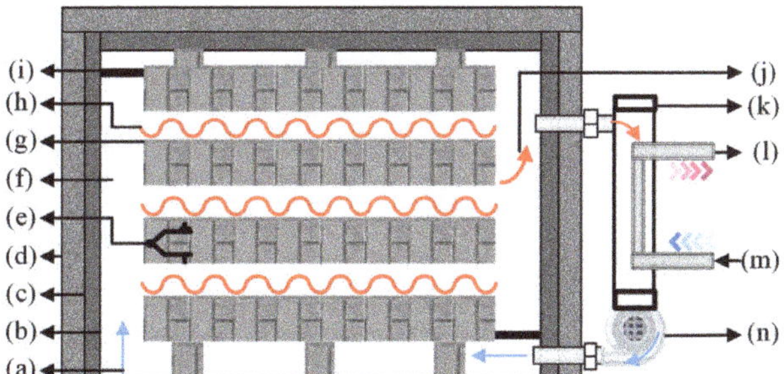

Figure 1. Schematic diagram of solid electric thermal storage (SETS). (**a**) Cold air. (**b**) Perlite. (**c**) Ceramic fiber. (**d**) Steel plates. (**e**) Thermocouple. (**f**) Flow channel. (**g**) Bricks. (**h**) Heating wires. (**i**) Partition wall. (**j**) Hot air. (**k**) Exchanger. (**l**) Supply water. (**m**) Return water. (**n**) Fan.

2.2. Principle

The basic assumption of the proposed PM is that the temperature of the bricks is spatially uniform in any transient process. This assumption implies that the temperature gradients within the bricks are negligible. The proposed model is based on the following assumptions: all the electric energy consumed by the electric heating wires is stored in the bricks in the form of thermal energy. The thermal energy stored in the SETS is mainly released by thermal radiation and convection. The isolation layer can prevent the thermal energy of hot air in the SETS from flowing outside. In heat transfer of the heat exchanger, the heat loss between the heat exchanger and the ambient environment, the pipe thermal resistance and fouling effect are ignored. In Figure 2, the PM of SETS in this paper is based on the basic principle of heat transfer instead of entransy dissipation-based thermal resistance theory presented in [22].

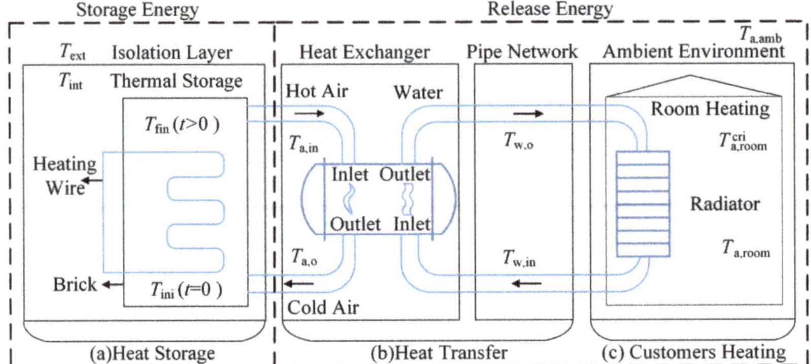

Figure 2. The structure of the temperature-based energy flow of the SETS. (**a**) Heat storage. Electric energy is consumed by the heating wires and stored in the thermal brick in the form of thermal energy. (**b**) Heat transfer. The stored thermal energy is transferred to the circulating water through the heat exchanger. (**c**) Customers heating. Uses the thermal energy to keep the room warm.

The initial and final temperatures of the bricks are assumed to be T_{ini} and T_{fin}, respectively. The internal and external temperatures of the isolation layer are T_{int} and T_{ext}, respectively. T_{ini} is equal to T_{int} before the electric heating wires are energized. The T_{ini} is increased as the heating wires are energized. In the heating process, thermal energy always transfers heat to the surroundings through the isolation layer, so T_{int} can be calculated by the average temperature of the bricks. According to a large number of survey and analysis of SETS operation, the T_{ext} approximation is 50 °C. The boundary and initial conditions of the PM are described in Equation (1).

$$\begin{cases} T_{ini} = T_{int} & (t = 0) \\ T_{fin} \geq T_{int} > T_{ini} & (t > 0) \\ T_{int} = (T_{fin} + T_{ini})/2 \\ T_{ext} = 50 \end{cases} \tag{1}$$

The inlet and outlet temperatures of the hot air through the heat exchanger are $T_{a,in}$ and $T_{a,o}$, respectively. The inlet and outlet temperatures of the circulating water through the heat exchanger are $T_{w,in}$ and $T_{w,o}$, respectively. $T_{a,room}^{cri}$ is the criterion minimum temperature of customers' room. The room temperature $T_{a,room}$ is influenced by the environmental conditions (e.g., ambient temperature $T_{a,amb}$. In Section 2.7. (Influencing Factors) explains why ambient temperature is selected). In order to keep the room temperature close to $T_{a,room}^{cri}$, the supply and return water temperatures $T_{w,in}$ and $T_{w,o}$ can be increased correspondingly. In the inlet of the heat exchanger, a high-temperature air is required to be input, so the stored thermal energy final temperature T_{fin} needs to be higher (about 700 °C). T_{fin} can be deduced by the change of ambient temperature $T_{a,amb}$, and the average power consumption of the SETS can be obtained by the T_{fin}. According to engineering experience, the variation ranges of T_{ini} and $T_{a,o}$ are very little that they are assumed to be constant. $T_{w,o}$ is affected by the ambient temperature $T_{a,amb}$, so the input and control variables of the heat exchanger $T_{w,o}$ and $T_{w,in}$ are deduced based on Equation (12).

Without considering the temperature gradients within the bricks, the initial temperature T_{ini} and final temperature T_{fin} changes in the SETS are used to predict the thermal energy consumption.

The proposed SETS PM prediction of the consumed thermal energy is equal to the electricity by the ambient temperature change.

2.3. Thermal Energy Storage

According to the above principle, the electric energy consumption $E_{\mathrm{pro},t}$ is equal to the sum of thermal energy $E_{\mathrm{sto},t}$ stored in the bricks and the thermal energy loss of the SETS $E_{\mathrm{los},t}$ as shown in Equation (2).

$$E_{\mathrm{pro},t} = E_{\mathrm{sto},t} + E_{\mathrm{los},t}. \tag{2}$$

The $E_{\mathrm{los},t}$ is not considered in the modeling. So the Equation (2) can be a simplify as $E_{\mathrm{pro},t} = E_{\mathrm{sto},t}$. The thermal energy $E_{\mathrm{sto},t}$ is transferred to the hot air by thermal convection ϕ_{conv}, thermal radiation ϕ_{rad}, and thermal conduction ϕ_{cond}. The three heat transfer formulas are introduced as follows:

2.3.1. Thermal Convection

Based on the initial temperature T_{ini} and final temperature T_{fin} of the thermal storage, the thermal convection from bricks into the isolation layer is given by Equation (3). The thermal convection equation is conveniently expressed by Newton's law of cooling [23].

$$\phi_{\mathrm{conv}} = h_{\mathrm{conv}} \left(T_{\mathrm{fin}} - T_{\mathrm{ini}}\right) A_{\mathrm{cha}} = N u \frac{\lambda}{D} \left(T_{\mathrm{fin}} - T_{\mathrm{ini}}\right) A_{\mathrm{cha}}, \tag{3}$$

where h_{conv} is the surface heat transfer coefficient of the bricks, A_{cha} is the surface area of the round holes in the bricks as shown in Figure 3. Many bricks are superimposed to form the core frame of the thermal storage. Two heat bricks are stacked together to form a circular hole in the center of the bricks as shown in the zoom figure. Electric heating wires are placed in the middle of the hole. L, W, H, and D are the length, width, height, and diameter of the brick, respectively. A_{cha} is the area of hot airflow channel which is stacked by two bricks together, $A_{\mathrm{cha}} = L \times D \times N$, where N is the number of bricks in a row. Based on the Nusselt–Number N_u, the heat transfer coefficient h_{conv} can be calculated by the Zukauskas formula [24]. λ is the thermal conductivity.

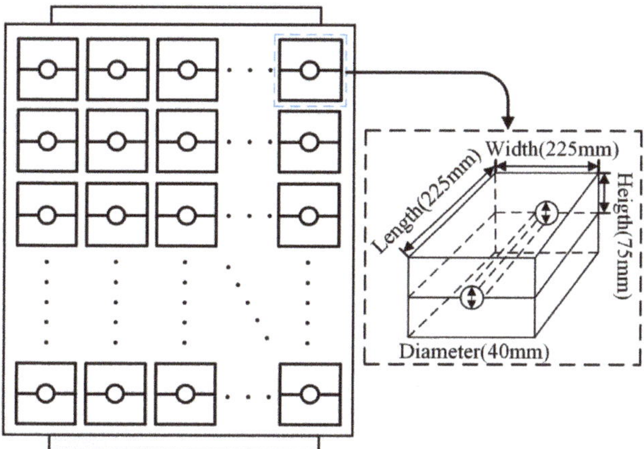

Figure 3. Schematic diagram of bricks placement.

In heat convection, multiple round holes can be approximately regarded as the tube bundle. The average heat transfer performance of the tube bundle is related to the flow Reynolds-number (R_e) [24].

$$R_e = u_\text{a} \times D / \nu_{T_\text{ave}} \qquad (4)$$

where u_a is the hot air flow rate, D is bricks round hole diameter, and ν_{T_ave} is the aerodynamic viscosity at the temperature of $T_\text{ave} = T_\text{fin}/2$. According to the range of R_e, the Nusselt-number (N_u) can be calculated by Zukauskas formula in Equation (5).

$$\begin{cases} N_u = 0.52 R_e^{0.5} P_{r_a}^{0.36} P_{r_c} & \text{if } 10^2 < R_e \leq 10^3 \\ N_u = 0.27 R_e^{0.63} P_{r_a}^{0.36} P_{r_c} & \text{if } 10^3 < R_e \leq 20^5 \\ N_u = 0.033 R_e^{0.8} P_{r_a}^{0.36} P_{r_c} & \text{if } 20^5 < R_e \leq 20^6 \end{cases} \qquad (5)$$

where P_{r_a} is the Prandtl-number (PN) of the fluid average temperature, PN reflects the contrast between momentum diffusion and thermal diffusion in fluid. P_{r_c} is a constant. $P_{r_c} = \left(P_{r_a} / P_{r_\text{bri}} \right)^{0.25}$, P_{r_bri} is the PN of the bricks average surface temperature. According to the temperature T_int and T_fin, the P_{r_a} and P_{r_bri} can be queried.

When $10^2 < R_e \leq 20^6$, the corresponding N_u can be selected. In most cases, the calculation results satisfy the condition that $10^3 < R_e \leq 20^5$. This paper uses $N_u = 0.27 R_e^{0.63} P_{r_g}^{0.36} P_{r_c}$ as an example. Substituting Equation (5) into Equation (3), the thermal energy ϕ_conv is obtained in Equation (6).

$$\phi_\text{conv} = 0.27 R_e^{0.63} P_{r_g}^{0.36} P_{r_c} \frac{\lambda}{D} \left(T_\text{fin} - T_\text{ini} \right) A_\text{cha}. \qquad (6)$$

2.3.2. Thermal Radiation

The thermal energy ϕ_rad released from the bricks to the flow channel hot air, is expressed by Equation (7), which is based on the Stefan–Boltzmann law [23].

$$\phi_\text{rad} = \varepsilon \sigma \left((T_\text{fin} + 273)^4 - (T_\text{ini} + 273)^4 \right) A_\text{bri}, \qquad (7)$$

where ε is the emissivity of the bricks' surface, σ is the blackbody radiation constant (Stefan–Boltzmann constant is 5.67×10^{-8} W/m^{2}K^4), and A_bri is the heat transfer surface area of the bricks.

2.3.3. Thermal Conduction

The isolation layer consists of fiberglass felt and perlite coated. The thermal conduction ϕ_cond between the internal hot air and the external ambient environment is given by the Fourier's law of heat conduction in (8) [23].

$$\phi_\text{cond} = \frac{T_\text{int} - T_\text{ext}}{\delta_f / \lambda_f + \delta_p / \lambda_p} A_\text{iso} = \frac{\left(T_\text{fin} - T_\text{ini} \right)/2 - 50}{\delta_f / \lambda_f + \delta_p / \lambda_p} A_\text{iso}, \qquad (8)$$

where $\delta_f \delta_p$ and $\lambda_f \lambda_p$ are the respective thickness, and thermal conductivity of fiberglass and perlite, and A_iso is the isolation layer area.

The prediction of average power consumption $\bar{Q}_\text{sto}$ can be obtained by Equations (6)–(8) as shown in Equation (9).

$$\bar{Q}_\text{sto} = \phi_\text{conv} + \phi_\text{rad} + \phi_\text{cond} \qquad (9)$$

$$= 0.27 R_e^{0.63} P_{r_g}^{0.36} P_{r_c} \frac{\lambda}{D} \left(T_\text{fin} - T_\text{ini} \right) A_\text{cha} + \varepsilon \sigma \left((T_\text{fin} + 273)^4 - (T_\text{ini} + 273)^4 \right) A_\text{bri}$$

$$+ \frac{\left(T_\text{fin} - T_\text{ini} \right)/2 - 50}{\delta_f / \lambda_f + \delta_p / \lambda_p} A_\text{iso},$$

where T_fin is unknown which can be obtained in Section 2.4 (Thermal Bricks Energy Release).

2.4. Thermal Bricks Energy Release

Thermal bricks energy release includes heat transfer and customers' heating.

2.4.1. Heat Transfer

When the SETS releases thermal energy, the inlet hot air temperature of the heat exchanger is regulated by the circulating wind speed of the fan. In the heat exchanger, according to the conservation of energy, the thermal energy released Q_a by the inlet hot air is equal to the thermal energy absorbed Q_w by the outlet circulating water. Therefore, the relationship between the inlet hot air temperature and the outlet circulating water temperature is given in (10) [25].

$$
\begin{cases}
Q_a = \dot{m}_a c_a \left(T_{a,in} - T_{a,o} \right) \\
Q_w = \dot{m}_w c_w \left(T_{w,o} - T_{w,in} \right)
\end{cases}
\tag{10}
$$

where, $\dot{m}_a$ and $\dot{m}_w$ are the flow rate of hot air and water through the heat exchanger, respectively. c_a and c_w are the specific heat of hot air and water through the heat exchanger, respectively. Since $Q_a = Q_w$, $T_{a,i}$ can be deduced as shown in Equation (11). Ignoring the heat loss from the hot air at the outlet of the SETS to the heat exchanger inlet, and assuming that $T_{a,i} \approx T_{fin}$, then, taking this into Equation (9), the average power consumption $\bar{Q}_{sto}$ can be obtained.

$$
T_{a,in} = T_{a,o} + \frac{\dot{m}_w c_w \left(T_{w,o} - T_{w,in} \right)}{\dot{m}_s c_s},
\tag{11}
$$

where, $T_{w,o} - T_{w,in}$ is determined by the Section 2.4.2 (Customers Heating).

2.4.2. Customers Heating

The thermal energy storage is intended to exchange heat with the circulating water to maintain the customers' room temperature greater than or equal to the designed heating temperature $T_{a,room}^{cri}$. The temperatures of supply and return water are obtained by the room and the ambient temperature in Equation (12).

$$
\begin{cases}
T_{w,o} = T_{a,room}^{cri} + \left(\dfrac{T_{w,o}^{cri} - T_{w,in}^{cri}}{2} \right) \left(\dfrac{T_{a,room}^{cri} - T_{a,amb}}{T_{a,room}^{cri} - T_{a,amb}^{cri}} \right) \\
\qquad + \left(\dfrac{T_{w,in}^{cri} + T_{w,o}^{cri}}{2} - T_{a,room}^{cri} \right) \left(\dfrac{T_{a,room}^{cri} - T_{a,amb}}{T_{a,room}^{cri} - T_{a,amb}^{cri}} \right)^{\frac{1}{1+\eta}} \\
T_{w,in} = T_{w,o} - \left(T_{w,o}^{cri} - T_{w,in}^{cri} \right) \left(\dfrac{T_{a,room}^{cri} - T_{a,amb}}{T_{a,room}^{cri} - T_{a,amb}^{cri}} \right)
\end{cases}
\tag{12}
$$

where $T_{w,o}^{cri}$ and $T_{w,in}^{cri}$ are the respective temperatures of water supply and return designed for SETS. η is the heat transfer coefficient of the radiator. According to the standard of heating in the north of China, the minimum value of $T_{a,room}^{cri}$ is 18 °C. During the winter heating period, the average ambient temperature of Shenyang is -16.9 °C.

The overshoot of water temperature and the frequent start of the fan are easily caused by the use of $T_{w,o}$ and $T_{w,in}$ owing to the inertia of temperature change. Therefore, in this paper, the highest supply water temperature $T_{w,o,max}$ and lowest return water temperature $T_{w,i,min}$ during the heating period are designed. Taking Shenyang city as an example, the minimum and maximum temperatures of $T_{a,amb}$ are taken into Equation (12) to obtain the temperature curves of supply and return water as shown in Figure 4. The minimum temperature of return water $T_{w,i,min}$ is selected as the benchmark so that the change of temperature $T_{w,o} - T_{w,in} = \gamma$ can be deduced as shown in Equation (13).

$$
\gamma = \left(1 + \frac{T_{w,o} - T_{w,i,min}}{T_{w,o,max} - T_{w,i,min}} \right) T_{w,i,min}.
\tag{13}
$$

Substituting γ into Equation (11), $T_{a,in}$ is obtained as shown in Equation (14).

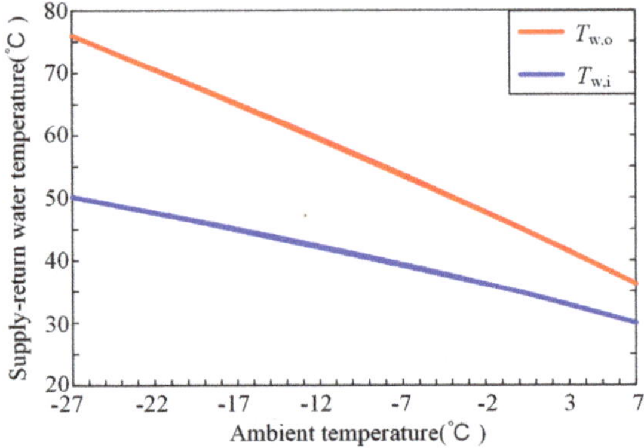

Figure 4. Influence curve of ambient temperature on circulating water temperature.

$$T_{a,in} = T_{a,o} + \frac{\dot{m}_w c_w \left(1 + \frac{T_{w,o} - T_{w,i,min}}{T_{w,o,max} - T_{w,i,min}}\right) T_{w,i,min}}{\dot{m}_s c_s}. \tag{14}$$

Based on the above analysis, the thermal energy store in bricks at time t can be obtained as shown in Equation (15).

$$E_{pro,t} = E_{sto,t} = \int_0^i \bar{Q}_{sto,t}\, dt \quad (i \in t), \tag{15}$$

where, $\bar{Q}_{sto,t}$ is the average power consumption of SETS at time t. $E_{pro,t}$ is the electricity consumption by the heating wires, and $E_{sto,t}$ is the stored heat energy in the bricks. The working time t is related to the behavior characteristics, and the superscript i is the time index.

2.5. Customers' Behavior Characteristics Extraction

The working time t of the SETS is determined by the different electricity consumption behavior characteristics of customers. For example, two types of behavior characteristics of SETS customer within one month is shown in Figure 5. One type starts from the previous night 22:00 and stops heating at 1:00. Then, SETS starts again at 3:00 and ends the reservoir at 5:00. The other type starts from the previous night 22:00 and then continues to work until the end at 5:00. The consumption time t of one type is $t = 3.5h$ on the left curve, and the other type is $t = 7h$ on the right curve.

Since the load curves of the same electrical behavior have regular working time, it is necessary to cluster the same electricity consumption behavior characteristics with the heat load data. Then, the customers' behavior types are used to determine the working time t of SETS. Also, the $E_{pro,t}$ daily electricity consumption of SETS is predicted by Equation (15). The method of K-means is used to identify customers' behavior types. After all SETS data clustering, the customers' behavior types are obtained. The load data with the same type has a similar working time, the same type is identified using the support vector machine that has been widely used to learn target class [26]. When the target types are larger than binary, the one-versus-one is one of the effective methods to find the discriminative binary classifiers to solve a multiclass problem [27]. Thus, the one-versus-one method is adopted in behavior extraction. The overall process of the behavior characteristics extraction is shown in Algorithm 1.

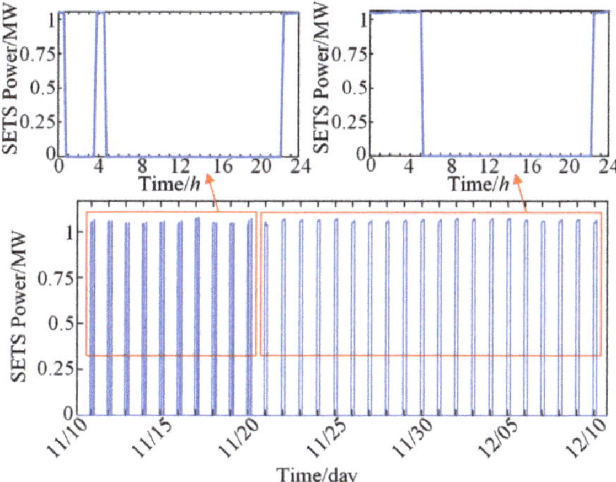

Figure 5. Customers' electricity consumption behavior curve.

Algorithm 1 The behavior characteristics extraction algorithm process.

1: Input SETS' load data.
2: Label the data of the same customer from **1, 2, 3** to **n**;
3: Using unsupervised clustering method (K-Means) to identify target data types c_1, c_2;
4: Statistics time labels $y_1, y_2, \cdots, y_m$ of the same type c_j;
5:     *if $(m > 2)$ then*
6:         Using the one-versus-one method, divide the time labels into binary pairs as the support

vector machine output time labels;
7:     *else*
8:         Using the support vector machine directly;
9:     *end if*
10: Output the working time *t*.

2.6. Summary of the PM Prediction

Based on the above analysis, the proposed PM of SETS can be used to predict the daily electricity consumption. The overall PM flowchart is shown in Figure 6. According to the initial values, the range of Reynolds-Number (R_e) is estimated. If $R_e > 20^6$ or $R_e \leq 10^2$, return the initialization parameters, otherwise, input the ambient temperature $T_{a,amb}$ to obtain temperatures of $T_{w,o}$ and $T_{w,in}$. Select the maximum value of supply water $T_{w,o,max}$ and the minimum value of return water $T_{w,in,min}$. Then, obtain the inlet temperature $T_{a,in}$. N_u is selected based on the range of R_e, so input the N_u and $T_{a,in} \approx T_{fin}$ to predict the average power consumption $\bar{Q}_{sto}$. The $\bar{Q}_{sto,t}$ is integrated over working hours t. According to the stored thermal energy $\bar{Q}_{sto,t}$ in the bricks, the predicted value of SETS heating wires electricity consumption $E_{pro,t}$ can be obtained.

2.7. Influencing Factors

Since influencing factors (including humidity, wind speed, and temperature) influence the prediction of the CPM, it is necessary to select the strong correlation factors as the input of the cyber and PMs. Thus, the Pearson correlation analysis in Equation (16) is adopted to analyze which one is strongly correlated. The influencing factors and the SETS load are represented as *a* and *b*, respectively.

$$\rho_{a,b} = \frac{\left(\sum ab - \frac{\sum a \sum b}{N} \right)}{\sqrt{\left(\sum a^2 - \frac{(\sum a)^2}{N} \right)\left(\sum b^2 - \frac{(\sum b)^2}{N} \right)}}. \tag{16}$$

The statistics on the correlation between the influencing factors and SETS load within 24 h is as shown in Figure 7. The histogram explains which influencing factors are strongly correlated with the SETS load. The influencing factors (including humidity, wind speed, and temperature) are negatively correlated with the load, and humidity is positively correlated with the SETS load only in the range of 8:00 to 10:00. The ranges of correlation coefficient of humidity (blue), wind speed (green) and temperature (yellow) are $[-0.24, 0.21]$, $[-0.47, 0.13]$, and $[-0.82, -0.66]$, respectively. It can be seen that the correlation between the temperature and SETS load is strongly negative, while humidity and wind speed are weakly correlated with SETS load. Thus, the temperature is selected as the input of the cyber and PMs.

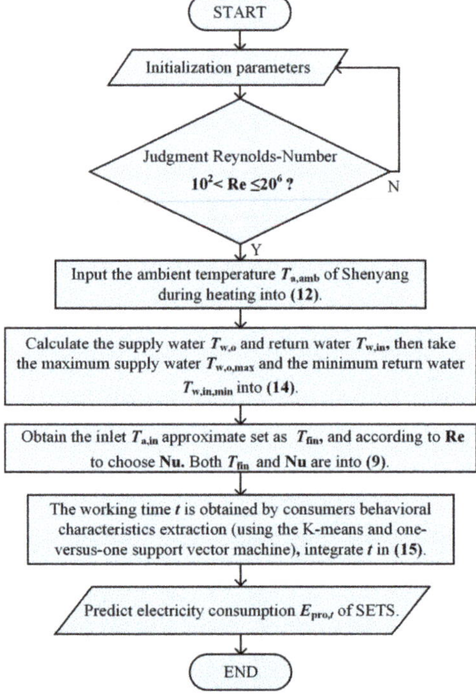

Figure 6. Flowchart for the physical model (PM).

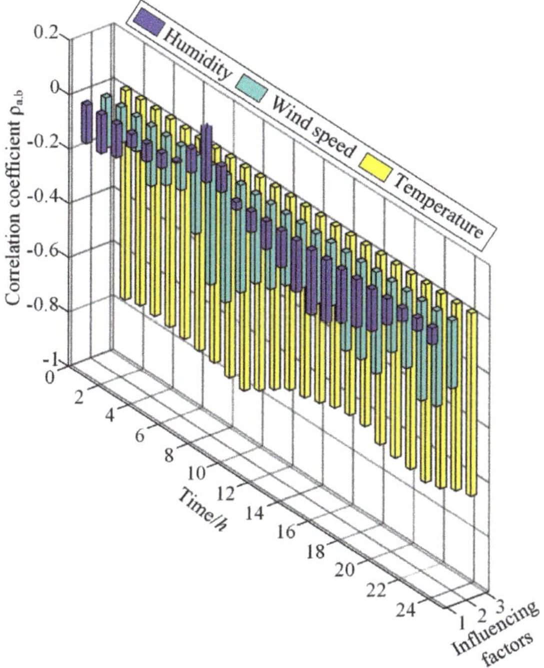

Figure 7. Correlation of SETS load influencing factors.

3. Cyber–Physical Approach

The parameters of SETS and ambient temperature are the inputs of the SETS PM. The PM prediction result is shown in Figure 8. It can be observed that January has the best fitting with the least error. The test result with the PM still has an irregular error as shown in Table 1. The root mean square error (RMSE), mean absolute error (MAE), and mean absolute percentage error (MAPE) are used to measure the prediction performance of the PM with a deviation between the predicted value and the real value.

$$\begin{cases} \text{RMSE} = \sqrt{\dfrac{1}{m} \sum_{i=1}^{m} (y_i - \hat{y}_i)^2} \\[2mm] \text{MAE} \ = \dfrac{1}{m} \sum_{i=1}^{m} |y_i - \hat{y}_i| \\[2mm] \text{MAPE} = \dfrac{100\%}{m} \sum_{i=1}^{m} \left| \dfrac{y_i - \hat{y}_i}{y_i} \right| \end{cases} \tag{17}$$

where m is the number of data. y_i is the real value of SETS load, $\hat{y}_i$ is the prediction value of SETS PM.

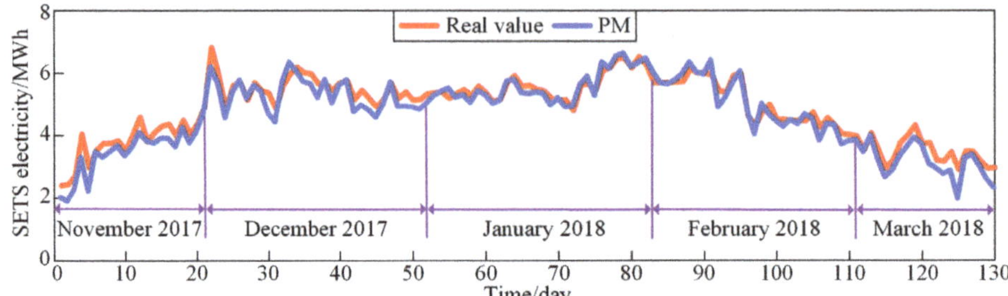

Figure 8. PM prediction of the load curve of a customer's SETS.

Table 1. Physical model (PM) prediction parameters.

Month	2017.11	2017.12	2018.1	2018.2	2018.3
Date	10–30	1–31	1–31	1–28	1–19
RMSE	372.05	310.29	167.36	227.1	389.45
MAE	323.36	264.99	139.33	183.92	318.54
MAPE%	9.08	4.82	2.49	3.75	9.53
Average $\bar{y}_i$(kWh)	3805	5526.9	5616	5035.1	3470.8
Average $\hat{\bar{y}}_i$(kWh)	3486.3	5323.3	5610	4990.2	3152.7

The PM predicts error according to the result of RMSE, MAE, and MAPE, and obtains a consistent evaluation. Selecting the month of January which has the smallest RMSE in five months, the predicted result RMSE error is 167.36 kWh. The electricity consumption of real value is 5616 kWh, the prediction accuracy is approximately 29.8%. Therefore, the prediction accuracy of the PM still needs to be further improved.

To improve the prediction accuracy, other prediction methods are considered. The CM is widely used in load prediction because it can accurately model the past. Hence, the application of the CM is considered to predict the electricity consumption of SETS. However, the electric behavior of SETS is different from the other conventional electric loads, it is difficult to predict a sudden increase and decrease in heat load, that is, the predictability of the CM is not good enough. On the contrary, the PM is sensitive to a sudden increase and decrease in the heat load. Therefore, the CM and PM can be combined to improve the prediction accuracy with mutual verification. CPS is widely used in many fields, for example, measurement recovery in false data injection attacks [28], smart home cyberattack detection [29], energy theft detection in multi-tenant data centers [30], and inter-well connectivity estimation in petroleum [31], etc. Hence, the cyber–physical approach is developed to predict electricity consumption of SETS. The CM established by the load data of SETS is used to calibrate the output error of the PM.

The cyber components (including influencing factors) and the physical components (including behavior characteristics and average power consumption) of SETS are integrated using the cyber–physical approach. The cyber–physical approach is used to predict the electricity consumption of SETS. The Pearson correlation analysis formula is used to solve the correlation coefficients $\rho_{a,b}$ between power consumption of SETS and influencing factors. The ambient temperature has the greatest impact on the power consumption of SETS, which is considered as an input of the PM flow.

Firstly, the PM of SETS is utilized to predict the daily average power consumption based on ambient temperature. Secondly, the integration of K-means and one-versus-one support vector machine is applied to extract the behavior characteristics of SETS to obtain the work time t. Then the electricity consumption of SETS is obtained by integrating average power consumption. The performance of the prediction is tested by the error calculation Equation (17). The error levels are divided according to the size of the errors.

Since RMSE is a good reflection of the precision of the measurement, it is used to design the error levels of the PM. According to the result of RMSE, which is represented by ϵ_s, the subscript indicates $s = 1, 2, 3, 4, 5$ which represents the months from November 2017 to March 2018, respectively. The number ϵ_1 to ϵ_5 are 372.05, 310.29, 167.36, 227.1, 389.45, respectively. Therefore, according to the order of ϵ_s ($\epsilon_5 > \epsilon_1 > \epsilon_2 > \epsilon_4 > \epsilon_3$), the error is divided into 5 levels, and then the data is input in the CM which is used in BP neural network.

Subsequently, the output of PM is calibrated by the error of CM output to obtain the electricity consumption of the SETS. The specific method of the CM is introduced as follows.

The error levels, ambient temperature, and PM prediction values are taken as input vectors x_n of BP neural network. Output vectors o_n is the error as shown in Figure 9. The number of hidden neurons is 13, the sigmoid function is selected as the activation function for the hidden layer and the output layer, and the gradient descent method is used to update the parameters. In the CM, the influencing factors and the power consumption of SETS are the input vectors x_n of BP neural network, and the electricity is the output vectors o_n. The number of hidden neurons is set as 25.

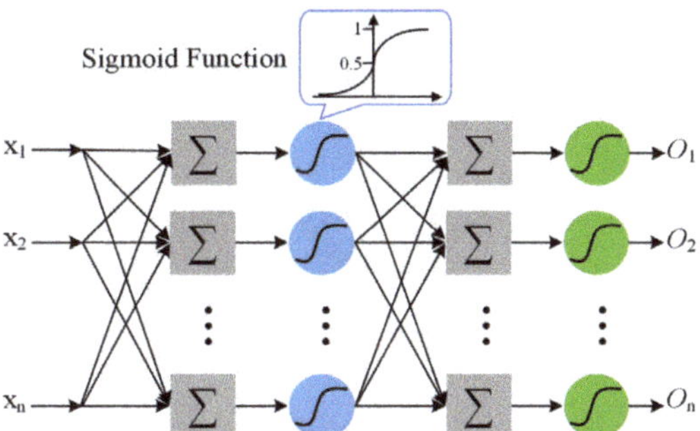

Figure 9. Cyber model (CM) structure of back propagation (BP) neural network.

During training, the CM of the BP neural network can automatically extract the "reasonable rules" between the input layer and output layer data and memorize the learning content in the weight of the network adaptively. The CM can predict the load deviation after training even if there is unknown noise-pollution. BP neural network is used to train the CM. The CM prediction result plus the PM predicted value is the electricity consumption of SETS.

4. Validation

In order to verify the effectiveness of the proposed CPM method, its prediction result is compared with PM, CM and real value. MRE is used to evaluate the performance of the prediction model.

The proposed method to predict the electricity consumption of SETS is verified using real operation data, the data is obtained from the power grid valley thermal storage system monitoring platform of State Grid Company. This paper uses the data of one customer from the monitoring platform. The SETS is located in the third printing plant of Heping District, Shenyang City, Liaoning Province. The scene device of SETS is shown in Figure 10. The nominal power of the SETS is 1 MW, and the maximum thermal storage capacity is 10 MWh.

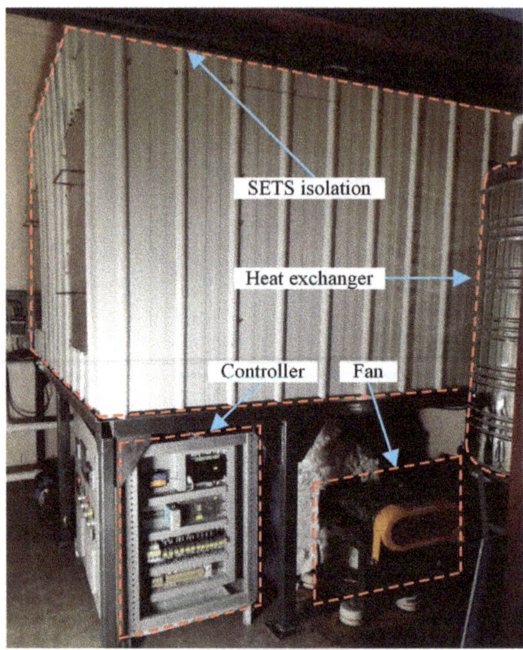

Figure 10. Scene device of SETS.

The load data of 1MW SETS during the heating period from 10 November 2017, to 19 March 2018, are taken. The PM parameters of SETS are selected as shown in Table 2. The data of 120 days are used as the training set, and the data of the later 10 days are used for the testing. The simulation results are shown in Figure 11. The four curves are the individual SETS real values, CPM, PM, and the CM predicted value. The trained model is used to test the load condition in the following 10 days. From the training results, the CPM predicted value coincides with the real value. In the testing results, the comparison data in the orange dotted frame is enlarged for 121–130 days as seen in Figures 12–14. The result after the application of the CPM is significantly better than that individually predicted by the PM and CM alone.

Table 2. PM parameters.

Parts	Symbol	Number	Unit	Symbol	Number	Unit
Bricks	T_{ini}	150	°C	$T_{a,o}$	105	°C
	L	0.225	m	W	0.225	m
	H	0.075	m	$D/2$	0.02	m
	L-row	7		W-row	16	
	H-row	43		ε	0.5	
Exchanger	$T_{w,in}^{cri}$	45	°C	$T_{w,o}^{cri}$	65	°C
	c_a	1.093	$\mathrm{kJ\,kg^{-1}{}^\circ C^{-1}}$	c_w	4.174	$\mathrm{kJ\,kg^{-1}{}^\circ C^{-1}}$
	$\dot{m}_a$	30,936	$\mathrm{m^3\,h^{-1}}$	$\dot{m}_w$	15	$\mathrm{kg\,s^{-1}}$
	u_a	1	$\mathrm{m\,s^{-1}}$			
Isolation	δ_g	0.5	m	λ_g	0.101	$\mathrm{W\,m^{-1}\,K^{-1}}$
	δ_f	0.05	m	λ_f	0.13	$\mathrm{W\,m^{-1}\,K^{-1}}$
	T_{ext}	50	°C			
Room	$T_{a,room}^{cri}$	18	°C	$T_{a,amb}^{cri}$	−16.9	°C
	λ	0.0574	$\mathrm{W\,m^{-1}\,K^{-1}}$	ν	79.38×10^{-6}	$\mathrm{m^2\,s^{-1}}$
	$P_{r_{bri}}$	0.706		P_{r_a}	0.687	
	η	0.287				

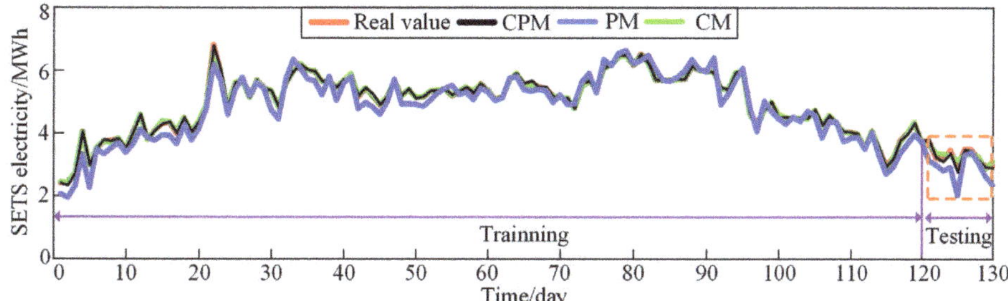

Figure 11. The simulation results of one customer's 130 days.

4.1. Comparison of CPM with Real Value

The simulation results are shown in Figure 12. The prediction of the CPM coincides with the real value v. The prediction data are shown in Table 3. Comparing the real value v with the CPM, the predicted absolute error is e_1. Selecting three days (123, 124, and 125) for calculation gives the CPM prediction with the MRE ($\Delta_1 = \frac{e_1}{v} \times 100\%$) of 2.4%, 3.3%, and 5%, respectively. Therefore, the MRE (Δ_1) of the prediction by the CPM is not more than 5%.

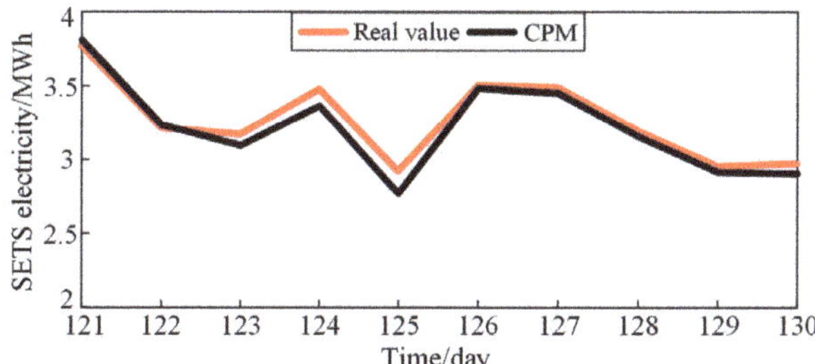

Figure 12. Comparison of CPM with real value testing curves.

Table 3. Performance comparison of simulation results (unit MWh).

Day	Real Value v	CPM e_1	MRE Δ_1	PM e_2	MRE Δ_2	CM e_3	MRE Δ_3
121	3.771	0.038	1	0.632	16.7	0.086	2.3
122	3.216	0.02	0.6	0.223	6.9	0.174	5.4
123	3.173	0.076	2.4	0.356	11.2	0.179	5.6
124	3.479	0.113	3.3	0.56	16.1	0.186	5.3
125	2.922	0.15	5	0.891	30.4	0.21	7.2
126	3.503	0.018	0.5	0.185	5.2	0.179	5.1
127	3.489	0.039	1.1	0.074	2.1	0.152	4.4
128	3.195	0.037	1.1	0.145	4.5	0.083	2.6
129	2.956	0.039	1.3	0.321	10.8	0.086	2.9
130	2.973	0.067	2.3	0.613	20.6	0.14	4.7

4.2. Comparison of CPM with PM

The simulation results are shown in Figure 13. Compared with the real value v, the prediction error of the PM is larger than in the CPM. The prediction data are shown in Table 3. Comparing the real value v with the PM, the predicted absolute error is e_2. Selecting three days (121, 125, and 130) for calculation gives the PM prediction with the MRE ($\Delta_2 = \frac{e_2}{v} \times 100\%$) of 16.7%, 30.4%, and 20.6%, respectively, while the CPM prediction gives the MRE (Δ_1) of 1%, 5%, and 2.3%, respectively. Therefore, the prediction result of the CPM when compared with the PM has the MRE reduced by 25.4%.

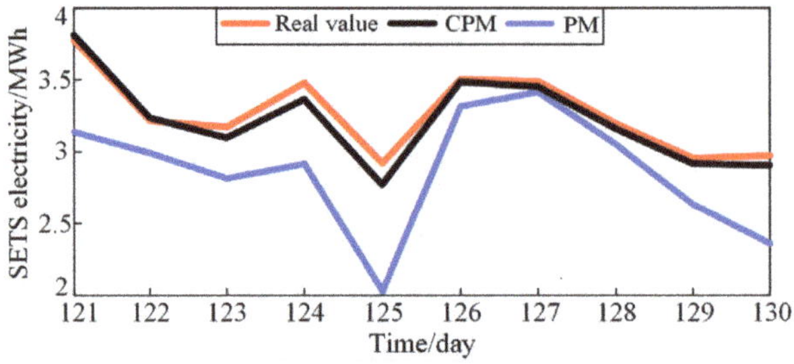

Figure 13. Comparison of cyber–physical model (CPM) with PM testing curves.

4.3. Comparison of CPM with CM

The simulation results are shown in Figure 14. Compared with the real value v, the prediction error of the CM is larger than in the CPM. The prediction data are shown in Table 3. Comparing the real value of v with the CM, the predicted absolute error is e_3. Selecting three days (122, 123, and 125) for calculation gives the CM prediction with the MRE ($\Delta_3 = \frac{e_3}{v} \times 100\%$) of 5.4%, 5.6%, and 7.2%, respectively, while the CPM prediction gives the MRE (Δ_1) of 0.6%, 2.4%, and 5%, respectively. Therefore, the prediction result of the CPM when compared with the CM has the MRE reduced by 4.8%.

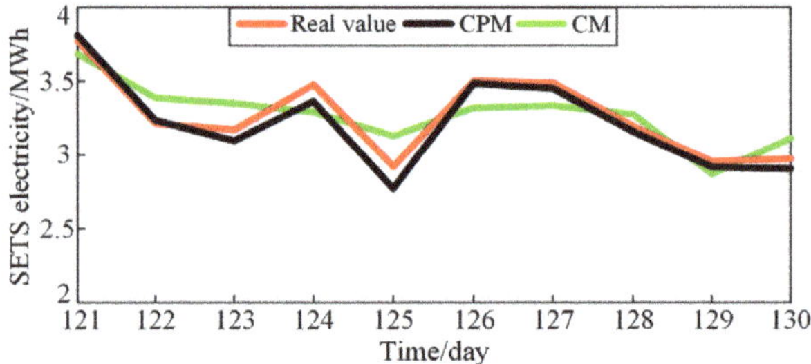

Figure 14. Comparison of CPM with CM testing curves.

5. Conclusions

Dispatching electricity consumption of SETS is an effective method for peak shaving in the power system. The prediction of thermal storage is a prerequisite for dynamic optimal dispatching, and the thermal storage can be used to shift times of resources. The CPS approach is proposed to predict power load, which promotes the application of the CPS in smart grids and ubiquitous power internet of things. This paper proposes a cyber–physical approach to predict the electricity consumption of SETS with consideration of the ambient temperature and electric behavior of the SETS into the PM. The CM adopts the BP neural network to calibrate the errors obtained in the PM. The 1MW SETS is established to validate the proposed cyber–physical approach. The simulation results show that when the CPM is compared with the PM, the MRE is reduced by 25.4%, and when compared with the CM is reduced by 4.8%. Using the CPM to calibrate the PM effectively improves the prediction accuracy. Conclusively, the prediction of SETS using the CPM is better than the individual PM and the CM alone.

The recommendations for future research are as follows.

- Thermal storage is an effective method for peak shaving and dispatching in power system. The electricity consumption prediction of SETS is worth to explore the combination with the heat and power generation unit.
- The application of the proposed cyber–physical model in other resources of the power system is recommended to be studied.

Author Contributions: Conceptualization, H.J., J.Y. and H.W.; methodology, H.J., J.Y. and H.W.; software, H.J.; validation, H.J., J.Y. and H.W.; formal analysis, H.J.; investigation, H.J.; resources, H.J.; data curation, H.J. and K.T.; writing—original draft preparation, H.J.; writing—review and editing, H.J., H.W. and M.O.O.; visualization, H.J. and J.F.; supervision, J.Y. and H.W.; project administration, J.Y. and H.W.; funding acquisition, H.W.

Funding: This work was supported in part by the China Postdoctoral Science Foundation under Grant 2019M651144, in part by the Liaoning Provincial Department of Education Research Funding under Grant LQGD2019005.

Conflicts of Interest: The authors declare no conflict of interest.

References

1. Ibrahim, D. On thermal energy storage systems and applications in buildings. *Energy Build.* **2002**, *34*, 377–388.
2. Cisek, P.; Taler, D. Numerical and experimental study of a solid matrix electric thermal storage unit dedicated to environmentally friendly residential heating system. *Energy Build.* **2016**, *130*, 747–760. [CrossRef]
3. Sauter, P.S.; Solanki, B.V.; Cañizares, C.A.; Bhattacharya, K.; Hohmann, S. Electric thermal storage system impact on northern communities' microgrids. *IEEE Trans. Smart Grid* **2019**, *10*, 852–863. [CrossRef]
4. Wong, S.; Pinard, J.P. Opportunities for smart electric thermal storage on electric grids with renewable energy. *IEEE Trans. Smart Grid.* **2017**, *8*, 1014–1022. [CrossRef]
5. Cooke, W.B.H.; Hardy, R.H.S.; Sulatisky, M.T. Thermal energy storage in forced-air electric furnaces. *IEEE Trans. Ind. Appl.* **1980**, *IA-16*, 127–133. [CrossRef]
6. Bedouani, B.Y.; Moreau, A.; Parent, M.; Labrecque, B. Central electric thermal storage (ETS) feasibility for residential applications: part 1. numerical and experimental study. *Int. J. Energy Res.* **2001**, *25*, 53–72. [CrossRef]
7. Wang, X.; Zhang, M.; Ren, F. Learning customer behaviors for effective load forecasting. *IEEE Trans. Knowl. Data Eng.* **2019**, *31*, 938–951. [CrossRef]
8. Quilumba, F.L.; Lee, W.J.; Huang, H.; Wang, D.Y.; Szabados, R.L. Using smart meter data to improve the accuracy of intraday load forecasting considering customer behavior similarities. *IEEE Trans. Smart Grid* **2015**, *6*, 911–918. [CrossRef]
9. Xu, G.; Hu, X.; Liao, Z.; Xu, C.; Yang, C.; Deng, Z. Experimental and numerical study of an electrical thermal storage device for space heating. *Energies* **2018**, *11*, 2180. [CrossRef]
10. Taylor, J.W. Short-term load forecasting with exponentially weighted methods. *IEEE Trans. Power Syst.* **2012**, *27*, 458–464. [CrossRef]
11. Wu, L.; Shahidehpour, M. A hybrid model for integrated day-ahead electricity price and load forecasting in smart grid. *IET Gener. Transm. Distrib.* **2014**, *8*, 1937–1950. [CrossRef]
12. Rejc, M.; Pantoš, M. Short-term transmission-loss forecast for the Slovenian transmission power system based on a fuzzy-logic decision approach. *IEEE Trans. Power Syst.* **2011**, *26*, 1511–1521. [CrossRef]
13. Jiang, H.; Zhang, Y.; Muljadi, E.; Zhang, J.J.; Gao, D.W. A short-term and high-resolution distribution system load forecasting approach using support vector regression with hybrid parameters optimization. *IEEE Trans. Smart Grid* **2018**, *9*, 3341–3350. [CrossRef]
14. Rafiei, M.; Niknam, T.; Aghaei, J.; Shafie-Khah, M.; Catalão, J.P.S. Probabilistic load forecasting using an improved wavelet neural network trained by generalized extreme learning machine. *IEEE Trans. Smart Grid* **2018**, *9*, 6961–6971. [CrossRef]
15. Hong, T.; Wilson, J.; Xie, J. Long term probabilistic load forecasting and normalization with hourly information. *IEEE Trans. Smart Grid* **2014**, *5*, 456–462. [CrossRef]
16. Thouvenot, V.; Pichavant, A.; Goude, Y.; Antoniadis, A.; Poggi, J.M. Electricity forecasting using multi-stage estimators of nonlinear additive models. *IEEE Trans. Power Syst.* **2016**, *31*, 3665–3673. [CrossRef]

17. Idowu, S.; Saguna, S.; Åhlund, C.; Schelén, O. Applied machine learning: Forecasting heat load in district heating system. *Energy Build.* **2016**, *133*, 478–488. [CrossRef]
18. Wang, Z.; Wang, J.; Zhao, L.; Jia, S. A thermal energy usage prediction method for electric thermal storage heaters based on deep learning. In Proceedings of the 2019 IEEE 4th International Conference on Cloud Computing and Big Data Analysis, Chengdu, China, 12–15 April 2019.
19. Nigitz, T.; Gölles, M. A generally applicable, simple and adaptive forecasting method for the short-term heat load of consumers. *Appl. Energy* **2019**, *241*, 73–81. [CrossRef]
20. Macana, C.A.; Quijano, N.; Mojica-Nava, E. A survey on cyber physical energy systems and their applications on smart grids. In Proceedings of the 2011 IEEE PES Conference on Innovative Smart Grid Technologies Latin America, Medellin, Colombia, 19–21 October 2011.
21. Seshia, S.A.; Hu, S.; Li, W.; Zhu, Q. Design automation of cyber–physical systems: Challenges, advances, and opportunities. *IEEE Trans. Comput-Aided Des. Integr. Circuits Syst.* **2017**, *36*, 1421–1434. [CrossRef]
22. Dai, Y.; Chen, L.; Min, Y.; Chen, Q.; Hao, J.; Hu, K. Modeling and analysis of electrical heating system based on entransy dissipation-based thermal resistance theory. In Proceedings of the 2016 IEEE Power and Energy Society General Meeting, Boston, MA, USA, 17–21 July 2016.
23. Alawadhi, E.M. Thermal analysis of a building brick containing phase change material. *Energy Build.* **2008**, *40*, 351–357. [CrossRef]
24. Gao, W.; Zhao, J.; Li, X.; Zhao, H.; Zhang, Y.; Wu, X. Heat transfer characteristics of carbon dioxide cross flow over tube bundles at supercritical pressures. *Appl. Therm. Eng.* **2019**, *158*, 1–16. [CrossRef]
25. Attonaty, K.; Stouffs, P.; Pouvreau, J.; Oriol, J.; Deydier, A. Thermodynamic analysis of a 200 MWh electricity storage system based on high temperature thermal energy storage. *Energy* **2019**, *172*, 1132–1143. [CrossRef]
26. Liu, L.; Huang, H.; Hu, S. Lorenz chaotic system-based carbon nanotube physical unclonable functions. *IEEE Trans. Comput-Aided Des. Integr. Circuits Syst.* **2018**, *37*, 1408–1421. [CrossRef]
27. Hashemi, S.; Yang, Y.; Mirzamomen, Z.; Kangavari, M. Adapted one-versus-all decision trees for data stream classification. *IEEE Trans. Knowl. Data Eng.* **2009**, *21*, 624–637. [CrossRef]
28. Li, Y.; Wang, Y.; Hu, S. Online generative adversary network based measurement recovery in false data injection attacks: a cyber–physical approach. *IEEE Trans. Ind. Inform.* **2019**. [CrossRef]
29. Liu, Y.; Zhou, Y.; Hu, S. Combating coordinated pricing cyberattack and energy theft in smart home cyber–physical systems. *IEEE Trans. Comput-Aided Des. Integr. Circuits Syst.* **2018**, *37*, 573–586. [CrossRef]
30. Zhou, Y.; Liu, Y.; Hu, S. Energy theft detection in multi-tenant data centers with digital protective relay deployment. *IEEE Trans. Sustain. Comput.* **2018**, *3*, 16–29. [CrossRef]
31. Chen, X.; Zhang, D.; Wang, L.; Jia, N.; Kang, Z.; Zhang, Y.; Hu, S. Design automation for interwell connectivity estimation in petroleum cyber–physical systems. *IEEE Trans. Comput-Aided Des. Integr. Circuits Syst.* **2017**, *36*, 255–264. [CrossRef]

Article

Load Profile Segmentation for Effective Residential Demand Response Program: Method and Evidence from Korean Pilot Study

Eunjung Lee [1], Jinho Kim [1,*] and Dongsik Jang [2]

[1]   School of Integrated Technology, Gwangju Institute of Science and Technology, 123 Cheomdangwagi-ro, Buk-gu, Gwangju 61005, Korea; jkl51149@gist.ac.kr

[2]   Korea Electric Power Research Institute, 105 Munji-ro, Yuseong-gu, Daejeon 34056, Korea; yojian2@kepco.co.kr

*   Correspondence: jeikim@gist.ac.kr; Tel.: +82-62-715-5322

Received: 26 February 2020; Accepted: 11 March 2020; Published: 13 March 2020

Abstract: Due to the heterogeneity of demand response behaviors among customers, selecting a suitable segment is one of the key factors for the efficient and stable operation of the demand response (DR) program. Most utilities recognize the importance of targeted enrollment. Customer targeting in DR programs is normally implemented based on customer segmentation. Residential customers are characterized by low electricity consumption and large variability across times of consumption. These factors are considered to be the primary challenges in household load profile segmentation. Existing customer segmentation methods have limitations in reflecting daily consumption of electricity, peak demand timings, and load patterns. In this study, we propose a new clustering method to segment customers more effectively in residential demand response programs and thereby, identify suitable customer targets in DR. The approach can be described as a two-stage k-means procedure including consumption features and load patterns. We provide evidence of the outstanding performance of the proposed method compared to existing k-means, Self-Organizing Map (SOM) and Fuzzy C-Means (FCM) models. Segmentation results are also analyzed to identify appropriate groups participating in DR, and the DR effect of targeted groups was estimated in comparison with customers without load profile segmentation. We applied the proposed method to residential customers who participated in a peak-time rebate pilot DR program in Korea. The result proves that the proposed method shows outstanding performance: demand reduction increased by 33.44% compared with the opt-in case and the utility saving cost in DR operation was 437,256 KRW. Furthermore, our study shows that organizations applying DR programs, such as retail utilities or independent system operators, can more economically manage incentive-based DR programs by selecting targeted customers.

Keywords: data analysis; demand response (DR), load profile clustering; k-means; targeting of customer

---

1. Introduction

Recently, distributed energy resources (DER) such as photovoltaic (PV), wind turbine (WT), energy storage system (ESS), and demand response (DR) have been rapidly expanded on the distribution system. Because of this trend, the power demand characteristics have been more complicated. In addition, various business models and policies as resources application increased have been created. However, the DER expansion leads to load fluctuation on the distribution system locally. The DR program has been regarded as one of the solutions to mitigate imbalance. For this reason, DR programs have recently received significant attention. Under a DR program, electricity consumers change their electricity consumption patterns in response to a time-based rate or incentive payments for the periods

Energies **2020**, *13*, 1348; doi:10.3390/en13061348   www.mdpi.com/journal/energies

when needed [1]. Utilities and/or independent system operators (ISO) manage DR programs to avoid peak demand, high prices, and variable generation of renewables.

DR programs can be divided into two types: price- and incentive-based. Price-based DR programs vary the electricity price depending on certain time conditions being met [1]. Time of use (TOU), critical peak pricing (CPP), and real time pricing (RTP) are examples of this type of DR. Meanwhile, incentive-based DR programs encourage customers to shed their load or sell back to the electricity market. In the case of incentive-based DR programs, targeting suitable customers takes priority before DR implementation [2]. According to the peak time rebate program implemented by San Diego Gas & Electric (SDG&E), targeted enrollment, which selects suitable customers to participate in incentive-based DR programs, is essential for efficient DR operation [3]. Before DR program introduction, the customer demand characteristics analysis is significant because of heterogeneous characteristics. Especially, the costs of recruiting DR customers may be considerable, as the process involves several activities such as marketing, education, and DR system support and operation. If utility companies or ISO do not select suitable customers for enrollment in the DR program, the losses caused by enrollment of inappropriate customers could be substantial. Therefore, to minimize losses, it is essential to secure a large DR capacity with a relatively small number of customers.

Before choosing suitable customers with potential in electricity consumption and similarity between peak time and event, analyzing the load profiles of customers is essential. We considered the customer targeting concept through analyzing several typical load profiles as a result of load profile segmentation. Therefore, load profile segmentation analysis should be conducted for selecting adequate customers. Various clustering methods are normally employed to perform electricity consumer segmentation. Residential electricity consumption is uncertain and variable due to various factors affecting demand, such as home appliance usage patterns, the number of family members, lifestyle patterns, customer occupations, and income levels. These factors cause residential demand to have far more variability than commercial and industrial demand [4], thus making the residential load profile segmentation problem relatively more difficult. When analyzing load profile clusters, their load patterns or characteristics are commonly applied as variables. However, in residential load profile clustering, only considering load patterns poses a number of problems such as an excessively broad spectrum of hourly consumption rates and different peak occurrence times within the same group, whereas the drawback of only considering load characteristics is that consumer patterns are not reflected accurately. To determine suitable DR participant groups, residential customers should therefore be segmented by both pattern and consumption scales.

This paper proposes a two-stage k-means model to address pattern and consumption scales. In the first stage, k-means clustering is conducted based on load characteristics, such as daily consumption and peak occurrence time. In the second stage, k-means clustering is performed based on hourly load profile of residential customers. This methodology is applied to over 800 Korean residential DR participants, for whom hourly electricity use data is available. The results reveal an appropriate segmentation methodology for DR participants. This paper contributes to the literature on load profile segmentation for targeting customers by:

- Extending the k-means clustering method to reflect all load patterns and characteristics, thus resulting in outstanding performance;
- Deriving home appliances and usage pattern data using only electricity consumption data and not any additional data such as customer information, thus making the analysis more efficient;
- Presenting load profile segmentation of Korean household electricity demand data; and
- Conducting data analysis to suitable select groups for DR.

The remainder of this paper is organized as follows. In Section 2, we illustrate the current state-of-the-art clustering methodology. In Section 3, we present the proposed two-stage k-means model, which ensures effective household load profile segmentation for targeting residential customers. In Section 4, we show the effect of targeting residential customers in the DR program and compare this effect to the effect of opt-in enrollment in Korea. Section 5 concludes and outlines ideas for further research in this area.

2. Literature Review

This section presents a review of the current state-of-the-art methodology for load profile segmentation. Many studies have been performed to segment load profile accurately by applying various clustering methods. K-means, self-organizing maps (SOM), mixture models, expectation maximization (EM), and spectral clustering have been widely used as clustering methods. Among the several methods available for clustering to address load pattern segmentation, the most commonly employed are standard k-means [5–10], adaptive k-means [11,12], fuzzy k-means [13,14], and g-means [15], which is an alternative clustering model to k-means. SOM [16,17] is commonly employed by itself but has also been combined with other clustering methods such as k-means and hierarchical clustering as a hybrid model [18]. Mixture models [19,20] and EM [21] are also popular as statistical clustering methods. For DR program operation, DR customer segmentation is commonly conducted for many reasons. Spectral clustering applying information entropy based piecewise aggregate approximation is proposed for commercial demand response application being able to reflect multiscale similarities [22]. Recently, deep learning based clustering such as deep embedded clustering has become a trend for use in residential baseline estimation [23]. Each of the existing clustering methods normally used for electricity consumer segmentation has its own characteristics and is summarized in Table 1 for each characteristic. As explained in Table 1, each clustering method has its advantages in terms of data type or separation process. Although there are a lot of existing clustering methods, k-means has great strength in that it is easier than other existing models and shows good performance in various problem solving cases.

Table 1. Characteristics of previous clustering methods.

Clustering Method	Characteristics
k-means	1. Easy to implement clustering model 2. Less computation compared with other clustering methods 3. Fast and applicable to a wide range of problems 4. Necessary to specify the number of initial clusters
SOM	1. Excellent clustering result 2. Easy evaluation grouped by visual inspection 3. Necessary to specify the number of initial clusters
Mixture models	1. Ability to model a mixture of both continuous and categorical data 2. Providing probability that a given point belongs to each of the possible clusters
Spectral clustering	1. Allowing more flexible distance metrics and performing well 2. Necessary to specialized machines with large memory to compute full graph Laplacian matrix (quadratic/super quadratic complexities in the number of data point)
Embedded clustering [24]	1. Able to simultaneously learn feature representations and clustering assignments using deep neural networks 2. Less sensitive to the choice of hyper parameters

It also can be used to increase accuracy of customer baseline and select appropriate customers for DR. Zhang et al. [7] proposed clustering by k-means before baseline estimation, and it demonstrated improved results. In regard to addressing clustering structure issues, some studies have employed two-stage clustering methods [14,15] which are similar with the proposed methodology in this study, showing that this structure could reflect all the load factors (i.e., voltage, residential type, consumption, and pattern) better than the structures prevalent in the literature. However, load profile segmentation for DR targeting enrollment was not performed in these studies, and they were just focused on similar patterns in groups, which has the limitation of large variation in customer daily consumption. It is hard to use the existing models as it is in this study. Therefore, we considered the two-stage methodology to reflect load characteristics affecting DR at the first stage.

Commonly, optimization methods are utilized for customer targeting in DR program, and there are some studies on this without load profile segmentation [25,26]. Kwac et al. [25] proposed solving the stochastic knapsack problem (SKP) as a means to recruit optimal customers for DR programs. Zhou et al. [26] designed an adaptive targeting method to estimate DR effects.

This paper describes a customer targeting and DR analysis model through a two-stage clustering analysis. The proposed methodology will enable the effective selection of customers for DR programs and illustrate a better DR effect than in opt-in enrollment.

3. Targeting Customers for Incentive DR Using a Two-stage Load Profile Clustering Method

Selecting and recruiting appropriate customers for DR programs is essential for the successful operation of incentive-based DR. DR potential can be estimated by analyzing customer load characteristics. In this study, we derived adequate customer groups for residential DR from demand data through the load profile segmentation. There are many methods for clustering such as k-means, SOM, fuzzy clustering, Gaussian Mixture Models (GMMs), and hierarchical clustering. We adopted k-means methods in view of simplicity and accuracy, and designed load profile segmentation framework as two-stage methodology considering load characteristics in the first step and load profile value in the second step.

3.1. k-means

k-means is a popular method for cluster analysis in data mining that is commonly employed to study electricity demand clustering. It is a simple and robust algorithm which aims to separate n observations into k clusters [15,27]. When a dataset $X = \{x_1, x_2, \ldots, x_N\}$ (with $x_i \in \mathbb{R}^n$) and K clusters $C = \{C_1, C_2, \ldots, C_K\}$ are given, each $x_i \in X$ is assigned to exactly one cluster $C_k \in C$, which is characterized by a cluster centroid μ_k. The classical k-means clustering method is performed as follows. First, the integer value K corresponding to the number of clusters is determined. Then, the initial cluster centroid set $\{\mu_1, \mu_2, \ldots, \mu_K\}$ is selected randomly. Data point $x_i \in X$ is assigned to the closest μ_k through distance comparison against $\{\mu_1, \mu_2, \ldots, \mu_K\}$ using the Euclidean distance. The formula for setting the data set in clusters is illustrated by Equation (1):

$$cluster(x_i) = \underset{k \in \{1,\ldots,K\}}{\mathrm{argmin}} \|x_i - \mu_k\|^2 \tag{1}$$

The clustering algorithm aims to minimize the sum of squares within the groups and maximize it between the groups. The cost function J to be minimized in k-means is therefore expressed by Equation (2):

$$J = \frac{1}{N} \sum_{k=1}^{K} \sum_{x_i \in C_k} \|x_i - \mu_k\|^2 \tag{2}$$

The cluster centroid set update is performed by calculating the mean data set belonging to cluster C_k as given by Equation (3):

$$\mu_k = \frac{1}{|C_k|} \sum_{x_i \in C_k} x_i \tag{3}$$

This process is repeated until the distribution of the dataset among the clusters no longer changes. In other words, cluster centroids do not change.

3.2. Methodology for Customer Targeting Based on Two-Stage Clustering Method in Efficient DR Operation

The framework used to segment customers into groups based on load profile and to determine appropriate groups for incentive-based DR program participation is depicted in Figure 1. First, load data is collected for load profile clustering. Subsequently, we perform data preprocessing comprising data selection (i.e., exclude weekends, holidays, and event days from the data) and cleansing (i.e., replace missing data and delete incomplete customer data). After data preprocessing, a two-stage load profile clustering is performed to segment residential DR customers in accordance with electricity consumption characteristics and their load profile.

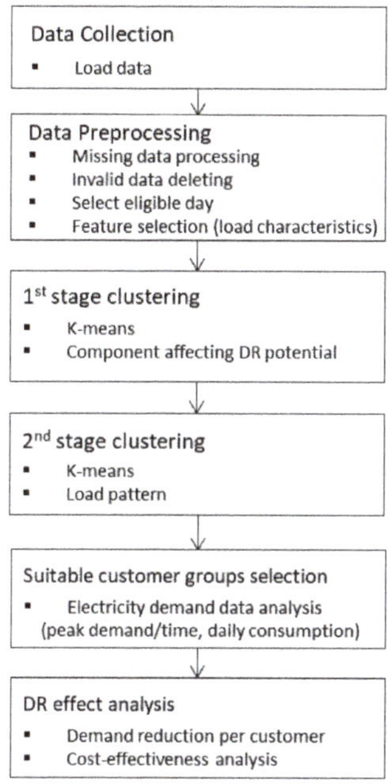

Figure 1. Flowchart of the proposed two-stage clustering methodology and customer targeting strategy.

Load profile including information such as peak time, duration, and electricity consumption can estimate approximately how much customers can reduce their capacity, so this information could be an important factor for determining which customers can reduce the most demand during the implementation of the DR program. These characteristics should be extracted from the load profile and treated as variables in the clustering method. Therefore, the characteristics (i.e., daily consumption, peak time) are considered in the first stage of clustering. In the second stage, the classification variable is the normalized load profile. Suitable DR participation groups are then derived by analyzing the segmentation results. Distributions of peak time, average consumption, and peak demand scale could be obtained from this analysis. After selecting the target groups, a DR effect analysis is conducted to verify the effect of targeted enrollment. This analysis shows the demand reduction capacity per customer of the targeted enrollment, and these results are compared with the results obtained assuming opt-in enrollment into the DR program. When the clustering method is applied, considering many variables does not always produce reliable results. Therefore, it is necessary to include the essential variables strategically. However, if there are too many variables to segment customers well, a method to deal with this problem should be devised. In this study, we improve load profile clustering performance by applying our proposed methodology. Figure 2 explains the proposed two-stage load profile clustering algorithm.

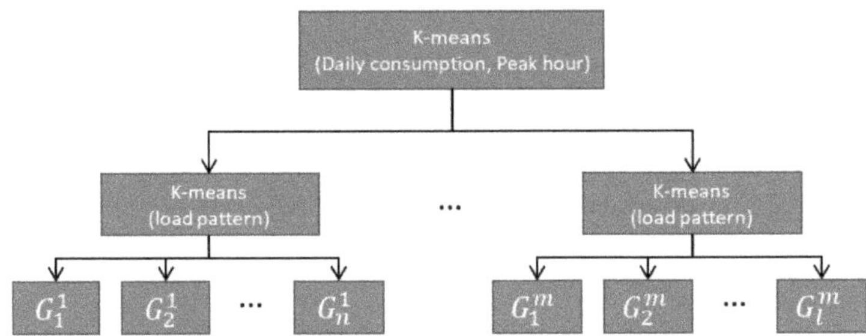

Figure 2. Two-stage clustering methodology for load profile segmentation.

Before load profile segmentation, load characteristics should be found from load profile by using feature selection (being the process of selection of a subset of relevant features). Features used for cluster input variables are selected through correlation analysis. When we derive relevant features from load profile, we consider factors (i.e., daily consumption, peak time, difference between peak demand and minimum demand) affecting effective DR operation.

The next step is normalization of load characteristics for 1st stage segmentation and load profile for 2nd stage segmentation instead of using raw data. Normalization transforms the load to a number from 0 to 1 and can provide better performance by changing the value of input data. The normalization about load characteristics was conducted on the basis of each variable. On the other hand, the normalization about load profile was used in accordance with each customer. Min–max normalization was used, as illustrated by Equation (4) in the case of load profile normalization:

$$\widetilde{d}_{i,t} = \frac{d_{i,t} - \min_{t \in T}(d_{i,t})}{\max_{t \in T}(d_{i,t}) - \min_{t \in T}(d_{i,t})} \tag{4}$$

where i, t, $d_{i,t}$, and $\widetilde{d}_{i,t}$ are customers, time, demand of customer i at time t, and the min–max normalization result, respectively.

After the normalization process, segmentation based on load characteristics is preceded by the k-means method before the load profile segmentation as explained in Figure 2. This process separates customers based on their consumption scale and peak times. In other words, it is a process to segment customers over a large range. The reason why these components are chosen is that consumption scale would be an indicator to estimate how much customers can reduce their demand, and the customer's peak time occurrence during an event indicates whether customers stay at home. The next step is customer segmentation based on load profiles, which is conducted for all members of each group following the first-stage clustering analysis. The effect of separation as two-stage k-means clustering is that features can be better reflected as compared to basic k-means.

The main goal of this analysis is to determine a way to produce the most significant effect with suitable customers enrolled in the DR program. To achieve this goal, we propose two standards to select customer groups with high DR potential. If a peak demand event occurs, the likelihood of customers staying in their homes is relatively high. It may be argued that the corresponding customers tend to be able to reduce their demand effectively. However, this is not an absolute indicator. In some cases, for instance, the demand of some customers could be high although peak demand time may not remain constant, or some customers may register an insignificant demand reduction although peak demand times remain constant. Therefore, we stipulate the following criteria to determine the target groups:

1. Customer groups with high demand consumption;
2. Customers groups who have similar peak demand times with an event.

After the load profile segmentation, result analysis through a boxplot chart is adopted as a method of excluding customer groups who are inappropriate customers in DR program participation.

3.3. Internal Evaluation of the Clustering Method

After performing customer segmentation via clustering, the accuracy the clustering result should be assessed. Evaluation methods are commonly divided into external and internal processes [28]. In external evaluation, the result is assessed by a comparison with the actual value. Internal evaluation is normally used when the data does not contain actual values; thus, the assessment is based on the idea that good results have minimum distance within clusters and maximum distance between clusters (i.e., high intracluster similarity and low intercluster similarity). Although there are many evaluation methods available, we consider only internal evaluations, since they are used to measure the goodness of clustering evaluation structure without respect to external information (i.e., labels or actual results). Among these, we use the Davies–Bouldin index (DBI) and Dunn index (DI). The DBI is an internal evaluation method to quantify clustering quality. It evaluates customer segmentation based on the similarity between clusters and is calculated as follows:

$$DBI = \frac{1}{n} \sum_{i=1}^{n} \max_{j \neq i} \frac{\sigma_i - \sigma_j}{d(\mu_i, \mu_j)} \tag{5}$$

where n, μ_i, σ_i, and $d(\mu_i, \mu_j)$ are the number of clusters, the centroid of cluster i, average distance between μ_i and all objects in cluster i, and the Euclidean distance between μ_i and μ_j, respectively. Its output is a single number, and clustering algorithms with lower output values indicate better performance.

The DI is another internal evaluation method to quantify clustering quality. The indicator measures how well clusters are separated and how dense they are. It can be formulated as follows:

$$DI = \frac{\min\limits_{1 \leq i < j \leq n} d(i, j)}{\max\limits_{1 \leq k \leq n} d'(k)} \tag{6}$$

where $d(i, j)$ and $d'(k)$ are the distances between centroids of cluster i and j and between objects within cluster k, respectively. Its output is a single number, and clustering algorithms with larger output values indicate improving performance.

3.4. Cost-effective Analysis

From the perspective of operation research (OR), cost-effective analysis is an important component. The effectiveness of customer targeting through load profile segmentation in DR operation is operation cost reduction. Thus, we need to identify the amount of cost variation compared with opt-in and targeting recruitment. We confirmed it by using the cost-effectiveness test which is one of the economic analysis methods usually performed before public project investment [29]. It is divided into Total Resource Cost (TRC), Program Administrator Cost (PAC), Ratepayer Impact Measure (RIM), and Participant Cost Test (PCT). We considered the PAC test to recognize the cost effect according to DR customer targeting in perspective of DR operator (i.e., utility or ISO). The cost and benefits list should be defined before economic effectiveness estimation. The list for analysis from the perspective of DR operators can be specified in Table 2.

To identify whether the utility project is appropriate for investment, each cost/benefit item should be calculated. If the cost-effectiveness test result has a positive value, it represents that the project has profit. The project is a nonprofitable business in the opposite case.

Avoided energy costs is the benefit of decreasing the amount of power purchased in accordance with electricity consumption reduction. It can be formulated as follows:

$$AEB = ER \times ARU \tag{7}$$

where *ER* and *ARU* are the amount of power reduction and the unit cost of energy avoidance (i.e., average system marginal price (SMP) during DR event), respectively.

Table 2. Benefits and cost list for cost-effectiveness analysis in DR operation.

Index	Benefit/Cost Lists
Benefits	Avoided energy costs
	Avoided transmission and distribution cost
	Revenue gain/loss from changes in sales
	Incentives paid
Cost	DR System operation cost
	Measure and evaluation cost
	Marketing and education cost

Avoided transmission and distribution cost is the benefit reducing demand for transmission and distribution construction as a result of decreasing annual peak demand. It can be formulated as follows:

$$ATDB = PR \times (ATU + ADU) \tag{8}$$

where *PR*, *ATU*, and *ADU* are peak reduction capacity in power system, unit cost of transmission construction avoidance, and unit cost of distribution construction avoidance, respectively.

In the case of the cost list, it contains the cost of revenue loss from changes in sales, incentives, DR system operation, measure, evaluation, marketing, and education. Revenue loss from changes in sales is a cost as the utility company cannot provide power to customers as an amount of DR reduction. Incentive paid cost is cost for utility companies to provide incentives to DR participants as a result of demand reduction. Measurement, evaluation, marketing, and education cost are included in DR operation cost and we assume that these costs are calculated proportionate to the number of DR customers.

4. Load Profile Segmentation for Effective DR Program Operation in Korea

DR options in Korea have mostly been unavailable to residential customers and have been implemented only for commercial and industrial customers. However, utility companies have recently attempted to attract residential customers by changing their policies and opening DR programs to them. The Korea Electric Power Corporation (KEPCO) which is a utility in Korea also conducted a peak-time rebate (PTR) pilot program from November 2017 to February 2018 in 10 events to develop an appropriate residential DR program in Korea [30]. It was performed with about 800 residential customers living in Seoul, Korea. The PTR program was designed based on incentive-based DR to mitigate peak demand by reducing participant demand in accordance with the utility's notification. After the DR event, the PTR provides incentives based on the amount of demand reduction achieved after participants receive a notification to reduce their demand. It does not have any penalty in the case of the PTR program and can make customers who pay a flat electricity price realize that the electricity price has a time-varying rate system.

Although this PTR program is designed for opt-in customers, targeted enrollment to select residential customers with high DR potential is necessary to improve the benefits of the DR program. Therefore, we analyze residential customer demand data from the PTR pilot program and apply the two-stage clustering methodology discussed in the previous section. From this study, we obtain customer clusters according to load pattern and consumption and select suitable groups for efficient DR operation through an analysis of group characteristics. Finally, we identify the actual demand reduction effect in the case of opt-in operation and targeted enrollment operation by applying residential customer data during an actual PTR event.

4.1. Input Data

This study was conducted using residential demand data. We obtained residential hourly demand data in the Korea Electric Power Cooperation (KEPCO) service area where residents live. This data covered 847 residential customers, all of whom participated in the PTR pilot program. The data covered the period from November 2017 through February 2018, during which time the PTR pilot program operated from mid-January to the end of February. The PTR events occurred throughout nine days from 17:00 to 20:00. The average hourly demand from residential PTR program customers in Seoul, Korea is illustrated in Figure 3.

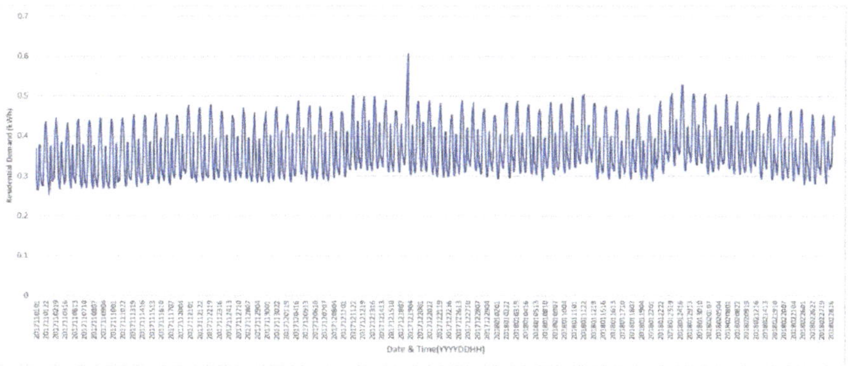

Figure 3. Average hourly demand for residential PTR program customers in Seoul, Korea, from November 2017 through February 2018.

4.2. Data Preprocessing: Feature Selection

In the data, preprocessing, missing data imputation, deleting invalid data, selecting eligible days, and reducing dimensions are performed to obtain reasonable results. We conducted a feature selection as part of the dimension reduction for DR potential. When we select features, we consider which factor affects to DR reduction as follows:

1. Is the customer living at home and contributing to peak reduction during the DR event?
2. Are there any incentives to reduce their demand due to large usage?
3. Is the large capacity that can be reduced compared to the base load?

Therefore, we selected three features (i.e., daily consumption, peak hour, difference between maximum and minimum demand) from the load profile as principal factors. Deleting features through correlation analysis between these features should be processed. As a result of the correlation analysis, we selected two features (daily consumption and peak hour) for 1st stage clustering based on demand characteristics. The correlation analysis between demand characteristics' features is illustrated in Figure 4.

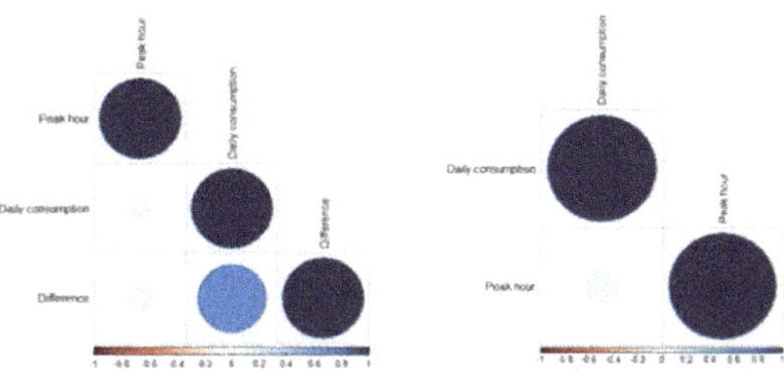

Figure 4. Feature selection from load profile characteristics of residential customers.

4.3. Load Profile Segmentation of Residential DR Customers

When load profile clustering is conducted for customer segmentation, it is essential to determine the optimal number of clusters in the data. We used the NbClust package in the R statistical software to estimate the number of clusters, following Charrad et al. [31]. This package provides 30 indices which determine the number of clusters in a data set and offers the best clustering scheme [32]. Hubert statistics values and Dunn index values are also provided by NbClust. These numbers provide a graphical method to determine the number of clusters. We can realize that the number of optimal clusters is situated in a peak point in their plots of second differences, indicating that the number of optimal clusters is six when first-stage clustering based on demand characteristics (i.e., daily consumption, peak demand time, and difference between peak and minimum demand) was conducted. Figure 5 shows the Hubert index and D index results.

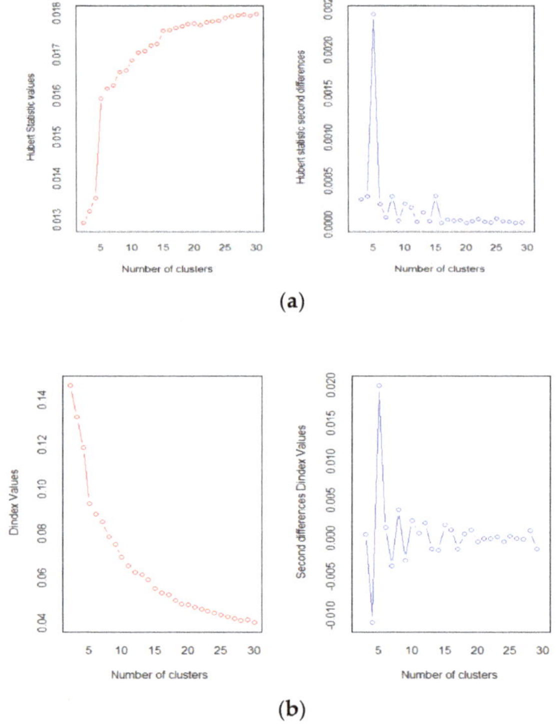

Figure 5. Examination methods to determine the number of clusters by (a) the Hubert index and (b) DI index.

After the first-stage clustering, second-stage clustering for customer segmentation based on demand patterns was conducted. The optimal number of clusters for each of the six groups separated by demand characteristics were 3, 2, 2, 2, 2, and 2. Therefore, we separated residential customers, who participate in the PTR pilot program into 13 groups according to load patterns and consumption.

We performed load profile segmentation through our proposed method and then compared the resulting customer segmentation according to different clustering models. We examined 12 methods, including our proposed method. The remaining clustering methods are based on the fundamental k-means, SOM [16,17,33], and FCM [34–36] methodology in which the classification variables are: (1) demand characteristics, (2) load patterns, and (3) both characteristics and load patterns.

To compare the results, the internal evaluation measures Davis-Bouldin index (DBI) and Dunn index (DI) were used. The DBI and DI result of clustering methods were presented as Table 3. The proposed methodology showed the best result according to Table 3, so we conclude that our proposed methodology is indeed appropriate. We judged that the reason why the proposed methodology has a better result can be explained as follows. It is at a point that separation as two-stage clustering framework can reflect each feature impact, considering that factors affecting DR reduction in 1st stage segmentation make it so that rough load profile clustering before 2nd stage segmentation separates each feature impact by its pattern. Generally, clustering methods separate data based on the distance of input variables. Therefore, the undesirable result would be presented if a lot of input variables which can make each variable effect difficult to verify are used unnecessarily. However, the proposed methodology considers all variables by separating the clustering method into two stages. It can make an outstanding result in two-stage k-means clustering. The proposed method separates residential customers into 13 groups, with the load profiles of each group illustrated in Figure 6. Groups 1 through 13 contain 14, 15, 25, 76, 56, 38, 85, 120, 88, 98, 68, 85, and 79 customers, respectively. The load profile of the 13 groups showed morning peak, evening peak, nighttime peak, and dual morning and night peaks. Residential customers do not usually consume electricity during daytime, so these peak characteristics were consistent with residential load profiles.

Table 3. DR operation clustering result evaluation according to variable selection and clustering structure.

Index	DBI	DI	Ranking
One-stage k-means (Characteristics)	0.5007	1.0084	2/10
One-stage k-means (Characteristics, Pattern)	1.2108	1.2065	3/9
One-stage k-means (Pattern)	1.3493	1.2481	10/7
Two-stage k-means (Characteristics, Pattern)	0.3637	1.8804	1/1
One-stage SOM (Characteristics)	1.2456	1.2298	7/8
One-stage SOM (Characteristics, Pattern)	1.2804	1.3270	9/6
One-stage SOM (Pattern)	1.2669	1.3330	8/5
Two-stage SOM (Characteristics, Pattern)	1.8948	0.7824	11/12
One-stage FCM (Characteristics)	1.2400	1.4659	5/3
One-stage FCM (Characteristics, Pattern)	1.2340	1.5159	4/2
One-stage FCM (Pattern)	1.2437	1.4554	6/4
Two-stage FCM (Characteristics, Pattern)	2.0115	0.7922	12/11

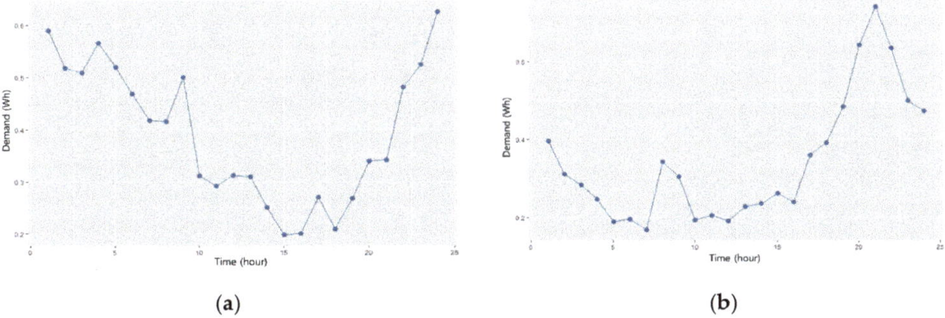

(a)   (b)

Figure 6. *Cont.*

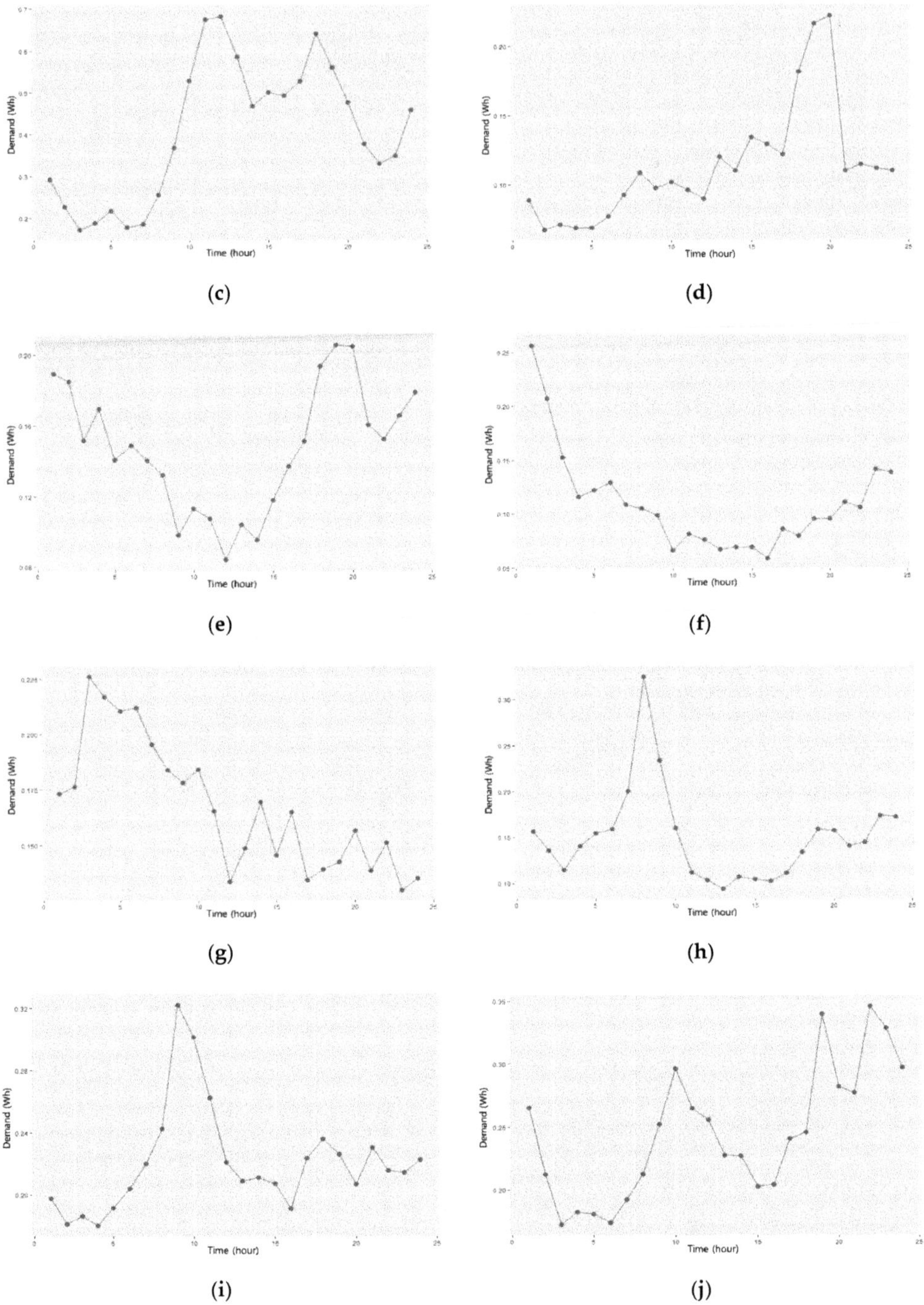

Figure 6. *Cont.*

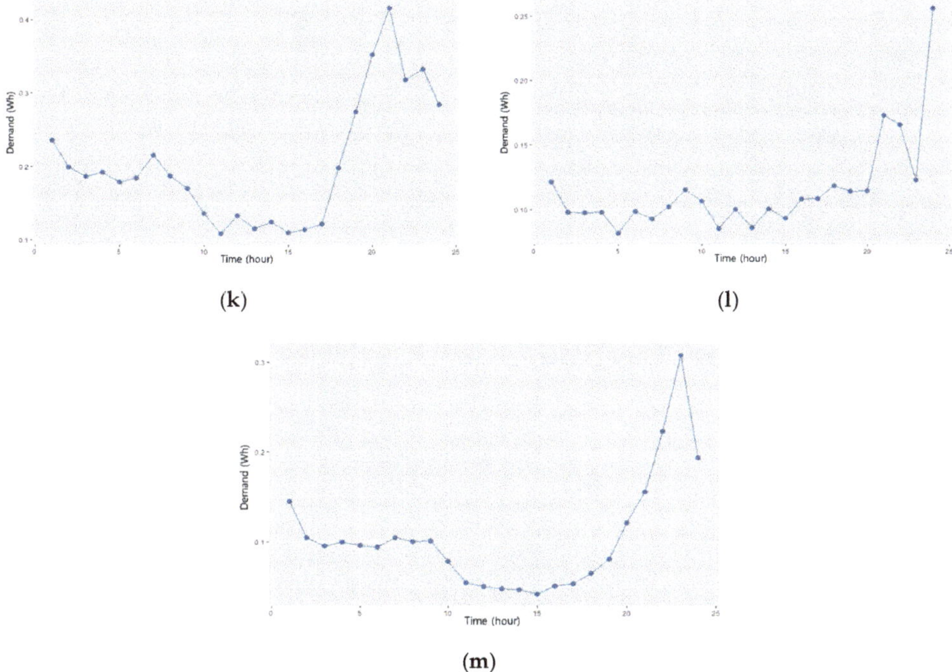

Figure 6. Load profiles of customer segmentation results by the proposed clustering methodology: (**a**) group 1; (**b**) group 2; (**c**) group 3; (**d**) group 4; (**e**) group 5; (**f**) group 6; (**g**) group 7; (**h**) group 8; (**i**) group 9; (**j**) group 10; (**k**) group 11; (**l**) group 12; (**m**) group 13.

4.4. Customer Targeting for DR Operation

Appropriate customer selection for DR participation by using the load profile segmentation result in the previous section is applied to study efficient DR operation. The 13 load patterns are shown in Figure 7. It can be possible to use load profile and the amount of consumption to estimate DR potential from the 13 load pattern. If customers consume little electricity, their DR participation would be inefficient to a DR operator, despite having a suitable pattern for DR (i.e., nighttime peak and dual morning and night peaks). Therefore, the DR operator considers both factors. To reflect these components, a boxplot analysis is conducted. Peak time (i.e., hour of the day) when maximum demand happens and average consumption boxplots for the 13 groups were analyzed, and they are illustrated in Figure 8.

First, we eliminated groups which experienced inconsistent peak demand occurrence times as the events. The groups not corresponding to this criterion were 6, 7, 8, and 9. Then, groups with low electricity consumption were also deleted, as they are inappropriate for economic purposes. The groups corresponding to little consumption were 4, 5, 12, and 13. We emphasize that utility companies should operate the PTR program using the remaining groups, namely, groups 1, 2, 3, 10, and 11. The total number of customers included in this targeted enrollment scenario was 220.

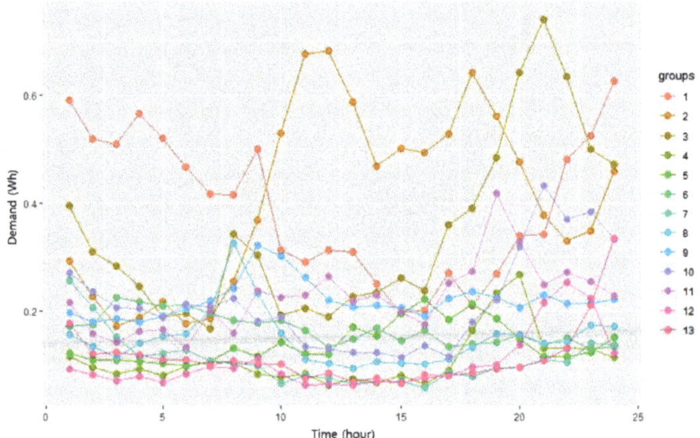

Figure 7. Load profile of customer clusters by the proposed method.

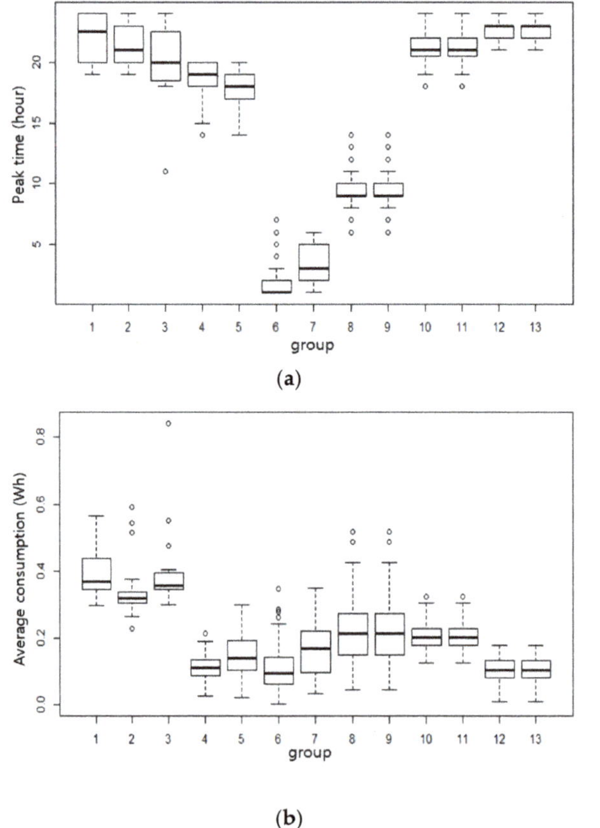

Figure 8. Boxplot of the proposed clustering method: (**a**) peak time, and (**b**) average consumption.

After finding targeting groups, calculating the amount of demand reduction was performed to identify the effect in accordance with targeting customers by using the actual PTR event data.

There were 847 DR participation customers and there were nine event days when the utility company notified residential PTR customers to reduce demand. We considered that all of residential customer (i.e., 847 customers) are participated in PTR pilot program in case of Opt-in enrollment, and targeted enrollment is attracted by a group of customers who are able to reduce their demand more than other groups during the event. There were 220 customers in the targeted enrollment group, which is different from the number of customers in the opt-in enrollment group. To compare the demand reduction for both types of enrollment, we calculated average demand reduction of event days per customer in both cases. As the customer baseline load (CBL) should be estimated for demand reduction capacity due to the DR event, we applied the Max 4 of 5 method, which has been used for the PTR program in Korea [30]. The Max 4 of 5 method estimates CBL by averaging high demand of four days among five eligible days, which means days excluding weekends, event days, and holidays. Average demand reduction for event days per customer in opt-in enrollment, targeted enrollment, and the 13 groups are illustrated in Table 4. Average demand reductions for the opt-in enrollment and targeted enrollment program were 0.2620 (kWh) and 0.3496 (kWh), and the difference between them was 0.0876 (kWh).

Table 4. Average demand reduction for event days per residential customers during the Peak Time Rebate (PTR) pilot program.

Groups	The Number of PTR Participants	Average Demand Reduction (kWh)
Opt-in enrollment	847	0.2620
Targeted enrollment	220	0.3496
Group 1	14	0.3646
Group 2	15	0.4177
Group 3	25	0.3402
Group 4	76	0.2690
Group 5	56	0.2365
Group 6	38	0.1914
Group 7	85	0.2362
Group 8	120	0.2862
Group 9	88	0.2544
Group 10	98	0.3129
Group 11	68	0.3126
Group 12	85	0.1985
Group 13	79	0.1990

The electricity consumption for 6~8PM was 1.3569 (kWh). The demand reduction ratio based on common demand during events was 19.31% and 25.76%, respectively. An improvement of 6.45% was observed, with targeted enrollment reduction increasing demand reduction by 33.44%, in comparison with opt-in enrollment. Thus, it is significantly more efficient to operate the DR program with customers who have larger DR potential, as defined in this study.

Additionally, we conducted a cost-effectiveness analysis for managing the DR program in two cases: residential customers who want to participate in the DR program and targeted residential customers who have large DR potential. We assume that the demand reduction of targeted customers is the same as the actual DR participants in identifying the cost-effectiveness of DR customer targeting. Economic analysis based on the California Standard Practice Manual is performed from the perspective of the DR operator [29]. There were 847 total households participating in the PTR pilot program whose total average reduction is 221.914 kWh, and 635 households (which comprise 75% of the total participants) that we determined as DR targeting participants.

Customer operation cost decreased due to the reduced number of customers, and the amount of increased benefit is 437.256 KRW, the exchange rate is 1100 KRW, marking a 108.58% benefit increase over the existing economic analysis result. The economic analysis changes by customer targeting is presented as Table 5.

Table 5. Cost-Effectiveness Analysis Changes by Incentive DR Targeting (Unit: KRW).

	Benefits and Cost List	No Targeting Customers (Before)	Targeting Customers (After)
Benefits	Avoided energy costs	861,043	861,043
	Avoided transmission and distribution cost	11,589,254	11,589,254
	Revenue gain/loss from changes in sales	1,335,866	1,335,866
	Incentives paid	4,271,845	4,271,845
Cost	DR System operation cost	1,131,856	848,245
	Measure and evaluation cost	437,926	328,194
	Marketing and education cost	175,248	131,335
	Utility company benefit result	5,097,557	5,534,813
	Difference	437,256	

5. Conclusions

We presented an appropriate DR customer selection methodology for a Korean residential DR program to maximize the DR effect with lower customer enrollment. The proposed method showed better performance than other methods. Our method is divided into two parts. The first is customer segmentation according to load profile and consumption, and the second is targeted group selection based on two standards for DR participation. When we conducted customer segmentation, a two-stage clustering method was introduced. Customers were clustered by demand characteristics as variables in the first stage, and then segmented based on load patterns in the second stage. It can reflect more features of residential demand data than existing clustering methods, that makes better result in customer segmentation. Customer groups were classified as having higher DR potential by peak time and consumption patterns to select adequate groups having large potential in PTR program. As a result, the targeted groups were 1, 2, 3, 10, and 11 in our sample of residential customers in Korea, and their average demand reduction was 0.3496 (kWh), for an improvement of approximately 0.0876 (kWh), which increased savings by 33.44% compared to demand reduction due to opt-in enrollment. The proposed method allowed identifying enhanced DR effects. After the DR targeting demand reduction, we also conducted the cost-effective analysis of the PTR program from the perspective of the DR operator.

As a result, we observed that targeted DR capacity may be achieved with a small number of customers if targeted enrollment is implemented, which can use infrastructure and operation costs effectively. These results provide insights into the efficient use of DR in Korea. The number of customers and total DR capacity of targeted enrollment decreased compared with opt-in enrollment. However, if the number of customers who would like to participate in the DR program is high enough when the official full-scale program starts, selecting optimal customers among them would be more highly important. Therefore, the proposed method would be of great help in ensuring an efficient and economically sensible DR program in Korea.

We considered the residential customer targeting based on customer segmentation in demand response in this paper. Customer segmentation focus on the model structure to reflect features affecting demand response well. Some researches consider clustering model with heuristic algorithm in other areas, so we will apply this concept in further study.

Author Contributions: Conceptualization, E.L. and J.K.; data analysis, simulation, and methodology framework development, E.L.; writing, review, and editing, J.K.; supporting data collection and comments for improving the article, D.J. All authors have read and agreed to the published version of the manuscript.

Funding: This work was supported by GIST Research Institute(GRI) grant funded by the GIST in 2020, and this work was supported by the Korea Institute of Energy Technology Evaluation and Planning(KETEP) and the Ministry of Trade, Industry & Energy(MOTIE) of the Republic of Korea (No. 20191210301930).

Conflicts of Interest: The authors declare no conflict of interest.

References

1. U.S. Department of Energy (DOE). *Benefits of Demand Response in Electricity Markets and Recommendations for Achieving Them: A Report to the United States Congress Pursuant to Section 1252 of the Energy Policy Act of 2005*; U.S. DOE: Washington, DC, USA, 2006.
2. Wang, Y.; Chen, Q.; Kang, C.; Zhang, M.; Wang, K.; Zhao, Y. Load profiling and its application to demand response: A review. *Tsinghua Sci. Technol.* **2015**, *20*, 117–129. [CrossRef]
3. George, S.; Bode, J.; Berghman, D. 2012 San Diego Gas & Electric Peak Time Rebate Baseline Evaluation; California Measurement Advisory Council. 2012. Available online: http://www.calmac.org/publications/SDGE_PTR_Baseline_Evaluation_Report_-_Final.pdf (accessed on 13 September 2018).
4. Mohajeryami, S.; Doostan, M.; Schwarz, P. The impact of Customer Baseline Load (CBL) calculation methods on Peak Time Rebate program offered to residential customers. *Electr. Power Syst. Res.* **2016**, *137*, 59–65. [CrossRef]
5. Rhodes, J.D.; Cole, W.J.; Upshaw, C.R.; Edgar, T.F.; Webber, M.E. Clustering analysis of residential electricity demand profiles. *Appl. Energy* **2014**, *135*, 461–471. [CrossRef]
6. Park, S.; Ryu, S.; Choi, Y.; Kim, J.; Kim, H. Data-driven baseline estimation of residential buildings for demand response. *Energies* **2015**, *8*, 10239–10259. [CrossRef]
7. Zhang, Y.; Chen, W.; Xu, R.; Black, J. A cluster-based method for calculating baselines for residential loads. *IEEE Trans. Smart Grid* **2016**, *7*, 2368–2377. [CrossRef]
8. Tureczek, A.; Nielsen, P.S.; Madsen, H. Electricity Consumption Clustering Using Smart Meter Data. *Energies* **2018**, *11*, 859. [CrossRef]
9. Cominola, A.; Spang, E.S.; Giuliani, M.; Castelletti, A.; Lund, J.R.; Loge, F.J. Segmentation analysis of residential water-electricity demand for customized demand-side management programs. *J. Clean. Prod.* **2018**, *172*, 1607–1619. [CrossRef]
10. Wang, F.; Li, K.; Liu, C.; Mi, Z.; Shafie-Khah, M.; Catalão, J.P. Synchronous Pattern Matching Principle-Based Residential Demand Response Baseline Estimation: Mechanism Analysis and Approach Description. *IEEE Trans. Smart Grid* **2018**, *9*, 6972–6985. [CrossRef]
11. Kwac, J.; Flora, J.; Rajagopal, R. Household energy consumption segmentation using hourly data. *IEEE Trans. Smart Grid* **2014**, *5*, 420–430. [CrossRef]
12. Kwac, J.; Tan, C.-W.; Sintov, N.; Flora, J.; Rajagopal, R. Utility customer segmentation based on smart meter data: Empirical study. In Proceedings of the 2013 IEEE International Conference on Smart Grid Communications (SmartGridComm), Vancouver, BC, Canada, 21–24 October 2013; pp. 720–725.
13. Ozawa, A.; Furusato, R.; Yoshida, Y. Determining the relationship between a household's lifestyle and its electricity consumption in Japan by analyzing measured electric load profiles. *Energy Build.* **2016**, *119*, 200–210. [CrossRef]
14. Tsekouras, G.J.; Hatziargyriou, N.D.; Dialynas, E.N. Two-stage pattern recognition of load curves for classification of electricity customers. *IEEE Trans. Power Syst.* **2007**, *22*, 1120–1128. [CrossRef]
15. Mets, K.; Depuydt, F.; Develder, C. Two-stage load pattern clustering using fast wavelet transformation. *IEEE Trans. Smart Grid* **2016**, *7*, 2250–2259. [CrossRef]
16. Verdú, S.V.; Garcia, M.O.; Senabre, C.; Marín, A.G.; Franco, F.G. Classification, filtering, and identification of electrical customer load patterns through the use of self-organizing maps. *IEEE Trans. Power Syst.* **2006**, *21*, 1672–1682. [CrossRef]
17. Räsänen, T.; Ruuskanen, J.; Kolehmainen, M. Reducing energy consumption by using self-organizing maps to create more personalized electricity use information. *Appl. Energy* **2008**, *85*, 830–840. [CrossRef]
18. Räsänen, T.; Voukantsis, D.; Niska, H.; Karatzas, K.; Kolehmainen, M. Data-based method for creating electricity use load profiles using large amount of customer-specific hourly measured electricity use data. *Appl. Energy* **2010**, *87*, 3538–3545. [CrossRef]
19. Haben, S.; Singleton, C.; Grindrod, P. Analysis and clustering of residential customers energy behavioral demand using smart meter data. *IEEE Trans. Smart Grid* **2016**, *7*, 136–144. [CrossRef]
20. Granell, R.; Axon, C.J.; Wallom, D.C. Impacts of raw data temporal resolution using selected clustering methods on residential electricity load profiles. *IEEE Trans. Power Syst.* **2015**, *30*, 3217–3224. [CrossRef]
21. Labeeuw, W.; Stragier, J.; Deconinck, G. Potential of active demand reduction with residential wet appliances: A case study for Belgium. *IEEE Trans. Smart Grid* **2015**, *6*, 315–323. [CrossRef]

22. Shunfu, L.; Fangxing, L.; Erwei, T.; Yang, F.; Dongdong, L. Clustering Load Profiles for Demand Response Applications. *IEEE Trans. Smart Grid* **2019**, *10*, 1599–1607.

23. Mingyang, S.; Yi, W.; Fei, T.; Yujian, Y.; Goran, S.; Chongqing, K. Clustering-Based Residential Baseline Estimation: A Probabilistic Perspective. *IEEE Trans. Smart Grid* **2019**, *10*, 6014–6028.

24. Junyuan, X.; Ross, G.; Ali, F. Unsupervised deep embedding for clustering analysis. *Int. Conf. Mach. Learn.* **2016**, 478–487.

25. Kwac, J.; Rajagopal, R. Data-driven targeting of customers for demand response. *IEEE Trans. Smart Grid* **2016**, *7*, 2199–2207. [CrossRef]

26. Zhou, D.P.; Balandat, M.; Tomlin, C.J. Estimation and Targeting of Residential Households for Hour-Ahead Demand Response Interventions–A Case Study in California. In Proceedings of the 2018 IEEE Conference on Control Technology and Applications (CCTA), Copenhagen, Denmark, 21–24 August 2018; pp. 18–23.

27. Hartigan, J.A.; Wong, M.A. Algorithm AS 136: A k-means clustering algorithm. *J. R. Stat. Soc. Ser. C (Appl. Stat.)* **1979**, *28*, 100–108. [CrossRef]

28. Aghabozorgi, S.; Shirkhorshidi, A.S.; Wah, T.Y. Time-series clustering–A decade review. *Inf. Syst.* **2015**, *53*, 16–38. [CrossRef]

29. California Public Utilities Commission (CPUC). Demand Response Cost-Effectiveness Protocols. Available online: https://www.cpuc.ca.gov/General.aspx?id=7023 (accessed on 7 August 2019).

30. Lee, E.; Jang, D.; Kim, J. A Two-Step Methodology for Free Rider Mitigation with an Improved Settlement Algorithm: Regression in CBL Estimation and New Incentive Payment Rule in Residential Demand Response. *Energies* **2018**, *11*, 3417. [CrossRef]

31. Charrad, M.; Ghazzali, N.; Boiteau, V.; Niknafs, A.; Charrad, M.M. Package 'NbClust'. *J. Stat. Softw.* **2014**, *61*, 1–36. [CrossRef]

32. Charrad, M.; Ghazzali, N.; Boiteux, V.; Niknafs, A. NbClust: An R Package for Determining the Relevant Number of Clusters in a Data Set. *J. Stat. Softw.* **2014**, *61*, 1–36. [CrossRef]

33. Wehrens, R.; Buydens, L.M. Self-and super-organizing maps in R: The Kohonen package. *J. Stat. Softw.* **2007**, *21*, 1–19. [CrossRef]

34. Bezdek, J.C.; Ehrlich, R.; Full, W. FCM: The fuzzy c-means clustering algorithm. *Comput. Geosci.* **1984**, *10*, 191–203. [CrossRef]

35. Cebeci, Z.; Yildiz, F.; Kavlak, A.T.; Cebeci, C.; Onder, H. Ppclust: Probabilistic and Possibilistic Cluster Analysis. 2019. Available online: https://CRAN.R-project.org/package=ppclust (accessed on 17 February 2020).

36. Pereira, R.; Fagundes, A.; Melício, R.; Mendes, V.M.F.; Figueiredo, J.; Martins, J.; Quadrado, J.C. A fuzzy clustering approach to a demand response model. *Int. J. Electr. Power Energy Syst.* **2016**, *81*, 184–192. [CrossRef]

Article

Evolutionary Analysis for Residential Consumer Participating in Demand Response Considering Irrational Behavior

Xiaofeng Liu [1], Qi Wang [1,*] and Wenting Wang [2]

[1] School of Electrical and Automation Engineering, Nanjing Normal University, Nanjing 210023, China; liuxiaofeng@seu.edu.cn

[2] School of Electrical Engineering and Information Engineering, Lanzhou University of Technology, Lanzhou 730050, China; wwt0706@126.com

* Correspondence: Wangqi@njnu.edu.cn; Tel.: +86-136-0146-9268

Received: 2 September 2019; Accepted: 23 September 2019; Published: 29 September 2019

Abstract: Demand response (DR) has been recognized as a powerful tool to relieve energy imbalance in the smart grid. Most previous works have ignored the irrational behavior of energy consumers in DR project implementation. Accordingly, in this paper, we focus on solving two questions during the execution of DR. Firstly, considering the bounded rationality of residential users, a population dynamic model is proposed to describe the decision behavior on whether to participate in the DR project, and then the evolutionary process of consumers participating in DR is analyzed. Secondly, for the DR participants, they have to compete dispatching amounts for maximal profit in a day-ahead bidding market, hence, a non-cooperative game model is proposed to describe the competition behavior, and the uniqueness of the Nash equilibrium is analyzed with mathematical proof. Then, the distributed algorithm is designed to search the evolutionary result and the Nash equilibrium. Finally, a case study is performed to show the effectiveness of the formulated models.

Keywords: demand response; population evolution; irrational behavior; Markov state; non-cooperative game

1. Introduction

As the development of the economy and society continues, energy demand is growing explosively in all walks of life. Consequently, the power grid has to face the serious challenge in balancing energy supply and demand. Especially in peak demand hours, the tense situation of supply and demand happens from time to time, which affects the stability of the grid. In order to relieve the pressure of energy supply and demand, the power grid can promote energy supply ability by building new power plants or reduce the energy demand of consumers. However, peak demand hours only take a tiny proportion in a whole year, and meanwhile, building new power plants needs a lot of manpower and material resources. Therefore, it is uneconomical to build new plants to solve the energy supply problem in such peak hours. For this reason, demand response (DR), which is one of the core technologies in the smart grid, is taking an increasingly important role in digging up demand-side resources and relieving the tension problem of supply and demand [1,2]. Generally, energy consumers include residential users, commercial users, and industrial users. In which, residential users have abundant flexible resources, hence, residential DR can effectively reduce energy demand in peak hours [3,4].

In recent years, there exists abundant research on the energy consumption scheduling or mechanism design of residential DR [5–7]. The authors in [8] proposed a reward mechanism for residential customers to shave peak loads, in which users' consumption characteristics were modeled by survey questionnaires. In order to aggregate a large number of households in the DR project, Mhanna et al. [9] designed a distributed algorithm from the perspective of the DR aggregator, through which households were aggregated and coordinated as a whole and then scheduled based on the objective of the aggregator. Moreover, Reference [10] studied residential DR with consideration of the power distribution network and the associated constraints, and proposed a distributed scheme where the load service entity and the households interactively communicate to compute an optimal demand schedule. However, the above research lacks consideration on the mutual effect of consumers' strategies and does not capture the dynamic property. To answer this issue, different game-theoretic frameworks have been proposed [11–15]. Authors in [16] formulated an energy consumption scheduling program with game theory, where players are residential users and their strategies are the daily schedules of household appliances. Authors in [17] proposed an event-triggered game-theoretic strategy for managing the power grid's demand side, capable of responding to changes in consumer preferences or the price parameters coming from the wholesale market. Reference [18] adopted a dynamic non-cooperative repeated game with Pareto-efficient pure strategies as the decentralized approach to optimize the energy consumption and energy trading amounts for the next day. Reference [19] focused on an hourly billing mechanism for DR management to solve several theoretical and practical questions, including the uniqueness of the consumption profile corresponding to the Nash equilibrium and the computational issue of the equilibrium profile. While in [20], the trading problem was formulated as a bargaining-based cooperative model, where DR aggregators and the generation company collaboratively decide the amounts of energy trade and the associated payments. Authors in [21] formulated a Stackelberg game among the DR aggregator and electricity generators, in which the DR aggregator plays as the leader to optimize the bidding strategy, and generators play as the followers to maximize their own profits.

By reviewing the above literature, it is found that the research hides a common assumption, that is, all DR participants are absolutely rational and their irrational behavior has been abandoned completely. However, in a real system, consumers on the demand side, residential users in particular, rarely have absolute rationality. Actually, a consumer's irrational behavior has a great influence on the decision-making process in the implementation of the DR project. One consumer may be affected to be in DR by its neighbors who have participated in the DR project. That is, the decision on whether consumers participate in DR is not only related with individual circumstances, but is also closely interrelated with the other group consumers. Although several papers have focused on the DR program considering consumers' irrationality [22,23], they mainly concentrated on the design of the DR mechanism to relieve the effect of irrationality and ignored the analysis of irrational behavior characteristics. For example, Reference [22] proposed a novel non-cooperative game among customers with prospect theory to incorporate the impact of customer irrational behavior, and Reference [23] put forward a dynamic pricing mechanism based on game theory, considering the existence of inexperienced or irrational users. Different from the existing research, this paper mainly concentrates on the analysis of irrational human behavior for residential users in the decision-making process of DR. In our proposed framework, a residential community is responsible for the load aggregation of internal users, and the DR project is divided into multiple stages. The residential community can independently decide whether it will participate in each stage of the DR project. In order to describe the irrationality of communities in the decision on whether to be in DR, a novel decision-making behavior model in each stage of the DR project is proposed based on the Markov chain. Accordingly, the population evolution result of being in DR can be obtained with the implementation of the DR project. Such an evolutionary process of the population is very significant for the grid to measure the feasibility of the designed DR mechanism. Furthermore, in each stage of the DR project, residential communities who are willing to be in DR have to participate in the day-ahead DR market to determine the bidding amount. To reduce the risk of participants' unreasonable bidding amount and price, a non-cooperative game approach is proposed to describe the competition behavior among communities in the day-ahead DR market. In brief, the contributions of this paper are as follows:

(1) A scenario is proposed for the multi-stage DR project to analyze the population evolution participating in DR considering the residential community's irrational behavior in the decision-making process, which can provide decision guidance in the design of the DR mechanism for dispatching the center of the smart grid.

(2) A novel decision-making behavior model is formulated with the Markov chain to forecast whether the residential community will participate in DR, which can provide a better understanding of the practical performance of the DR project.

(3) A non-cooperative game approach is formulated to search the bidding equilibrium among residential communities willing to participate in DR, which can contribute to the stability of the DR bidding market.

The rest of this paper is organized as follows. The system model is introduced in Section 2. In Section 3, we formulate the non-cooperative game approach and prove the existence of the Nash equilibrium. And then, the Markov model for the population evolution is given in Section 4. The case study and simulation results are presented in Section 5. Finally, this paper is concluded in Section 6.

2. System Model

A DR framework for residential community is proposed in Figure 1. Assume that there are total I residential communities with the set $I = N \cup M$, in which group N contains N residential communities who are willing to participate in DR, while group M contains M communities who are unwilling to participate. Each community contains many residential users with a rich flexible load, such as air condition and an electrical vehicle. The dispatching center of the smart grid is mainly responsible for DR transaction with the residential communities in group N [24,25]. In the day-ahead DR market, since only the bidding price and amount are exchanged between the dispatching center and community, each community does not reveal the details about the energy consumption of native users' appliances. Therefore, privacy can be protected from the residential community level [26]. Furthermore, the members in each group do not always remain unchanged. After a period of time, each community obtains the opportunity to choose to participate in DR or not. In the paper, such a time slot is set to one week. Since the residential community in the scenario has bounded rationality, we assume that a community can be infected with probability β by each neighboring community in group N, while a community in group N will transfer to group M with probability α. That is, at the beginning of each week, each community will make a new choice with the corresponding probability. Accordingly, in the proposed framework, there exists two time scales: one is a short-time scale for daily energy consumption scheduling with the set $T = [1, 2, ..., T]$; the other is long-time scale for residential community decision-making with the set $H = [1, 2, ..., H]$.

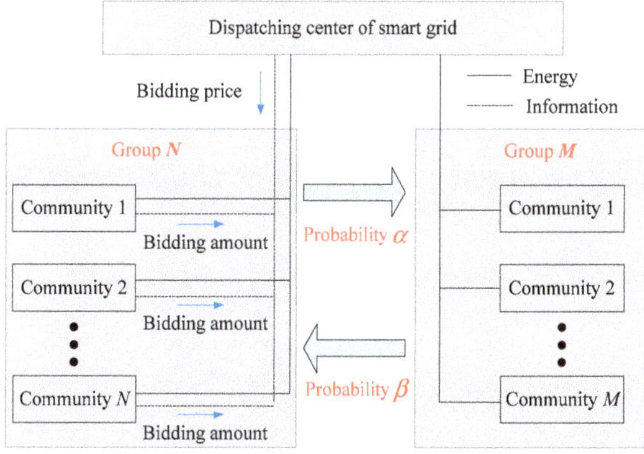

Figure 1. Framework for residential community participating in DR.

2.1. Energy Dispatching Model

Assume that residential community $n \in N$ chooses to participate in DR in week $h \in H$, then its bidding amount is $L_n^{h,t}$ in time slot $t \in T$. Considering the limitation of flexible DR resources on the residential side, the bidding amount has to satisfy the following constraint:

$$L_n^{h,\min} \leq L_n^{h,t} \leq L_n^{h,\max} \tag{1}$$

where $L_n^{h,\min}$ and $L_n^{h,\max}$ represent the minimal and maximal bidding amount of community n in week h, respectively. Therefore, the individual feasible bidding amount set of community n in week h can be expressed as follows:

$$\mathcal{L}_n^h = \left\{ \boldsymbol{L}_n^h : L_n^{h,\min} \leq L_n^{h,t} \leq L_n^{h,\max}, \forall t \in T \right\} \tag{2}$$

where $\boldsymbol{L}_n^h = \left[L_n^{h,1}, L_n^{h,2}, \cdots, L_n^{h,T} \right]$ is the bidding amount set of community n in all dispatching slots. Accordingly, the feasible bidding amount set of all residential communities in group N can be expressed as follows:

$$\mathcal{L}^h = \mathcal{L}_1^h \times \mathcal{L}_2^h \times \cdots \times \mathcal{L}_N^h \tag{3}$$

Note that this paper mainly concentrates on peak load shaving, hence, bidding amount refers to the dispatching amount that will be cut down in real time.

2.2. Bidding Price Model

When a residential community agrees to be scheduled by the dispatching center of the smart grid, it can obtain an economic benefit from the grid, but it firstly has to take part in the day-ahead bidding market. In order to maintain bidding market stability, it is necessary for the dispatching center to design a reasonable bidding price model. In the paper, we assume that the bidding price mechanism must satisfy the following conditions:

(1)	The bidding price model should be smooth or at least piecewise smooth.
(2)	The bidding price in a certain time slot should be a decreasing model with respect to the total bidding amount of all communities in group N.

Accordingly, a linear function is employed as the bidding price model. Since the bidding amount of community n in week h is $L_n^{h,t}$, the total bidding amount of all communities in time slot t can be expressed as:

$$L_h^t = \sum_{n=1}^{N} L_n^{h,t} \tag{4}$$

Therefore, the bidding price in the market can be expressed as follows:

$$p_h^t = a_h^t L_h^t + b_h^t \tag{5}$$

where $a_h^t < 0$ and $b_h^t > 0$ are constants correlated with time slot t and week h. Parameter $a_h^t < 0$ can guarantee the bidding price decreases with the increase of the bidding amount and, at the same time, can also effectively reduce the implementation cost of the DR project for the dispatching center.

2.3. Utility Model of Energy Consumption

A residential community receives utility when it consumes energy in its own ways. When the energy consumption of the community is scheduled by the dispatching center of the smart grid, consumption utility will be affected. In order to quantitatively measure the utility, a utility model needs to be formulated. In many DR studies [27,28], quadratic and logarithmic utility functions are frequently used, because they are non-decreasing and their marginal benefits are non-decreasing. In this paper, without loss of generality, the quadratic function is adopted as the utility model. That is:

$$u_n^{h,t} = c_h^t\left(L_n^{h,t}\right)^2 + d_h^t L_n^{h,t} \tag{6}$$

where $c_h^t > 0$ and $d_h^t > 0$ are time-varying parameters. Utility Equation (6) shows that, when a residential community shaves $L_n^{h,t}$ energy, then the utility will lose $u_n^{h,t}$. Therefore, the whole utility of community n in all time slots T can be calculated as:

$$u_n^h = \sum_{t=1}^{T} u_n^{h,t} \tag{7}$$

Utility Equation (7) shows that community n will lose utility u_n^h when it shaves $\sum_{t=1}^{T} L_n^{h,t}$ energy in the daily dispatching period.

3. Non-Cooperative Game for Group N Participating in Day-Ahead Bidding

In this section, we focus on the model formulation for residential communities who participate in DR. According to Figure 1, all communities in group N have to take part in the day-ahead bidding market. Generally, the game-theoretic approach can be divided into a cooperative game and non-cooperative game [18,29]. In this paper, we assume that each community is only concerned about self-interest, and the non-cooperative game approach is employed to optimize communities' bidding strategy.

3.1. Day-Ahead Bidding Optimization Problem

Load aggregators taking part in DR usually have different optimization targets, such as energy cost or load factor [30]. In the proposed scenario, the main purpose of residential community participating in DR is to obtain extra economic profit. Considering energy consumption scheduling affects residents' satisfaction, the dispatching center will give corresponding economic compensation to communities in group N. Based on the bidding price Equation (5), economic compensation of community n in all time slots T can be calculated as:

$$e_n^h = \sum_{t=1}^{T} p_h^t L_n^{h,t} = \sum_{t=1}^{T} \left(a_h^t L_h^t + b_h^t\right) L_n^{h,t} \tag{8}$$

Since a community's utility will be reduced in DR participation, each community will take the maximization of the comprehensive income as the target to compete with other communities in the bidding market. That is:

$$\begin{aligned} \text{maximize} \, & e_n^h\left(L_n^{h,t}\right) - u_n^h\left(L_n^{h,t}\right) \\ s.t. \, & L_n^{h,\min} \le L_n^{h,t} \le L_n^{h,\max} \end{aligned} \tag{9}$$

Each community will obtain the optimal bidding strategy in each time slot $t \in T$ by solving the optimization Equation (9).

3.2. Non-Cooperative Game Formulation

The bidding price in the market is determined by the total bidding amount of all communities in group N. Therefore, the economic compensation of community n is determined not only by its own bidding strategy, but also the bidding strategies of other communities in group N. That is, community n has to take other communities' strategies into consideration when it makes its bidding strategy. Hence, the bidding strategy problem for residential communities belongs to the typical non-cooperative game. Based on the objective Equation (9), the non-cooperative game can be formulated as follows [31]:

- Players: all residential communities in group N;
- Strategies: the bidding amount $L_n^{h,t}$;
- Payoffs: comprehensive income of community n:

$$R_n^h\left(L_n^h, L_{-n}^h\right) = \sum_{t=1}^{T}\left[\left(a_h^t - c_h^t\right)L_n^{h,t} + \left(b_h^t - d_h^t\right) + a_h^t L_{-n}^{h,t}\right]L_n^{h,t} \tag{10}$$

where $L_{-n}^h = \left[L_1^h, \cdots, L_{n-1}^h, L_{n+1}^h, \cdots, L_N^h\right]$ represents the bidding strategy set of other communities except n in group N; $L_{-n}^{h,t}$ represents the total bidding amount of $N-1$ communities with:

$$L_{-n}^{h,t} = \sum_{i=1, i\neq n}^{N} L_i^{h,t} \tag{11}$$

All residential communities will constantly update their own strategy for the higher economic compensation based on payoff Equation (10). Once all communities in group N obtain their own maximal profit, no one will change the strategy. Such an equilibrium state is called the Nash equilibrium, which can be expressed as:

$$R_n^h\left(L_n^{h*}, L_{-n}^{h*}\right) \geq R_n^h\left(L_n^h, L_{-n}^{h*}\right) \tag{12}$$

where $\left(L_n^{h*}, L_{-n}^{h*}\right)$ represents the Nash equilibrium of a formulated non-cooperative game.

3.3. Nash Equilibrium

According to the above definition of the Nash equilibrium, this section focuses on the mathematical proof of the existence and uniqueness of the Nash equilibrium.

Lemma 1. *For each residential community $n \in N$, the function $R_n^h\left(L_n^h, L_{-n}^h\right)$ is continuously differentiable in L_n^h. For the fixed value of $L_{-n}^{h,t}$, the function $R_n^h\left(L_n^h, L_{-n}^h\right)$ is concave about L_n^h.*

Proof. It is obvious that, $R_n^h\left(L_n^h, L_{-n}^h\right)$ is continuously differentiable in L_n^h. As for the concavity of $R_n^h\left(L_n^h, L_{-n}^h\right)$, we just need to prove the Hessian matrix of $R_n\left(L_n, L_{-n}\right)$ is negative definite. The Hessian matrix of $R_n^h\left(L_n^h, L_{-n}^h\right)$ is:

$$\nabla_{L_n^h}^2 R_n^h\left(L_n^h, L_{-n}^h\right) = \mathrm{diag}\left[2\left(a_h^t - c_h^t\right)\right]_{t=1}^{T} \tag{13}$$

Due to the negative value of $2\left(a_h^t - c_h^t\right)$, Equation (13) is a diagonal matrix with all diagonal elements being negative. Hence, the Hessian matrix of $R_n^h\left(L_n^h, L_{-n}^h\right)$ is negative definite. Consequently, function $R_n^h\left(L_n^h, L_{-n}^h\right)$ is concave about L_n^h. □

Definition 1. *The variational inequality (VI), denoted by $VI(\mathcal{L}, F)$, is to find a vector $x^* \in \mathcal{L}$ such that:*

$$\left(x - x^*\right)^T F(x^*) \leq 0 \; \forall x \in \mathcal{L} \tag{14}$$

According to Lemma 1 and Definition 1, we can obtain the following lemma:

Lemma 2. *The optimization problem of the non-cooperative model (10) is equivalent to the VI problem $VI(\mathcal{L}, F)$ where:*

$$F\left(L^h\right) = \left[F_n\left(L_n^h, L_{-n}^h\right)\right]_{n=1}^{N} \tag{15}$$

where $L^h = \left(L^h_n, L^h_{-n}\right)$ and $F^h_n\left(L^h_n, L^h_{-n}\right)$ are expressed as follows:

$$F^h_n\left(L^h_n, L^h_{-n}\right) = \nabla_{L^h_n} R^h_n\left(L^h_n, L^h_{-n}\right) \tag{16}$$

Proof. The proof can be found in [32]. $\square$

Based on Lemma 2, the following proposition can be obtained.

Proposition 1. *In the formulated non-cooperative model (10), its Nash equilibrium is unique.*

Proof. According to Lemma 2, VI problem $VI(\mathcal{L}, F)$ has the same solution with the solution of Equation (10). That is, we just need to prove the uniqueness of $VI(\mathcal{L}, F)$'s solution, then **Proposition 1** can be proved. According to [33], we know that $VI(\mathcal{L}, F)$ will have a unique solution when $F\left(L^h\right)$ is strictly monotone about feasible set $\mathcal{L}$.

To prove the strict monotone of $F\left(L^h\right)$ is to prove:

$$\sum_{t=1}^{T}\sum_{n=1}^{N}\left[\left(x^{h,t}_n - y^{h,t}_n\right)\left(\nabla_{x^{h,t}_n}R^h_n\left(x^h\right) - \nabla_{y^{h,t}_n}R^h_n\left(y^h\right)\right)\right] > 0 \tag{17}$$

where $x^h = \left\{x^h_n\right\}^N_{n=1} \in \mathcal{L}^h, y^h = \left\{y^h_n\right\}^N_{n=1} \in \mathcal{L}^h$.

Let $l^{h,t} = \left\{x^{h,t}_1, x^{h,t}_2, \cdots, x^{h,t}_N\right\}$ and $j^{h,t} = \left\{y^{h,t}_1, y^{h,t}_2, \cdots, y^{h,t}_N\right\}$, then Equation (17) can be rewritten as follows:

$$\sum_{t=1}^{T}\left[\left(l^{h,t} - j^{h,t}\right)\left(\nabla_{l^{h,t}}R^{h,t}_n\left(l^{h,t}\right) - \nabla_{j^{h,t}}R^{h,t}_n\left(j^{h,t}\right)\right)\right] > 0 \tag{18}$$

where:

$$R^{h,t}_n\left(l^{h,t}\right) = p^t_h x^{h,t}_n - u^{h,t}_n$$

and:

$$\nabla_{l^{h,t}}R^{h,t}_n\left(l^{h,t}\right) = \left[\nabla_{x^{h,t}_1}R^{h,t}_n\left(l^{h,t}\right), \nabla_{x^{h,t}_2}R^{h,t}_n\left(l^{h,t}\right), \cdots \nabla_{x^{h,t}_N}R^{h,t}_n\left(l^{h,t}\right)\right]^T$$

If the following condition is satisfied, then Equation (18) will hold:

$$\left(l^{h,t} - j^{h,t}\right)\left(g_{h,t}\left(l^{h,t}\right) - g_{h,t}\left(j^{h,t}\right)\right) > 0 \; \forall t \in T \tag{19}$$

where $g_{h,t}\left(l^{h,t}\right) = \nabla_{l^{h,t}}R^{h,t}_n\left(l^{h,t}\right)$.

If the Jacobian matrix of $g_{h,t}\left(l^{h,t}\right)$ is negative definite, then Equation (19) will be satisfied. Assume that $G_{h,t}\left(l^{h,t}\right) = \nabla_{l^{h,t}}g_{h,t}\left(l^{h,t}\right)$, then:

$$G_{h,t}\left(l^{h,t}\right) + G_{h,t}\left(l^{h,t}\right)^T = 2\left(a^t_h - c^t_h\right)\left(11^T + I\right) \tag{20}$$

where I is a unit matrix and 1 is a $N \times 1$ matrix where all elements are 1. Since characteristic values of $11^T + I$ are 1 and $N + 1$, the characteristic values of $G_{h,t}\left(l^{h,t}\right) + G_{h,t}\left(l^{h,t}\right)^T$ are $2\left(a^t_h - c^t_h\right)$ and $2(N+1)\left(a^t_h - c^t_h\right)$. Consequently, $G_{h,t}\left(l^{h,t}\right) + G_{h,t}\left(l^{h,t}\right)^T$ is negative definite. That is, $F(L)$ is a strictly monotone function. Therefore, the Nash equilibrium of the formulated non-cooperative game is unique. $\square$

Based on the above analysis, the Nash equilibrium for the game can be solved. Equation (9) is to search the maximal value of the comprehensive income in all dispatching slots T. But, when the comprehensive income achieves the maximal value in each time slot $t \in T$, then the comprehensive income in all dispatching slots will also achieve the maximal value. That is, Equation (9) can be translated into the following optimization problems:

$$\begin{cases} \text{maximize} w_n^{h,t} = \left[\left(a_h^t - c_h^t\right)L_n^{h,t} + \left(b_h^t - d_h^t\right) + a_h^t L_{-n}^{h,t}\right]L_n^{h,t} \\ s.t.\ L_n^{h,\min} \le L_n^{h,t} \le L_n^{h,\max} \end{cases} \quad \forall t \in T \tag{21}$$

Furthermore, when the bidding strategies of other communities $L_{-n}^{h,t}$ are regarded as fixed values, then the optimal bidding strategy of community n can be expressed as follows:

$$\varphi_n\left(L_{-n}^{h,t}\right) = \underset{L_n^{h,t}}{\text{argmaximize}}\ w_n^{h,t}\left(L_n^{h,t}, L_{-n}^{h,t}\right) \forall n \in N \tag{22}$$

where $\varphi_n\left(L_{-n}^{h,t}\right)$ represents the optimal bidding strategy of community n corresponding to strategies $L_{-n}^{h,t}$; $L_{-n}^{h,t} = \left[L_1^{h,t}, \cdots, L_{n-1}^{h,t}, L_{n+1}^{h,t}, \cdots, L_N^{h,t}\right]$ represents the bidding strategies of other communities in time slot t.

4. Evolution Analysis between Groups N and M

According to the above analysis, the economic compensation of each residential community in group N is correlated not only with the bidding price parameters, but also with the total bidding amount in the market. Therefore, a community's economic compensation will be influenced when the population of group N changes. In the initial period of the DR project, communities will obtain high economic compensation for participating in DR. Hence, the neighboring communities may be infected to participate in DR for the high economic compensation. However, when the population of group N has consistent growth, the economic compensation of each community will be reduced gradually. Consequently, those residential communities who care more about energy consumption satisfaction will not choose to participate in DR anymore. Finally, the population of group N and group M will reach a dynamic balance. In this section, communities' transition probability model between group N and group M is formulated, and then the group population is analyzed.

4.1. Transition Probability Model

To describe the population evolution briefly, we define the state of residential community $i \in I$ at week h as $S_i(h)$, in which $S_i(h) = 0$ represents community i belonging to group M is unwilling to participate in DR, $S_i(h) = 1$ represents community i belonging to group N adopts the DR project. Then, the transition probability can be calculated following four cases.

(1) Case 1: $S_i(h) = 0 \rightarrow S_i(h + 1) = 1$

When the state of community i is $S_i(h) = 0$ at week h, then it may participate in DR if some of its neighboring communities have adopted it. That is, community i can be infected by each neighboring community with probability β, where $0 \le \beta \le 1$. Such probability shows the effect of social networking or mutual imitation among DR communities. Note that the probability β is a networking-related parameter. Therefore, the greater the number of neighboring communities who participate in DR, the higher the probability community i will adopt the DR project. Consequently, the corresponding transition probability can be expressed as follows:

$$P\left(S_i(h + 1) = 1 \middle| S_i(h) = 0\right) = 1 - (1 - \beta)^{N_i^h} \tag{23}$$

where N_i^h is the total number of neighboring communities in group N that have connection to the community i; $1 - \beta$ represents the probability that community i is not infected by one neighboring community; $(1 - \beta)^{N_i^h}$ represents the probability that community i is not infected by N_i^h neighboring communities.

(2) Case 2: $S_i(h) = 0 \rightarrow S_i(h + 1) = 0$

Since community i with $S_i(h) = 0$ can only choose $S_i(h + 1) = 1$ or $S_i(h + 1) = 0$, the probability that community i still remains in group M can be obtained according to Equation (23). That is:

$$P\Big(S_i(h+1) = 0 \big| S_i(h) = 0\Big) = 1 - P\Big(S_i(h+1) = 1 \big| S_i(h) = 0\Big) = (1 - \beta)^{N_i^h} \tag{24}$$

(3) Case 3: $S_i(h) = 1 \rightarrow S_i(h + 1) = 0$

When the state of community i is $S_i(h) = 1$ at week h, it may switch back to $S_i(h + 1) = 0$ at week $h + 1$. For example, a community may find the DR project is inconvenient or uneconomical and thus abandon it. In this paper, we assume that the probability from $S_i(h) = 1$ to $S_i(h + 1) = 0$ are correlated with economic compensation and energy consumption utility. If a community in the group N switches from state 1 to state 0, it will lose economic compensation, but will obtain the corresponding utility. Hence, when economic compensation decreases due to the increasing of group N's population, the probability from $S_i(h) = 1$ to $S_i(h + 1) = 0$ will increase gradually. To quantitatively measure such probability, the average comprehensive income for all communities in group N is defined as:

$$\overline{w}^h = \frac{1}{N}\sum_{i \in N}\Big(e_i^h - u_i^h\Big) \tag{25}$$

Basically, the probability from $S_i(h) = 1$ to $S_i(h + 1) = 0$ is defined as:

$$\alpha = \eta\left(1 - \frac{\overline{w}^h}{\overline{w}^{h,\max}}\right) \tag{26}$$

where η is a constant parameter; $\overline{w}^{h,\max} = \text{maximize}\big[\overline{w}^1, \overline{w}^2, \ldots, \overline{w}^h\big]$ represents the maximal value of group N's average income in the preceding h weeks. Equation (26) shows that the lower income group N receives, the higher probability the community switches from state 1 to state 0. Specifically, when the average income in group N reaches the maximal value, the corresponding probability will reach the minimal value. According to Equations (25) and (26), the transition probability from state 1 to state 0 can be expressed as:

$$P\Big(S_i(h+1) = 0 \big| S_i(h) = 1\Big) = \alpha = \eta\left(1 - \frac{\overline{w}^h}{\overline{w}^{h,\max}}\right) \tag{27}$$

(4) Case 4: $S_i(h) = 1 \rightarrow S_i(h + 1) = 1$

Similarly, since community i with $S_i(h) = 1$ can only choose $S_i(h + 1) = 1$ or $S_i(h + 1) = 0$, the probability that community i still remains in group N can be obtained according to Equation (27). That is:

$$P\Big(S_i(h+1) = 1 \big| S_i(h) = 1\Big) = 1 - P\Big(S_i(h+1) = 0 \big| S_i(h) = 1\Big) = 1 - \alpha \tag{28}$$

In summary, the transition probability model from $S_i(h)$ to $S_i(h + 1)$ can be expressed as follows:

$$P\Big(S_i(h+1) \big| S_i(h)\Big) = \begin{cases} 1 - (1 - \beta)^{N_i^h} & (S_i(h), S_i(h+1)) = (0,1) \\ (1 - \beta)^{N_i^h} & (S_i(h), S_i(h+1)) = (0,0) \\ \alpha & (S_i(h), S_i(h+1)) = (1,0) \\ 1 - \alpha & (S_i(h), S_i(h+1)) = (1,1) \end{cases} \tag{29}$$

4.2. Markov Model for Group Population

In reality, the state of a residential community in past weeks has no effect on the decision in future weeks, and the decision result in week $h + 1$ is only correlated with a community's state in week h. Therefore, this paper adopts the Markov chain to describe the decision-making process of residential community. The Markov chain mainly indicates that the future decision-making state is independent of the past state and is only dependent on the current state [34]. In our proposed scenario, the Markov chain can be expressed as:

$$P\Big(S_i(h+1)\big|S_i(1),\cdots,S_i(h)\Big) = P\Big(S_i(h+1)\big|S_i(h)\Big) \tag{30}$$

where $S_i(1),\cdots,S_i(h)$ are decision-making states from week 1 to week h. Assume that $\mathrm{Pr}_N^i(h)$ and $\mathrm{Pr}_M^i(h)$ are the probabilities of residential community i in group N and group M in week h. Then, the Markov state evolution can be described as:

$$\begin{aligned}
\mathrm{Pr}_N^i(h+1) &= \mathrm{Pr}_N^i(h)P\Big(S_i(h+1)=1\big|S_i(h)=1\Big) + \mathrm{Pr}_M^i(h)P\Big(S_i(h+1)=1\big|S_i(h)=0\Big)\\
&= (1-\alpha)\mathrm{Pr}_N^i(h) + \Big(1-(1-\beta)^{N_i^h}\Big)\mathrm{Pr}_M^i(h)
\end{aligned} \tag{31}$$

and:

$$\begin{aligned}
\mathrm{Pr}_M^i(h+1) &= \mathrm{Pr}_M^i(h)P\Big(S_i(h+1)=0\big|S_i(h)=0\Big) + \mathrm{Pr}_N^i(h)P\Big(S_i(h+1)=0\big|S_i(h)=1\Big)\\
&= (1-\beta)^{N_i^h}\mathrm{Pr}_M^i(h) + \alpha\mathrm{Pr}_N^i(h)
\end{aligned} \tag{32}$$

In Equations (31) and (32), $(1-\beta)^{N_i^h}$ can be rewritten as:

$$\begin{aligned}
(1-\beta)^{N_i^h} &= \prod_{i\in I_i}\Big(\mathrm{Pr}_N^i(h)(1-\beta)+\mathrm{Pr}_M^i(h)\Big) = \prod_{i\in I_i}\Big(\mathrm{Pr}_N^i(h)(1-\beta)+\big(1-\mathrm{Pr}_N^i(h)\big)\Big)\\
&= \prod_{i\in I_i}\Big(1-\beta\mathrm{Pr}_N^i(h)\Big) \approx 1-\beta\sum_{i\in I_i}\mathrm{Pr}_N^i(h)
\end{aligned} \tag{33}$$

where I_i represents the set of community i's neighbors in set I. Therefore, Equations (31) and (32) can also be expressed as follows:

$$\mathrm{Pr}_N^i(h+1) = (1-\alpha)\mathrm{Pr}_N^i(h) + \left(1-\left(1-\beta\sum_{i\in I_i}\mathrm{Pr}_N^i(h)\right)\right)\mathrm{Pr}_M^i(h) \tag{34}$$

and:

$$\mathrm{Pr}_M^i(h+1) = \left(1-\beta\sum_{i\in I_i}\mathrm{Pr}_N^i(h)\right)\mathrm{Pr}_M^i(h) + \alpha\mathrm{Pr}_N^i(h) \tag{35}$$

According to the Markov state Equations (34) and (35), the probability for each community i in groups N or M can be calculated. However, the decision of a single community is not our concern and the main purpose in this section is to analyze the group population. For simplification, assume that I residential communities have good communication and each community can be infected by any other $I-1$ communities. Then, each residential community in I has equal probability to participate in DR. Consequently, we have:

$$\overline{\mathrm{Pr}}_N = \mathrm{Pr}_N^i, \overline{\mathrm{Pr}}_M = \mathrm{Pr}_M^i, I_i = I \tag{36}$$

where $\overline{\mathrm{Pr}}_N$ and $\overline{\mathrm{Pr}}_M$ represent the probability of any community being in group N or M. Then, Equations (34) and (35) are simplified to:

$$\overline{\mathrm{Pr}}_N(h+1) = (1-\alpha)\overline{\mathrm{Pr}}_N(h) + \beta I\overline{\mathrm{Pr}}_N(h)\overline{\mathrm{Pr}}_M(h) \tag{37}$$

and:

$$\overline{\mathrm{Pr}}_M(h+1) = \Big(1-\beta I\overline{\mathrm{Pr}}_N(h)\Big)\mathrm{Pr}_M^i(h) + \alpha\overline{\mathrm{Pr}}_N(h) \tag{38}$$

Therefore, the average number of residential communities in groups N and M in week h can be expressed as:

$$\begin{cases} N(h) = \overline{\mathrm{Pr}}_N(h)I \\ M(h) = \overline{\mathrm{Pr}}_M(h)I \end{cases} \tag{39}$$

According to Equations (37) and (38), the average number of residential communities in groups N and M in week $h + 1$ can be calculated as:

$$\begin{cases} N(h+1) = (1-\alpha)N(h) + \beta N(h)M(h) \\ M(h+1) = (1-\beta N(h))M(h) + \alpha N(h) \end{cases} \tag{40}$$

Equation (40) is used to describe the population evolution in groups N and M.

4.3. Distributed Algorithm

To search the Nash equilibrium of the formulated non-cooperative game and population evolution result of the residential communities participating in DR, a distributed algorithm is proposed which is shown in Algorithm 1. In the algorithm, an interior point method is employed to solve Equation (22), which has superior performance in solving convex optimization problems. In addition, steps 2–10 are designed to search the Nash equilibrium among communities in the day-ahead bidding market, and steps 11–14 are designed to obtain the population evolution result of residential communities participating in DR.

Algorithm 1: Searching for Nash equilibrium and population evolution result

---

Input: Parameters a_h^t, b_h^t, $L_n^{h,\min}$, $L_n^{h,\max}$, etc.
Output: Nash equilibrium, population evolution result.
Initialization: Number of group N in week 1.
1   $h = 1$;
2   **Repeat**
3       $n = 1$;
4   **for** $n \le N(h)$ **do**
5   Initialize bidding strategy vector L_n^h;
6   Each community $n \in N$ updates L_n^h by solving Equation (22);
7           $n = n + 1$;
8   **end**
9   **Until** No community changes its strategy;
10   **Return** Nash equilibrium $\left(L_n^{h*}, L_{-n}^{h*} \right)$;
11   Calculate transition probability α and β;
12   **Update** group population with Equation (40);
13   $h = h + 1$;
14   Go back to step **2** until $|N(h+1) - N(h)|$ changes in a termination criterion.

---

5. Case Study

The performance of the proposed approach is evaluated in this section. In the simulation, assume that there are $|I| = 50$ residential communities. At the beginning of the DR project (i.e., the 1st week), there are $|N| =$ four residential communities who are willing to participate in DR, and other $|M| = 46$ communities are in a waiting state. For these communities participating in DR, they have to take part in the day-ahead bidding market to obtain the dispatching amount. Generally, the daily peak hours appear in 10:00–14:00 and 18:00–21:00. Here, energy consumption scheduling during 18:00–21:00 in one day of each week is taken as an example. Accordingly, suppose that peak shaving hours are 18:00–21:00 and a scheduling interval is 15 minutes. That is, residential communities in group N will bid for a load shaving amount in time slots $T = [1, 2, ..., 12]$. Furthermore, bidding price parameters are shown as (unit: 10^3 dollars/MWh): $a_h^t = -0.068$, $b_h^t = 0.553$ ($t = 1$–5 and 11–12); $a_h^t = -0.058$, $b_h^t = 0.774$ ($t = 6$–10). Utility model parameters are shown as (unit: $10^{\wedge}3$ dollars/MWh): $c_h^t = 0.012$, $d_h^t = 0.117$ ($t = 1$–5 and 11–12); $c_h^t = 0.013$, $d_h^t = 0.126$ ($t = 6$–10). Transition probability model parameters are shown as: β is in [0.02,0.04], η is in [0.3,0.4]. As for the available DR resource in the community, we assume that the maximal value and minimal value of the DR resource are set as in Figure 2. And each community's DR resource is given with a random value among the range.

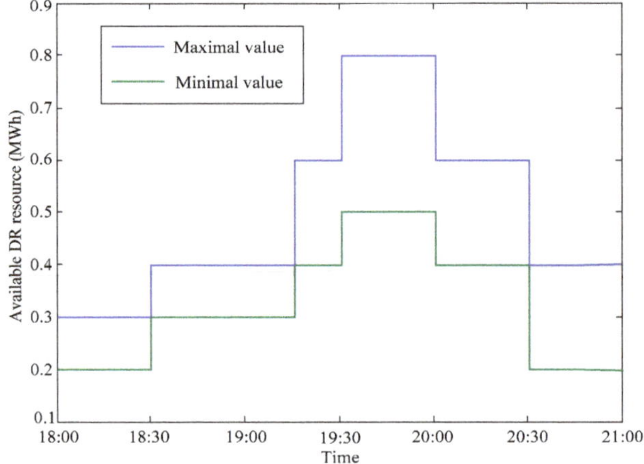

Figure 2. Range of available DR resource of each residential community.

5.1. Evolution Result in Groups *N* and *M*

Based on the above simulation parameters, the population evolution result and the Nash equilibrium was obtained by performing Algorithm 1. Note that Algorithm 1 is set to 70 iterations and each iteration means one week. Accordingly, Figure 3 is the convergence process of the population in groups *N* and *M*. It depicts that the population in each group gradually converged to the evolutionary equilibrium in 42 weeks. Specifically, in the first 20 weeks, group *N*'s population increased rapidly from four communities to 36 communities, while between the 21st week and 42nd week, group *N*'s population increased slowly from 37 communities to 44 communities. Note that, since the average number of communities in a group is deduced from Equation (45), in which the probability $\overline{\mathrm{Pr}}_N$ and $\overline{\mathrm{Pr}}_M$ are non-integral values, the group's population in Figure 2 was also a non-integral value. But the group's population was rounded as an integral value when searching the Nash equilibrium in the day-ahead bidding market.

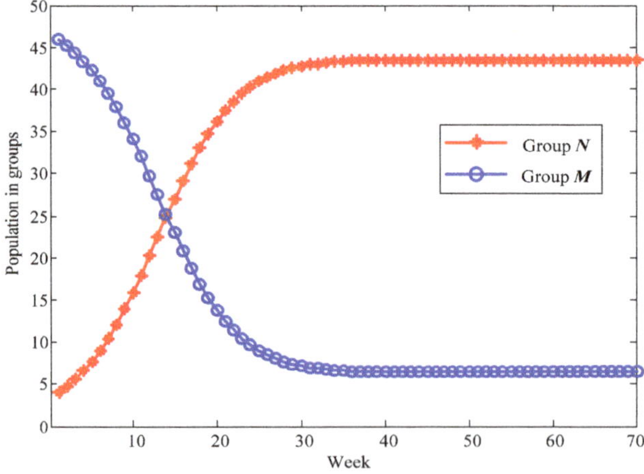

Figure 3. Convergence process of population in groups *N* and *M*.

Figure 4 is the convergence process of the total bidding amount in group N in all dispatching time slots T. From the figure, one can see that, at the beginning of the DR project, the total bidding amount of four communities was only about 20 MWh, while the bidding amount reached 80 MWh after performing the DR project for about 10 weeks. The main reason for such rapid growth was due to the increase of the population in group N. However, the bidding amount tends to reach saturation after 10 weeks. It was because the bidding price was reduced gradually with the further increase of group N's population and the bidding amount of each community was declined as well. Moreover, the comprehensive income of group N in all dispatching time slots T is shown in Figure 5. It demonstrates that the total income of group N increased firstly and then decreased dramatically in 10 weeks, and finally converged to the fixed value in the weeks after 10 weeks. The reason for such a trend was mainly related with the variation of economic compensation and utility. In the first five weeks, the bidding amount in the market was not very large, then the bidding price was relatively high and the economic compensation increased dramatically with the increase of group N's population. Therefore, the income in the first five weeks was mainly dependent on the economic compensation. However, with the increase of the bidding amount, the utility that the community lost in DR increased gradually but the economic compensation increased slowly. Therefore, the comprehensive income in the 6th weeks and 10th weeks decreased rapidly. Since the bidding amount tends to be stable after 10 weeks, the total income of group N was also convergent.

From the above evolution result, it is clear that most of the residential communities will be attracted to participate in DR with the implementation of the DR project. However, not all communities in the set I will be involved in DR. When residential communities in the bidding market become saturated, the average economic compensation of each community will be decreased to the minimal value. Consequently, some members in group N will be unwilling to participate in DR and switch from group N to group M with a high probability. At last, the population of groups N and M will reach the dynamic balance.

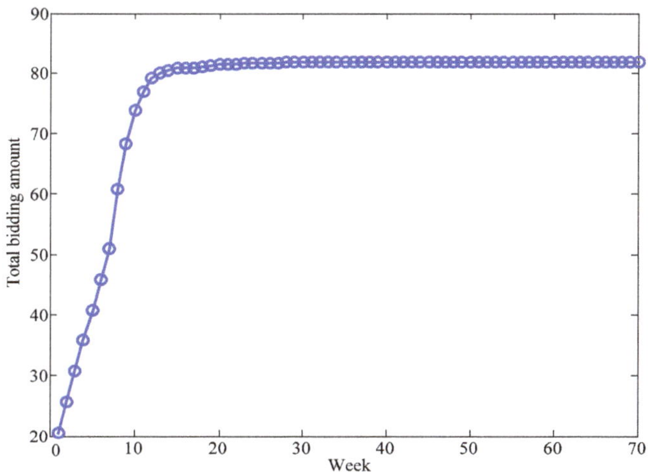

Figure 4. Convergence process of total bidding amount in group N.

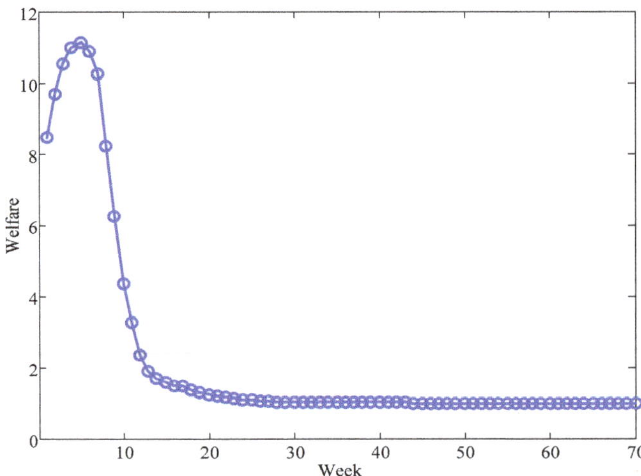

Figure 5. Convergence process of total comprehensive income in group N.

5.2. Equilibrium Result for Group N

This section shows the Nash equilibrium among the residential communities in group N. In our proposed scenario, at the beginning of each week, the community will choose to participate in DR or not, and then, communities participating in DR have to bid for the dispatching amount in the day-ahead market. Therefore, in each week exists one equilibrium solution and there are altogether 70 equilibrium solutions, considering the DR project was conducted in 70 weeks. For the limitation of paper space, we took the equilibrium solutions of two special weeks (i.e., 1st week and 70th week) as examples to illustrate the optimal bidding strategy.

Figures 6 and 7 are the optimal bidding amounts of each community in the 1st week and 70th week, respectively. In which, communities 1–4 are the communities who were willing to participate in DR at the beginning of the DR project. From the figures, we can see that the bidding amount in the 1st week was much more than that in the 70th week in all 12 time slots. Additionally, the bidding strategies of four communities in each time slot were all different in the 1st week, while four communities had the same bidding strategies in the 70th week. The main reason was that, at the beginning of the DR project, the bidding market needed a large amount of DR resources, hence the community with more DR resources will compete for more dispatching amount. However, when the DR project was conducted for 70 weeks, the DR resource in the market was saturated and the bidding price was also saturated with the lowest value. Consequently, the bidding amount of the community was reduced to the minimal value, even for those communities with abundant DR resources. Actually, in this case study, the bidding amount of each community was only related with bidding price parameters and utility model parameters after the bidding market reached saturation. The bidding amount changed with the change of bidding price parameters. For example, we see that from Figure 7, the values of bidding price parameters were different between $t = 1$–5 and $t = 6$–10, then the bidding amount between $t = 1$–5 and $t = 6$–10 were also different. Concretely, the average bidding amount of all communities in group N is presented in Table 1. From the table, it shows that the maximal bidding amount was 0.739 MWh between 19:45–20:00 in the 1st week, while the maximal bidding amount was only 0.203 MWh between 19:15–20:30 in the 70th week.

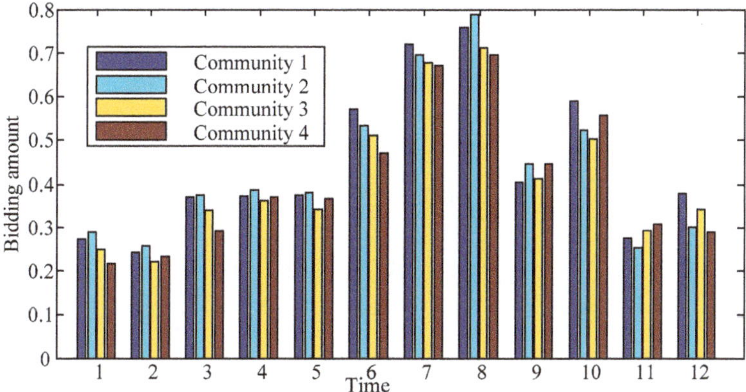

Figure 6. Optimal bidding amount of each community in the 1st week.

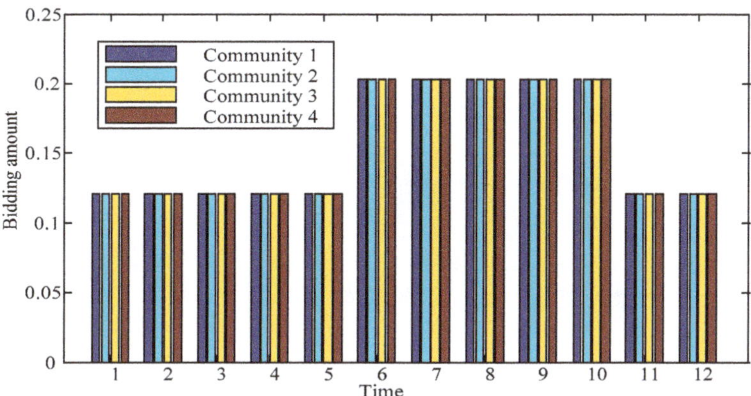

Figure 7. Optimal bidding amount of each community in the 70th week.

Table 1. Average bidding amount of all communities in group N.

Time	Average Bidding Amount (1st Week)	Average Bidding Amount (70th Week)
18:00–18:15	0.256	0.121
18:15–18:30	0.238	0.121
18:30–18:45	0.344	0.121
18:45–19:00	0.371	0.121
19:00–19:15	0.366	0.121
19:15–19:30	0.522	0.203
19:30–19:45	0.692	0.203
19:45–20:00	0.739	0.203
20:00–20:15	0.428	0.203
20:15–20:30	0.543	0.203
20:30–20:45	0.283	0.121
20:45–21:00	0.327	0.121

In addition, the bidding price in the 1st week and 70th week is shown in Figure 8. It depicts that the bidding price in the 1st week was much higher than the price in 70th week. For example, the highest bidding price in the 1st week was 675 dollars/MWh during 20:00–20:15, while the highest price in the 70th week was only 203 dollars/MWh. Therefore, from the aspect of the dispatching center of the smart grid, the designed bidding price mechanism can effectively reduce the dispatching cost of the grid. Of course, the dispatching center of the smart grid can also control the DR resource amount in the market by regulating price parameters. The next section analyzes the influence of bidding price parameters on DR.

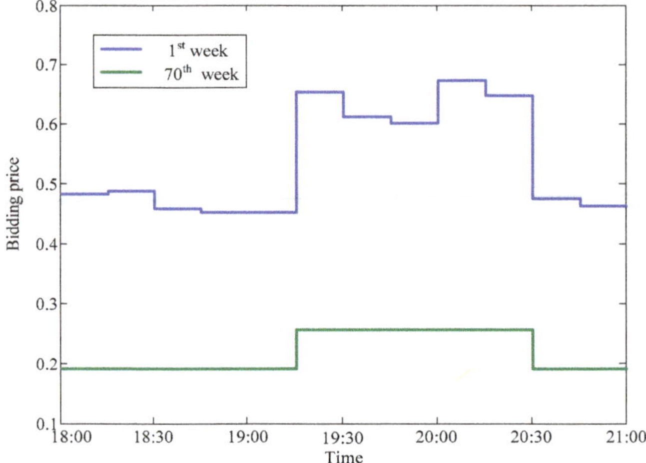

Figure 8. Bidding price in the 1st week and 70th week.

5.3. Impact of Bidding Price Parameters

According to the above analysis, it is clear that bidding price plays an important role in the DR project. It is necessary to analyze how the bidding price affects the DR. Therefore, this section analyzes the influence of bidding price parameters on the population evolution result and the Nash equilibrium. Here, we took initial price parameters (i.e., $a_h^t = -0.068$, $b_h^t = 0.553$ ($t = 1$–5 and 11–12); $a_h^t = -0.058$, $b_h^t = 0.774$ ($t = 6$–10)) as the benchmark, then varied the price parameters by 1.1–2.0 times the benchmark. The corresponding result is presented as follows.

Figure 9 shows the bidding amount of each community and population in group N for different bidding price parameters. Note that the optimal bidding strategy in the figure was the Nash equilibrium in the week that evolution equilibrium was reached. Since the bidding strategies of different communities were all the same under the same price parameters, each community had the same bidding amount in dispatching slots T. From the figure, it depicts that the bidding amount increased gradually with 1.1–2.0 times the benchmark. Specifically, the population in group N increased from 44 to 45 communities when the parameters reached 1.3 times the benchmark. However, the population in group N still had only 45 communities even when price parameters reached 2 times the benchmark. Figure 10 shows the total bidding amount of all communities for different bidding price parameters. It is clear that, although the total bidding amount achieved consistent growth, the increment was declined gradually. By analyzing the above result, it demonstrates that raising the bidding price can improve the growth of the bidding amount, but with the market saturation, the dispatching cost increased for the same increment. Therefore, for the dispatching center of the smart grid, it was very necessary to optimize the bidding price parameters to make a balance between the DR amount and the dispatching cost.

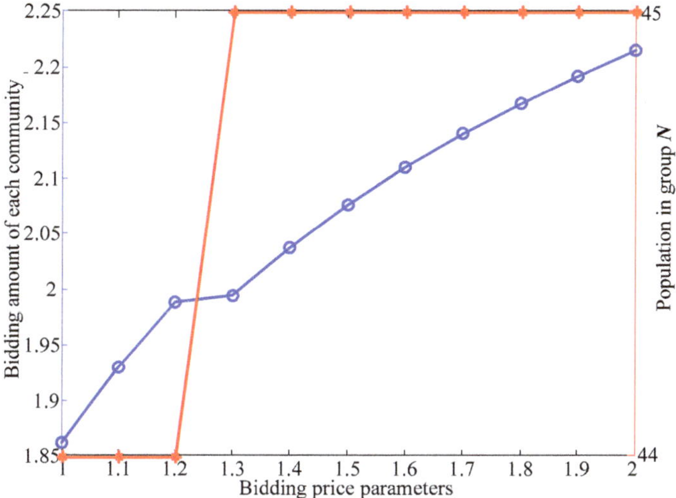

Figure 9. Bidding amount and population for different bidding price parameters.

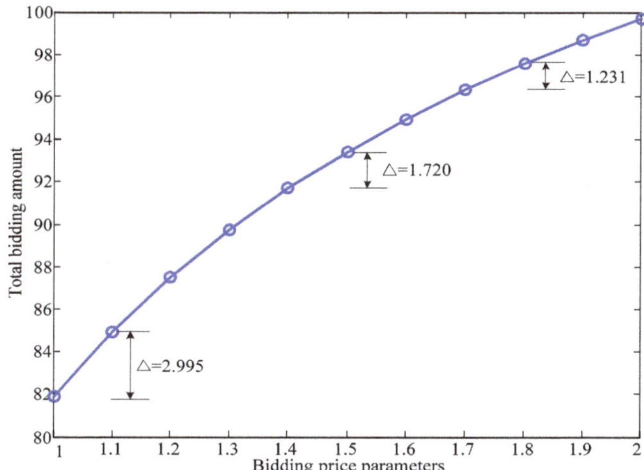

Figure 10. Total bidding amount for different bidding price parameters.

6. Conclusions

This paper not only focused on the evolutionary analysis of group population participating in the DR project, but also the optimization of the bidding strategy in a day-ahead bidding market. In the proposed scenario, a residential community's irrationality was considered in the decision-making process. In particular, the dynamic evolutionary process of group population was described with the Markov model, and the bidding strategy for communities participating in DR was optimized with the non-cooperative game approach. Furthermore, the uniqueness of the Nash equilibrium was proved with the mathematical method. Finally, a case study was performed to verify the effectiveness of formulated models. It showed that the group population in DR gradually converges to the fixed value with the implementation of the DR project. In addition, by analyzing the influence of bidding price parameters on DR, it showed that raising the bidding price can improve the growth of the group population in DR and the bidding amount, but the smart grid had to pay for the high dispatching cost with market saturation.

Author Contributions: Writing—original draft preparation, X.L.; methodology, Q.W.; validation, W.W. All authors were involved in preparing this manuscript.

Funding: This research received no external funding.

Conflicts of Interest: The authors declare no conflict of interest.

References

1. Gyamfi, S.; Krumdieck, S. Price, environment and security: Exploring multi-modal motivation in voluntary residential peak demand response. *Energy Policy* **2011**, *39*, 2993–3004. [CrossRef]
2. Yu, M.; Hong, S.H.; Ding, Y.; Ye, X. An incentive-based demand response (DR) model considering composited DR resources. *IEEE Trans. Ind. Electron.* **2018**, *66*, 1488–1498. [CrossRef]
3. Muratori, M.; Rizzoni, G. Residential demand response: Dynamic energy management and time-varying electricity pricing. *IEEE Trans. Power Syst.* **2015**, *31*, 1108–1117. [CrossRef]
4. Elghitani, F.; Zhuang, W. Aggregating a large number of residential appliances for demand response applications. *IEEE Trans. Smart Grid* **2017**, *9*, 5092–5100. [CrossRef]
5. Moradi, M.H.; Reisi, A.R.; Hosseinian, S.M. An optimal collaborative congestion management based on implementing DR. *IEEE Trans. Smart Grid* **2017**, *9*, 5323–5334. [CrossRef]
6. Gkatzikis, L.; Koutsopoulos, I.; Salonidis, T. The role of aggregators in smart grid demand response markets. *IEEE J. Sel. Areas Commun.* **2013**, *31*, 1247–1257. [CrossRef]
7. Ghazvini, M.A.F.; Soares, J.; Horta, N.; Neves, R.; Castro, R.; Vale, Z. A multi-objective model for scheduling of short-term incentive-based demand response programs offered by electricity retailers. *Appl. Energy* **2015**, *151*, 102–118. [CrossRef]
8. Vivekananthan, C.; Mishra, Y.; Ledwich, G.; Li, F. Demand response for residential appliances via customer reward scheme. *IEEE Trans. Smart Grid* **2014**, *5*, 809–820. [CrossRef]
9. Mhanna, S.; Chapman, A.; Verbic, G. A fast distributed algorithm for large-scale demand response aggregation. *IEEE Trans. Smart Grid* **2016**, *7*, 2094–2107. [CrossRef]
10. Shi, W.; Li, N.; Xie, X.; Chu, C.-C.; Gadh, R. Optimal residential demand response in distribution networks. *IEEE J. Sel. Areas Commun.* **2014**, *32*, 1441–1450. [CrossRef]
11. Yu, M.; Hong, S.H. Supply–demand balancing for power management in smart grid: A Stackelberg game approach. *Appl. Energy* **2016**, *164*, 702–710. [CrossRef]
12. Chen, H.; Li, Y.; Louie, R.H.Y.; Vucetic, B. Autonomous demand side management based on energy consumption scheduling and instantaneous load billing: An aggregative game approach. *IEEE Trans. Smart Grid* **2014**, *5*, 1744–1754. [CrossRef]
13. Chakraborty, P.; Khargonekar, P.P. A demand response game and its robust price of anarchy. In Proceedings of the IEEE International Conference on Smart Grid Communications, Venice, Italy, 3–6 November 2014; pp. 644–649.
14. Baharlouei, Z.; Hashemi, M.; Mohsenian-Rad, H.; Narimani, H. Achieving optimality and fairness in autonomous demand response: Benchmarks and billing mechanisms. *IEEE Trans. Smart Grid* **2013**, *4*, 968–975. [CrossRef]
15. Misra, S.; Bera, S.; Ojha, T.; Mouftah, H.T.; Anpalagan, A. ENTRUST: Energy trading under uncertainty in smart grid systems. *Comput. Netw.* **2016**, *110*, 232–242. [CrossRef]
16. Mohsenian-Rad, A.-H.; Wong, V.W.S.; Jatskevich, J.; Schober, R.; Leon-Garcia, A. Autonomous demand-side management based on game-theoretic energy consumption scheduling for the future smart grid. *IEEE Trans. Smart Grid* **2010**, *1*, 320–331. [CrossRef]
17. Latifi, M.; Rastegarnia, A.; Khalili, A.; Vahidpour, V.; Sanei, S. A distributed game-theoretic demand response with multi-class appliance control in smart grid. *Electr. Power Syst. Res.* **2019**, *176*, 105946. [CrossRef]
18. Mediwaththe, C.P.; Stephens, E.R.; Smith, D.B.; Mahanti, A. A dynamic game for electricity load management in neighborhood area networks. *IEEE Trans. Smart Grid* **2015**, *7*, 1329–1336. [CrossRef]
19. Jacquot, P.; Beaude, O.; Gaubert, S.; Oudjane, N. Analysis and implementation of an hourly billing mechanism for demand response management. *IEEE Trans. Smart Grid* **2019**, *10*, 4265–4278. [CrossRef]

20. Fan, S.; Ai, Q.; Piao, L. Bargaining-based cooperative energy trading for distribution company and demand response. *Appl. Energy* **2018**, *226*, 469–482. [CrossRef]

21. Nekouei, E.; Alpcan, T.; Chattopadhyay, D. Game-theoretic frameworks for demand response in electricity markets. *IEEE Trans. Smart Grid* **2014**, *6*, 748–758. [CrossRef]

22. Wang, Y.; Saad, W.; Mandayam, N.B.; Poor, H.V. Load shifting in the smart grid: To participate or not? *IEEE Trans. Smart Grid* **2015**, *7*, 2604–2614. [CrossRef]

23. Barabadi, B.; Yaghmaee, M.H. A new pricing mechanism for optimal load scheduling in smart grid. *IEEE Syst. J.* **2019**, *13*, 1737–1746. [CrossRef]

24. Ning, J.; Tang, Y.; Chen, Q.; Wang, J.; Zhou, J.; Gao, B. A Bi-level coordinated optimization strategy for smart appliances considering online demand response potential. *Energies* **2017**, *10*, 525. [CrossRef]

25. Gao, C.; Li, Q.; Li, Y. Bi-level optimal dispatch and control strategy for air-conditioning load based on direct load control. *Proc. CSEE* **2014**, *34*, 1546–1555.

26. Chen, S.; Cheng, R.S. Operating reserves provision from residential users through load aggregators in smart grid: A game theoretic approach. *IEEE Trans. Smart Grid* **2017**, *10*, 1588–1598. [CrossRef]

27. Samadi, P.; Mohsenian-Rad, A.H.; Schober, R.; Wong, V.W.; Jatskevich, J. Optimal real-time pricing algorithm based on utility maximization for smart grid. In Proceedings of the IEEE International Conference on Smart Grid Communications, Gaithersburg, MD, USA, 4–6 October 2010; pp. 415–420.

28. Fan, Z. A distributed demand response algorithm and its application to PHEV charging in smart grids. *IEEE Trans. Smart Grid* **2012**, *3*, 1280–1290. [CrossRef]

29. Gensollen, N.; Gauthier, V.; Becker, M.; Marot, M. Stability and performance of coalitions of prosumers through diversification in the smart grid. *IEEE Trans. Smart Grid* **2016**, *9*, 963–970. [CrossRef]

30. Chiu, W.-Y.; Hsieh, J.-T.; Chen, C.-M. Pareto optimal demand response based on energy costs and load factor in smart grid. *IEEE Trans. Ind. Inform.* **2019**. [CrossRef]

31. Soliman, H.M.; Leon-Garcia, A. Game-theoretic demand-side management with storage devices for the future smart grid. *IEEE Trans. Smart Grid* **2014**, *5*, 1475–1485. [CrossRef]

32. Facchinei, F.; Pang, J.S. *Finite-Dimensional Variational Inequalities and Complementarity Problems: Volume I and II*; Springer: New York, NY, USA, 2007.

33. Scutari, G.; Palomar, D.P.; Facchinei, F.; Pang, J.-S. Monotone games for cognitive radio systems. In *Nonlinear Systems*; Springer Science and Business Media: London, UK, 2012; Volume 417, pp. 83–112.

34. Marzband, M.; Azarinejadian, F.; Savaghebi, M.; Guerrero, J.M. An optimal energy management system for islanded microgrids based on multiperiod artificial bee colony combined with Markov chain. *IEEE Syst. J.* **2015**, *11*, 1712–1722. [CrossRef]

Article

Ramping of Demand Response Event with Deploying Distinct Programs by an Aggregator

Omid Abrishambaf [1], Pedro Faria [1,*] and Zita Vale [2]

[1] GECAD—Research Group on Intelligent Engineering and Computing for Advanced Innovation and Development, IPP—Polytechnic Institute of Porto, Rua DR. Antonio Bernardino de Almeida, 431, 4200-072 Porto, Portugal; ombaf@isep.ipp.pt

[2] IPP—Polytechnic Institute of Porto, Rua DR. Antonio Bernardino de Almeida, 431, 4200-072 Porto, Portugal; zav@isep.ipp.pt

* Correspondence: pnf@isep.ipp.pt; Tel.: +351-228-340-511; Fax: +351-228-321-159

Received: 7 February 2020; Accepted: 10 March 2020; Published: 16 March 2020

Abstract: System operators have moved towards the integration of renewable resources. However, these resources make network management unstable as they have variations in produced energy. Thus, some strategic plans, like demand response programs, are required to overcome these concerns. This paper develops an aggregator model with a precise vision of the demand response timeline. The model at first discusses the role of an aggregator, and thereafter is presented an innovative approach to how the aggregator deals with short and real-time demand response programs. A case study is developed for the model using real-time simulator and laboratory resources to survey the performance of the model under practical challenges. The real-time simulation uses an OP5600 machine that controls six laboratory resistive loads. Furthermore, the actual consumption profiles are adapted from the loads with a small-time step to precisely survey the behavior of each load. Also, remuneration costs of the event during the case study have been calculated and compared using both actual and simulated demand reduction profiles in the periods prior to event, such as the ramp period.

Keywords: aggregator; demand response; ramp period; real-time simulation

---

1. Introduction

Electrical energy demand has significantly increased in the last decades. This has led to huge peak of greenhouse gas emissions in order to provide and supply the required demand [1]. In the past decades, fossil fuels were the raw materials for electricity production [2]. However, nowadays several low carbon technologies and renewable energy resources (RERs) have been utilized to produce electricity [3]. Smart grids and microgrids are new some new concepts for the future distribution networks to eliminate the hierarchical structure of the grid and convert them to a fully decentralized and transactive energy system [4]. To do this, the process of energy production should be also placed in the demand side, among all electricity consumers. Therefore, the concept of prosumer (a consumer who also produces electricity) has been raised [5]. However, this makes the network instabilities more tangible than before, as a significant number of small and medium scale consumers and producers will be involved in the network management scenarios [6].

Distributed generation (DG) and RERs are considered as one of the bases of smart grids and microgrids implementation [7]. However, these paradigms would be fully addressed while they have been integrated with demand response (DR) programs [8].

In fact, the DR program is a feature in the upcoming distribution network to connect low carbon technologies without the need for reinforcement [9]. There are various definitions in the literature for DR program. While each definition has its own strengths and weak points, the most completed one is defined by Federal Energy Regulatory Commission (FERC) [10] as:

"Changes in electric use by demand-side resources from their normal consumption patterns in response to changes in the price of electricity, or to incentive payments designed to induce lower electricity use at times of high wholesale market prices or when system reliability is jeopardized."

According to the definition mentioned above, in a simple word, DR can be described as the reaction of electricity consumers to the price signals considered as incentives to reduce/modify the electricity use pattern. There are several types of DR programs: price-based and incentive-based [11]. In Price-based category, there are several programs such as real-time pricing (RTP), time-of-use (TOU), critical peak pricing (CPP), etc. [12]. Also, in the incentive-based DR programs, various strategies are proposed, namely direct load control (DLC), emergency demand response service (EDRS) capacity market programs (CMP), interruptible demand response program (IDRP), etc. [13]. Each program has its own specifications that are applicable depending on technical or economic reasons of the electricity network. As an example, the ERCOT market in Texas utilizes EDRS to maintain network stability in emergency conditions to reduce power outages. In this market, EDRS participants can provide DR reduction within a 10 to 30 min ramp period in advance to the event [14]. However, there is a minimum reduction capacity for the DR event and its participants in order to directly participate in the electricity markets. According to the surveyed references [15–17], this minimum reduction capacity for a consumer who intends to have an active role in the electricity market negotiations, is various depending on the DR type, typically from a few kilowatts to megawatts. In other words, this makes small scale consumers almost incapable to directly participate in electricity markets [18]. To overcome this barrier, an aggregator can be considered as a third-party entity between the upstream and downstream sides of the network [19,20]. In fact, this entity aggregates all small and medium scales DR participants and contribute them as a unique DR resource in the electricity market negotiations [21].

By a simple look on the current trends, a lot of papers and research projects can be found that are focused on the concepts of aggregator, as the role of the aggregator is being legalized in several European countries (e.g., France, Finland, Austria, Denmark, etc.) [22]. However, one of the aspects of aggregator that has not been discussed widely in the literature is the way that aggregator deals with ramp period before the DR event is started. Incentive calculations and payments between aggregator and DR participants during the ramp period, required information exchange between aggregator and network operators (e.g., independent system operator–ISO) during the ramp period, and several other issues need to be discussed before the models move from theoretical phase towards implementation level. Besides this, both scientific and practical features of any model should be scrutinized, learning from past experiences to estimate and prevent probable future issues. To do this, adequate real models and laboratory tools are essential to test and verify the functionalities of any developed model under practical challenges.

To address these issues, this paper proposes an aggregator model with a precise vision of DR timeline and ramp period. The model includes introducing the roles of aggregator in the electricity system as a third party, and how it deals with DR programs implementation and remuneration payments. Furthermore, a case study is developed in this paper to validate the model under practical challenges and technical issues. This has been done by a real-time simulator machine (OP5600) and a set of laboratory equipment as hardware-in-the-loop (HIL). In this way, the behavior of aggregator is surveyed in each moment of DR implementation, especially in the ramp period before the event being started. In the end, the remuneration costs are compared using both experimental and numerical results to reveal the importance of experimental tests and validations.

The rest of the paper is organized as follows: A literature review is presented in Section 2 to identify the challenges and gaps in the current trend of aggregator and DR programs. Section 3 presents the proposed methodology in four different subsections including the role of the aggregator in the power system, DR timeline, and ramp period evaluation, key points of aggregator to define DR programs, and a linear programming optimization for the aggregator to minimize DR remuneration costs. Subsequently, Section 4 proposes a case study and real-time simulation model developed for the aggregator to validate it using practical challenges, and its results are shown in Section 5. Finally, the main conclusions of the work are presented in Section 6.

2. Literature Review and Paper Contributions

There are plenty of research works focused on the context of the aggregator model and DR implementation. In [8], a methodology has been described to use multitype DR programs to smooth the uncertainty of RERs. The method utilizes a multi-objective scheduling algorithm for smoothing the fluctuations in RERs. Several constraints for DR programs have been considered in the same paper in order to maximize the user of RERs and maintain the balance in the network.

A real-time simulation model has been proposed in [12] for an aggregator entity, which uses several real hardware resources to emulate the consumption and generation profiles. Also, an optimization problem has been developed in the same work to optimally schedule the DR resources and RERs aiming at minimizing the aggregator operational costs. Although an actual infrastructure has been proposed in [12] for testing aggregator's concepts, there is no discussion about the ramping of DR programs prior to the event, and how aggregator deals with remunerations in this period. A realistic model of an aggregator in the scope of curtailment service provider has been presented in [17]. The authors proposed a real-time simulation model that supports decision making for DR validation in real-time. Also, a preliminary discussion has been proposed in the paper regarding the ramp period before DR events proposing the use of RERs and DR programs. The presented results proved the performance of the curtailment service provider using real measured data from the laboratory equipment. However, there is no discussion about how curtailment service provider behaves with incentive payments, and also a precise vision to the DR program information received from system operator, such as notification deadline, program duration, etc.

In [23], the authors proposed an assessment of a DR program for consumers who are equipped with a smart meter. A load-serving entity plays the role of the aggregator to offer incentives to the participant relying on near real-time information. Moreover, a timeline has been proposed in the same work regarding the information exchange between the load-serving entity and ISO regarding DR programs. Although the paper presented an interesting model, the authors validated their model through a numerical case study, and there is a lack of an experimental test and validation.

A bottom-up model has been proposed in [24] for an aggregator dealing with DR programs. Load shifting, load recovery, and load curtailment are considered as three types of DR programs available in the aggregator network. Also, through this model, the aggregator participates in the day-ahead markets by trading these DR flexibilities. In the end, the authors validated their model by a numerical case study using Nordic electricity market.

In [25], the authors discussed a short-term decision-making model for an electricity retailer that included RERs. Also, short-term DR trading methodology has been proposed somehow the retailer submits this flexibility to the markets in each hour. Through the simulation performed in the case study, the authors validated that the financial profits of the retailer will be increased if it participates in both real-time market and short-term DR trading mechanism. A short-term self-scheduling model has been developed in [26], which is used by the DR aggregators. It also addressed the uncertainties of the electricity customers participating in the market. Two types of DR programs have been used in this model, which are reward-based DR and time-of-use. The proposed approach has been validated through a case study with realistic data from electricity markets.

Focusing on communication infrastructures in the aggregator network, the work presented in [27] focused on an energy quality aware bandwidth aggregation scheme. The authors firstly modelled the delay-constrained energy quality tradeoff for multipath video communications using wireless networks, and then, they present an approach to merge the rate adaptation. They surveyed the performance of the system using real wireless networks and emulations test beds. In addition to this work, in [28], the authors provided an energy efficient and quality guaranteed video transmission solution. In fact, they proposed an approach to characterize energy distortion tradeoff for video transmission in wireless networks. Furthermore, the authors of the same work developed an algorithm to optimize the energy consumption to achieve video quality target. Their experimental results demonstrated that the developed solution has performance advantages comparing to the schemes in term of energy conversion and video quality.

The main contribution of the present paper, according to the reviews presented above, where only a few of them were focused on the behavior of aggregator during ramp period of DR implementation, is the development of a model that considers short and real-time DR programs ramping, using real-time simulator and laboratory equipment. In fact, most of the previous works considered a simple period prior to the event showing and mainly focusing on the aggregator scheduling process or DR programs itself. Furthermore, a lot of interesting models and research works are available in the literature focusing on the concepts of aggregator under short and real-time DR programs. However, those lack adequate testing on real infrastructures under practical challenges and technical issues.

The following topics are addressed, supporting the main contribution of the paper:

- Evaluating the performance of aggregator during DR implementation timeline, especially the ramp period, in term of scheduling and remuneration;
- Improving aggregator resources management in short and real-time DR;
- Developing a real-time simulation model using a set of laboratory equipment to evaluate the aggregator's performance under practical and technical challenges;
- Remunerating consumers by comparing costs between the actual and simulated demand reduction profiles.

3. Proposed Methodology

This section presents the entire developed model for the aggregator and DR programs. The structure of this section consists of proposing the aggregator model focusing on its role in the electricity systems and wholesale markets. Then, it focuses directly on the DR timeline and the ramp period before the event being started. Later, DR programs specifications presented for the aggregator are demonstrated, and in the last subsection, a linear programming with the objective of DR cost minimization is shown for the aggregator model.

3.1. Aggregator Architecture

This part describes the architecture of the presented aggregator model for DR programs. In fact, the responsibility of the aggregator in this paper is to gather all small and medium scale DR participants located in the same geographical area and present this flexibility to the electricity market negotiation as a unique resource. To do this, the aggregator has to make bidirectional contracts with the electricity consumers who intend to participate in one or more DR programs. This enables the aggregator accordingly to control and monitor the consumption of the end-users. Figure 1 shows the role of aggregator as a network player in electricity systems and smart grid technology.

As Figure 1 illustrates, the aggregator has a transactional role between the downstream (medium and small-scale DR participants) and upstream (electricity markets and network operators) sides of the network. Indeed, the aggregator has two layers, namely communication-based controlling and monitoring sides. In the upper layer, the aggregator is in touch with network operators or electricity market operators. While some technical or economic instabilities occur in the network, and it is required to reduce network consumption, the aggregator will be notified by the upstream level players to apply DR programs. Subsequently, in the lower layer, the aggregator has a multi-round communication with the downstream level of the network (i.e., consumers), as it may have some voluntary DR programs. In other words, the aggregator cannot forecast the response of consumers to each DR event. This leads to having several iterations of DR requests from aggregator to the consumers, and in response, the consumers reply with their preferences, demand bids, reduction capacity, price signal, etc. This procedure will be continued until the aggregator reaches the reduction baseline, which is, in fact, one of the responsibilities of the aggregator during the ramp period. In the last step, the aggregator presents DR bids in the electricity markets with a certain rate based on the real-time price of the market. Generally, the aggregator has day-ahead information of DR bids, such as forecasted reduction rate. Therefore, it should consider the reduction rate a bit higher than the forecasted baseline. This is due to preventing possible failures in the case of some consumers opting out during the event.

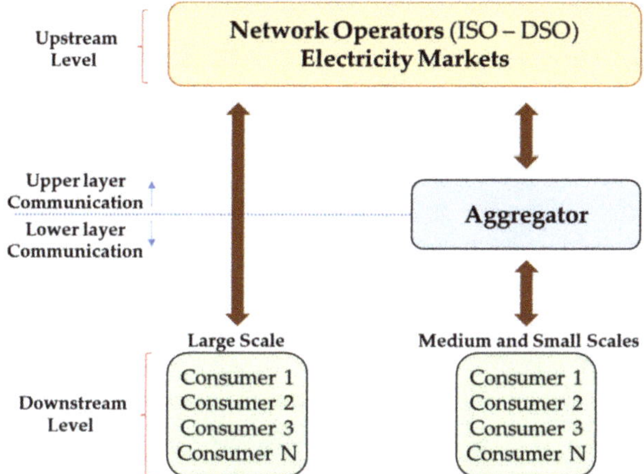

Figure 1. Participation of aggregator in the electricity system as a third-party entity.

In the practical phase, communication infrastructures are very important as they play as a base for the network management strategies, as all network players must exchange data continuously in real-time. This is more visible in critical moments, such as the ramp period, as the aggregator should have real-time information of the DR participants. Furthermore, there are instances during the DR event or ramp period whereby aggregator needs to verify that the consumers have followed the DR programs and contractual reduction correctly, according to the request of the operator (in the case of incentive-based DR). Therefore, all consumers should be equipped with a local energy management system, smart metering, and advanced metering infrastructure (AMI) to monitor real-time data, especially power consumption.

3.2. DR Timeline and Ramp Period

The ramp period is the time that the DR program manager gives to consumers in order to reach the contractual DR event baseline. The duration of the ramp period can be various based on the type of consumers, type of loads, the geographical location of the electricity market and system operator, etc. Also, all consumers will be notified in advance (from several months to 5 min) prior to the ramp period. Consequently, in short, and real-time DR programs, the tasks of aggregator are more complex as it should process the advance notification time, ramp period, and response duration before starting the event.

While the DR program manager specifies a DR event to be implemented by the aggregator, lots of information and setpoints will be dispatched and transmitted between these two entities. Figure 2. illustrates the timeline and information specified for a DR program. In fact, most of the parameters in the following timeline, such as the duration of assessment and ramp period, are defined by the DR program manager and transmitted to the aggregator using the upper communication layer shown in Section 3.1.

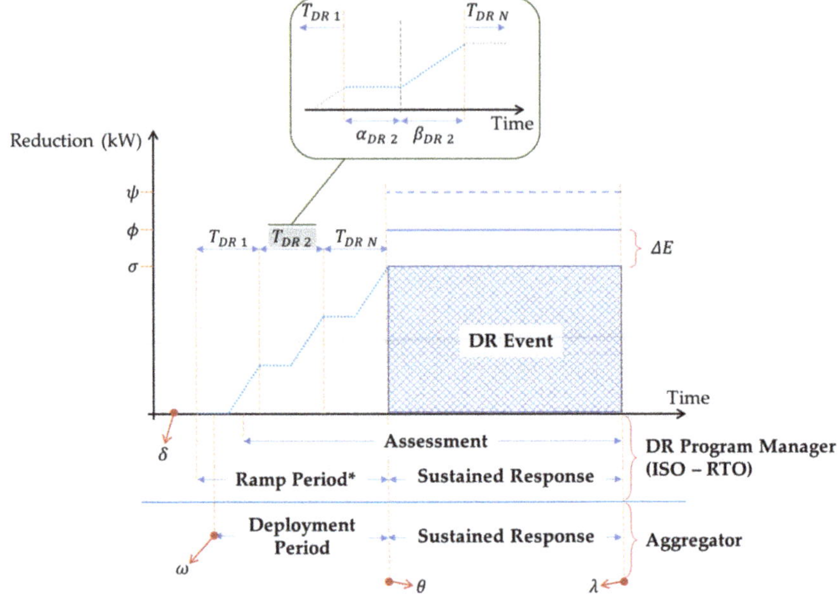

Figure 2. Timeline for demand response implementation by the aggregator.

The first point in the timeline shown in Figure 2 is the announcement deadline (δ), which is the last opportunity for the DR program manager to notify the aggregator for the DR event. After then, there is a ramp period considered as the time that aggregator is allowed to reach the desired amount of reduction. In the meanwhile, the deployment starting point (ω) is a moment during the ramp period that aggregator can take a risk to start the event by relying on forecasting the availability of DR resources. The main difference between deployment and assessment periods is that assessment is a paid period. The deployment period is the period that aggregator collects all the DR amount by different programs. In fact, the aggregator is able to collect all DR amounts within the ramp period, but it is free to start the event a bit later than the starting point of the ramp period. However, the aggregator can operate cautiously and notify DR resources for the event, during the ramp period, and wait for their response. In the last stage, the reduction deadline (θ) is the point at which the aggregator evaluates the available reduction capacity and verifies that their capacity is above the forecasted reduction baseline (ϕ). However, the event could also be started if the available reduction capacity is in the margin of forecast error (ΔE), above the reduction baseline (σ) defined by the DR program manager. While the DR event has been started, the timeline enters a sustained response period, which is the time that DR participants have to maintain their committed level of reduction until the end of the event (λ). During the sustained response period, both communication layers shown in Figure 1 are involved. In the lower layer, consumers transmit the related information to the aggregator, namely real-time consumption, and in the upper layer, aggregator conveys the consumer's information to the DR program manager.

Focusing on the ramp period in the lower communication layer of aggregator, Figure 3 demonstrates a cascade communication process during the ramp period between aggregator and consumers. By comparing Figures 2 and 3, in the first point of the ramp period, aggregator notifies consumers associated with demand response program (DRP) 1. This leads to having an activation notification period, indicated by α_{DR} in the timeline illustrated in Figure 2. After that, consumers reply with OPT IN or OPT OUT, and then if they are OPT IN, they will start the load reduction process. This has been indicated by β_{DR} in the timeline shown in Figure 2 that stands for the actual response period.

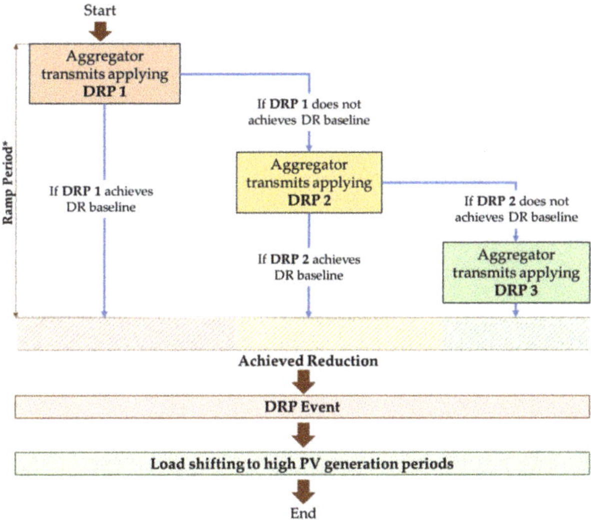

Figure 3. Ramp period evaluation process by the aggregator.

While consumers are replying with the actual reduction, the aggregator assesses the DRP 1 to check if it is sufficient for achieving the reduction baseline. If DRP 1 was insufficient, the aggregator evaluates the use of DRP 2 and notifies the consumers associated with DRP 2. This procedure is continued until the aggregator achieves to the forecasted reduction baseline, so it can inform the DR program manager and start the event.

3.3. Demand Response Programs

The aggregator is able to implement various types of DR programs. However, each program has its specific timescale. Therefore, the aggregator should select the most appropriate program according to the available timescale and objective, from long-term to real-time (Figure 4). In this paper, the main focus is given to short and real-time programs, as the ramp period is more critical in such programs. Therefore, the programs with long-term timescale (months or years planning) will be ignored.

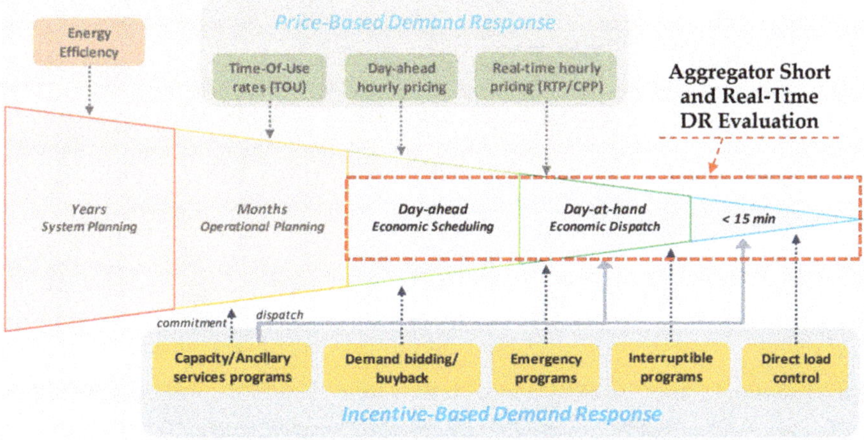

Figure 4. The timescale of demand response implementation. Adapted from [22].

Furthermore, short and real-time DR programs are more applicable comparing to other types of programs. The reason is they are usually implemented for improving or maintaining power quality as well as security of the power distribution network (e.g., voltage and frequency instabilities issues). In this context, incentive-based DR programs are the ones that can be implemented in short-term to real-time timescale, especially less than 15 min. If the aggregator intends to define a DR program, there are plenty of parameters and specifications that have to be considered. In below, there is a list of the most relevant ones to the model proposed in this paper:

- Program Type: Depending on the specification of the market, the DR program can be mandatory or voluntary. In voluntary programs, consumers can decide to participate in the event or not. In mandatory programs (e.g., DLC), the aggregator has full control over the contractual equipment as there is no need for consumer's permission. However, all DR participants will be notified before events whether in mandatory or voluntary programs;
- Remuneration: Most of DR programs has a remuneration rate that can be a power tariff discount, or an incentive paid for reduced kWh. However, some DR programs can be without remuneration as they are price-based;
- Activation signal: The aggregator can activate the event by transmitting a signal to the local controller on the consumer side. This signal can be a reduction notification, actual consumption level, electricity price notification, DLC control signal, etc.;
- Measure/Contract: The aggregator should specify in the program that what kinds of information DR participants must convey to the aggregator during the sustained response period. As an example, it can be the actual kWh reduction or real-time consumption data.

While a lot of DR programs with various specifications and parameters are accessible for the aggregator to implement, a question would be raised that which program is the most economical and optimal solution for the network. Therefore, DR program dispatch can make the use of a linear programming approach as will be explained in the next subsection.

3.4. Optimization

In this part, a set of the mathematical formulation is proposed for the aggregator model to choose the most optimal DR solution. The formulation is related to linear programming with the objective of DR cost minimization from the aggregator standpoint. As mentioned before, all consumers participating in DR programs have a contractual reduction limit as well as a remuneration tariffs associated with each program.

Equation (1) shows the objective function of the proposed linear programming, which aims to minimize the costs related to the DR programs. In this model, technical specifications of the grid, such as load balance, voltage control, etc. are not considered as it is assumed that the network operator is accountable for them. Furthermore, it is presumed aggregator will not sell/buy electricity to/from consumers, and it is only responsible for DR program implementation and provide these flexibilities to the market negotiations. So, the focus of these formulations is only given to economic aspects of DR programs from aggregator standpoint.

$$Minimize$$
$$DR\ COST = \sum_{t=\theta}^{\lambda} \sum_{c=1}^{C} \left[P_{DR\ S_{(t,c)}} \times I_{DR_{(t,c)}} \right] \tag{1}$$

The proposed objective function is modeled as a linear programming optimization problem using Rstudio® tool (www.rstudio.com), using a computer with Intel® Xeon® CPU @2.10 GHz, and 16 GB RAM. The linear and convex problem implemented, which includes in the present case study 4860 variables, can be solved by brute-force, heuristics, and others. There are several constraints that are applied to this objective function. Equation (2) shows the limitation of each DR resource in terms of minimum and the maximum capacity of them. Also, Equation (3) presents that the sum of capacity in available schedulable and non-schedulable DR resources ($P_{DR\ S} - P_{DR\ N}$) during the event should be higher than the reduction baseline in addition to the forecast margin error. This means that aggregator is always counting on an extra reduction capacity higher than the defined baseline to prevent the possible failures if some consumers opted out during the event. However, there is a limit for this extra capacity, and if the reduction goes higher than this limit, the additional capacity is not being paid. This is shown by Equation (4).

$$0 \leq P_{DR_{(t,c)}} \leq P^{max}_{DR_{(t,c)}} \qquad \forall t \in [\theta : \lambda], \ \forall c \in \{1, \ldots, C\}, \tag{2}$$

$$\sum_{c=1}^{C} P_{DR\ S_{(t,c)}} + \sum_{c=1}^{C} P_{DR\ N_{(t,c)}} \geq \sigma + \Delta E \qquad \forall t \in [\theta : \lambda] \tag{3}$$

$$\sum_{c=1}^{C} P_{DR\ S_{(t,c)}} + \sum_{c=1}^{C} P_{DR\ N_{(t,c)}} \leq \psi \qquad \forall t \in [\theta : \lambda] \tag{4}$$

In sum, this section presented the developed aggregator model with a focus on the DR timeline and aggregator's responsibility during the ramp period before the DR event is started. In the next section, a case study is proposed in order to validate and survey the functionalities of the presented model using an actual methodology and real infrastructures.

4. Case Study and Real-Time Simulation

This part explains a case study for validating and surveying the performance of the developed model under different challenges. To do this, it is considered that there is a small village with a lot of residential and commercial consumers, and only 27 of the residential consumers have direct DR contract with the aggregator. This means that the aggregator has no interaction with other consumers that are not participating in the DR programs. In the aggregator network, there are 10 consumers equipped with Photovoltaic (PV) panels as RERs, and five consumers with energy storage system (ESS). The use of RERs makes the aggregator capable to shift the load in the high PV generation periods, as the energy produced by this resource is uncontrollable. Figure 5 shows the village and the related aggregator network. In this case study, a part of the DR participants in the aggregator is being emulated by a set of laboratory equipment, so-called resistive loads bench in this paper. In this way, the aggregator's performance can be surveyed in both simulation and experimental aspects.

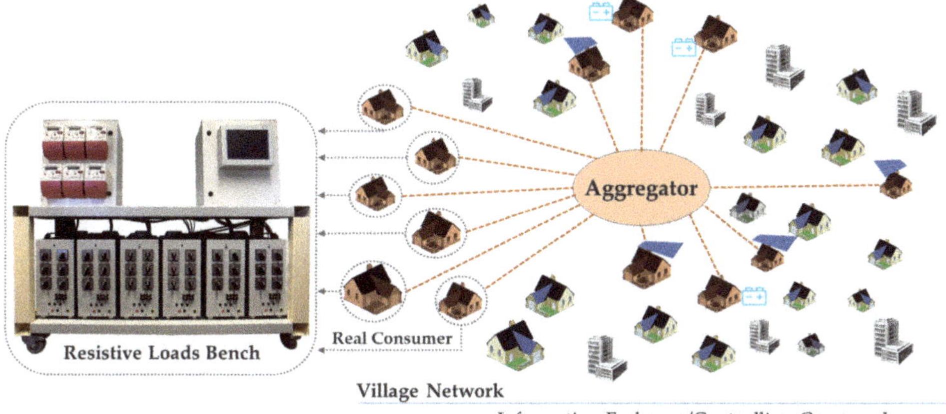

Figure 5. Schematic of the aggregator network in the case study.

The resistive loads bench consists of six resistive consumer loads that each of which has a nominal capacity of 4 kW. The bench is equipped with a set of controlling and monitoring devices, including a programmable logic controller (PLC) as a central controller of the bench, a set of relays to control the rate of consumption in each load, and a set of energy meters and commercial smart meters in order to monitor the real-time consumption of the loads. To survey both numerical and experimental features of the aggregator, a MATLAB™/Simulink model has been developed representing the electrical network of the 27 consumers in the aggregator model, as Figure 6 shows. Moreover, the OP5600 real-time simulator has been utilized to integrate the resistive loads bench in the Simulink model as hardware-in-the-loop (HIL). In fact, OP5600 executes the Simulink model in real-time enabling the user to control and monitor real hardware resources outside of simulation environments. This is done through several communication protocols as well as Digital and Analog slots in OP5600. More information about the performance of OP5600 and HIL methodology are available in [12].

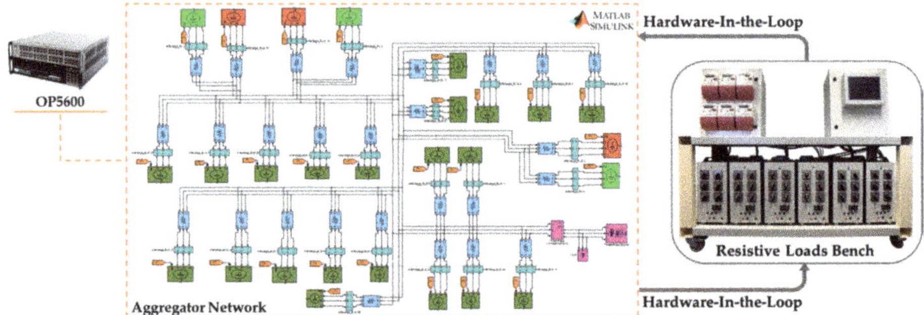

Figure 6. Real-Time simulation model of aggregator integrated with HIL methodology.

Regarding the DR programs, three types of programs are considered that consumers are able to establish with the aggregator, as Table 1 shows. Each program has its own features and specifications, which the aggregator defined according to the instructions and regulations of the upstream network players, such as market operator and DR program manager.

Table 1. DR contracts available for consumers to establish with aggregator.

DR Type	Mandatory/ Voluntary	Activation/ Signal	Remuneration	Measure/ Contract	Associated Device	Notification Time
DLC T1	Mandatory	Directly to the device	0.05 EUR/kWh	Actual kWh reduction	Washing Machine, Dishwasher	5 Min
DLC T2	Mandatory	Directly to the device	0.30 EUR/Event	Actual positive reply	Air Conditioner	5 Min
IDRP	Voluntary	Reduction Notification	0.03 EUR/kWh	Actual kWh reduction	Water Heater, Fan Heater	30 Min

The DR programs and the related remuneration rates shown in Table 1, has been developed by the authors in the scope of their previous works, and only the most relevant description has been mentioned in this part. More detailed information about these DR programs is available in [29]. According to Table 1, if a consumer establishes a contract with the aggregator for DLC T1, it will give permission to the aggregator to directly control its air conditioner, and in exchange, the consumer receives an incentive of 0.05 EUR/kWh for the reduction. In the DLC T2 contract, the aggregator is able to directly control the washing machine and dishwasher of the DR participant for a contractual number of events per month. Consumers who participate in DLC T2 receive an incentive of 0.06 EUR/kWh for the reduction. The last DR contract that consumers can establish with aggregator is IDRP. As this program is voluntary, the consumer can decide to participate or not. Therefore, a longer notification time is considered for this program, so the consumer has an adequate amount of time to make a decision. Table 2 demonstrates the capacity of the associated devices in all 27 consumers related to the aggregator network. The capacities shown on the same table are in kiloWatt (kW). As is clear in Table 2, each consumer has at least one DR contract with the aggregator. This means that the associated devices dedicated in Table 2, are being controlled by aggregator in DLC T1 and T2 programs, and by consumer itself in IDRP. The capacities shown in Table 2 are an average between the minimum and maximum active power consumption of each during an entire day.

Table 2. Controllable devices involved in the DR programs (AC = Air Conditioner, WM = Washing Machine, DW= Dishwasher, WH= Water Heater, FH = Fan Heater).

| Consumer ID | DLC T1 | DLC T2 | | IDRP | | Consumer ID | DLC T1 | DLC T2 | | IDRP | |
	AC (kW)	WM (kW)	DW (kW)	WH (kW)	FH (kW)		AC (kW)	WM (kW)	DW (kW)	WH (kW)	FH (kW)
1	0.54		0.62	0.33		15					
2	0.24		0.19	0.29		16	0.24				
3	0.39	0.16	0.16	0.20		17					0.17
4	0.23	0.18	0.50	0.31		18	0.51	0.27	0.45		
5	0.55	0.15	0.50			19			0.09	0.19	
6	0.17		0.50			20	0.05				
7	0.05		0.07	0.35		21	0.39	0.34	0.66	0.46	
8	0.76		0.07	0.23		22	0.33	0.25	0.63		
9	0.39	0.19	0.22	0.30		23	0.20				
10	0.05					24	6.90				0.27
11	0.43					25	1.50				0.17
12					0.27	26	4.71				0.17
13	0.05		0.19			27	0.52				0.12
14	0.02					-	-	-	-	-	-

Furthermore, as Table 2 shows, 85% of consumers are equipped with air conditioner, 26% of them have washing machine, 52% include dish washer, 33% have water heater, and 22% of them have fan heater. Figure 7 illustrates a day-ahead consumption and generation profile of the entire aggregator network as well as available DR capacities presumed for the case study based on Table 2. The data shown in Figure 7 are for a random winter day with a 15-min time interval, and adapted from a research project [30] related to the implementation of an intelligent energy management system in two small cities in Portugal.

The consumption shown in Figure 7 is only related to the aggregator network (27 consumers) and not to the entire village. Moreover, the uncontrollable part of consumption is related to the devices on the consumer side that aggregator has no interaction with them. In this case study, it is considered that the aggregator receives a DR event from the DR program manager with 10 kW as the reduction baseline, starting at 12:00 PM for two hours. Also, the aggregator is notified one hour in advance. The reason for this DR event could be a technical fault or any economic causes in the main grid. Figure 8 shows the DR event applied in the aggregator consumption profile.

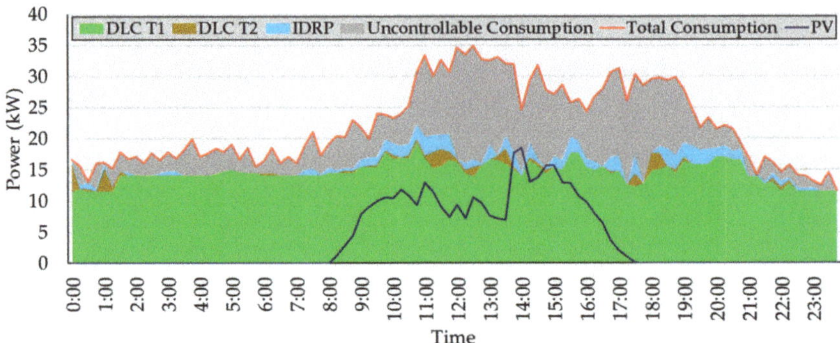

Figure 7. Aggregator consumption profile and stacked available DR capacities.

As Figure 8 shows, the reduced consumption during the DR event has been shifted to some periods after the event that have a high rate of PV generation. This enables the aggregator to use the energy produced by the local resources. Since the notification time of the event is one hour in advance, the aggregator should reach the reduction baseline for one hour (i.e., ramp period). Therefore, the aggregator starts announcing the consumers one by one to participate in the event. In the meantime, if some consumers have delay on replying to the DR announcement, the aggregator is able to use and discharge the available ESS in order to compensate the response time of the consumers. Also, the aggregator adjusts its internal reduction baseline to around 15 kW for keeping the consumption rate at 20 kW. This is due to overcoming the possible issues during the event, namely some consumers opting out.

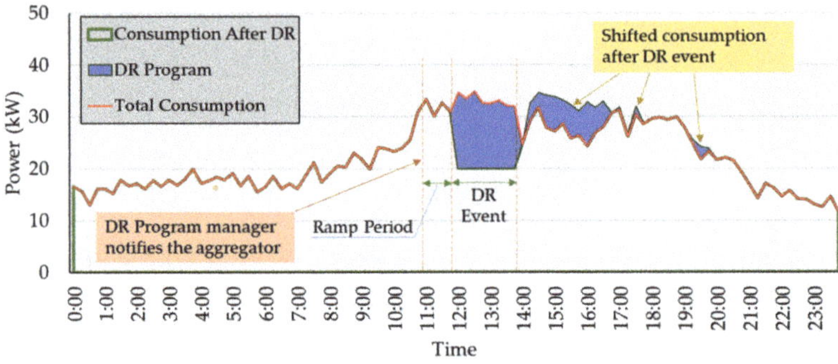

Figure 8. Proposed DR event applied in the aggregator consumption profile.

In practice, a consumption profile with a 15-min time interval is not that applicable for remuneration and scheduling purposes, as a lot of changes could happen during this time. In some cases, it is possible that the DR program manager pays incentives to the aggregator with a 15-min time interval. However, in the downstream side of the network, as the aggregator is dealing with every single consumer, the 15-min time interval is a long period. Consequently, in order to have a clear vision of the consumption profile during the ramp period and DR event, Figure 9 illustrates the aggregator consumption curve between 11:00 to 14:00, with a 1-min time interval. In the same figure, the uncontrollable part of aggregator consumption is not shown as the focus is given to the controllable part of consumption.

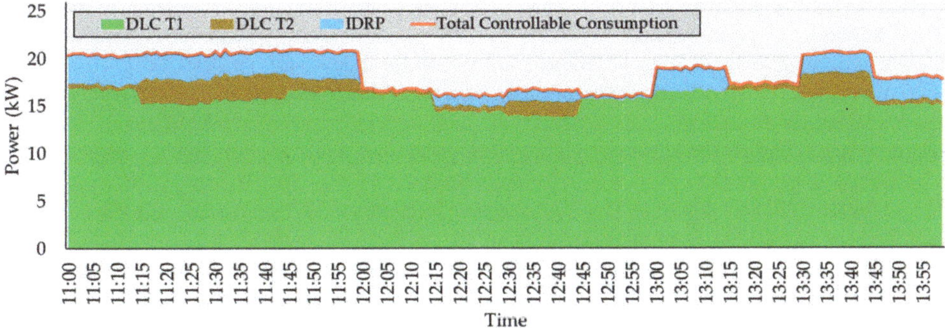

Figure 9. Available Demand response programs during ramp period and event.

As is clear in both Figures 7 and 9, almost half of the consumption during ramp period and the event itself is dedicated to DLC T1 (air conditioner), and only a few parts are devoted to DCL T2 (washing machine and dishwasher) and IDRP (water heater and fan heater). As the fan heaters in the IDRP program are resistive loads, they are a suitable target for emulating by resistive loads bench. Thus, in the next section (results) all fan heaters in the aggregator that are associated in the IDRP program are being emulated by the resistive loads bench and the behavior of each device as well as the aggregator facing an actual profile will be scrutinized.

5. Results

This section presents all the gained results from the aggregator standpoint. All the results provided in this section are for surveying the performance of the aggregator model during the ramp period and the DR event itself. Figure 10 shows the consumption reduction profiles after applying DR programs. The results shown in the same figure are with the one-minute time interval between 11:00 where the DR program manager notified the aggregator for the event, until the end of the event (14:00).

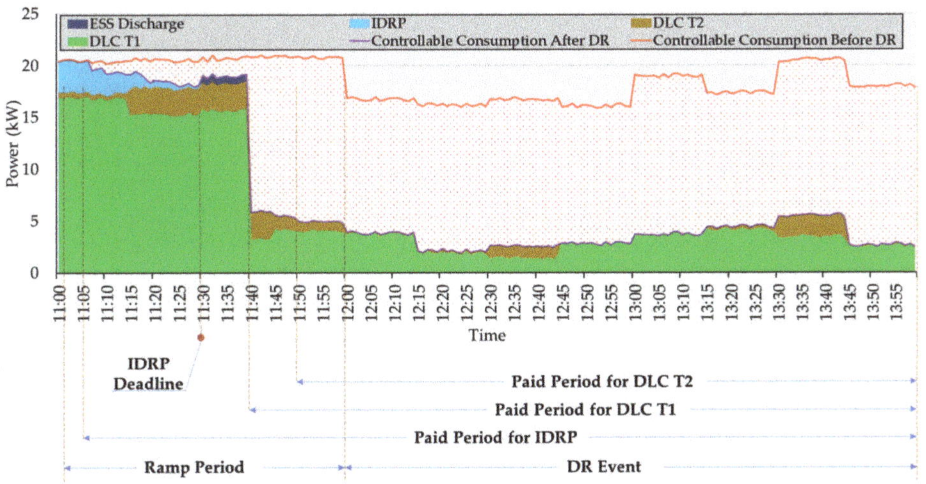

Figure 10. Results of applying demand response programs by the aggregator.

As it is clear in Figure 10, the aggregator firstly started to apply IDRP since it has the lowest remuneration rate from the aggregator point of view. To do this, the aggregator announced all the IDRP resources one by one and waited for their response as the program is voluntary. Then, all IDRP participants replied with their desired responses (OPT IN or OPT OUT) until 11:30 (program deadline). After that, the aggregator evaluates the use of the IDRP program and as it did not reach the reduction baseline, it decided to apply DLC T1 program. Therefore, the aggregator notifies DLC T1 participants 5 min in advance (11:35) to the starting point of the event (11:40). During this 5 min, the aggregator takes advantage of the available ESSs and start discharging them, so there would be a reduction in the consumption. While all resources in DLC T1 have participated in the event and the reduction rate of DLC T1 has been reached, the aggregator stops discharging the ESSs. The same procedure is also applied for DLC T2, and finally, at 11:50, when the aggregator reached the desired reduction baseline and it is ready to start the event at 12:00.

Moreover, as Figure 10 shows, the starting point of the paid period for each program is the moment that the first participant reduced its consumption, so the aggregator has to pay the contractual remuneration according to the reduced power. In other words, the aggregator receives the remuneration from the DR program manager only for event duration (i.e., in this case, two hours between 12:00 to 14:00). However, the aggregator must start paying the remuneration before the event during ramp period as the DR participants started the consumption reduction. That's why aggregator should pay remuneration to the DR participant with a lower rate than the one that it receives from the DR program manager, so it would be able to manage all the paid periods without a financial downturn.

In order to have a more precise and technical vision to the model, Figure 11 illustrates the experimental results adapted from the real-time simulation model and Resistive Loads Bench as HIL. The results shown Figure 11 are related to the 6 DR participants that own Fan Heater and they are involved in the IDRP (indicated in Table 2). In fact, each consumer load in the Resistive Loads Bench emulates a Fan Heater in each DR participant. The results demonstrated in Figure 11 are adapted from MATLAB™/Simulink and OP5600 in 3600 periods of 0.5 s, which is in total 30 min, between 11:00 to 11:30 while all IDRP resources are announced to participate. In other words, the time step of this model in real-time simulation is set at 0.5 s. This means OP5600 conveys the reference signal (power reference in Figure 11) to the resistive load bench with one-minute time interval, and then, it acquires real-time consumption data with 0.5 s time interval. The actual power measurement curve in Figure 11 shows the real behaviors and reactions of resistive consumer loads, and it is only shown until the IDRP deadline, as after this moment all their consumption was cut.

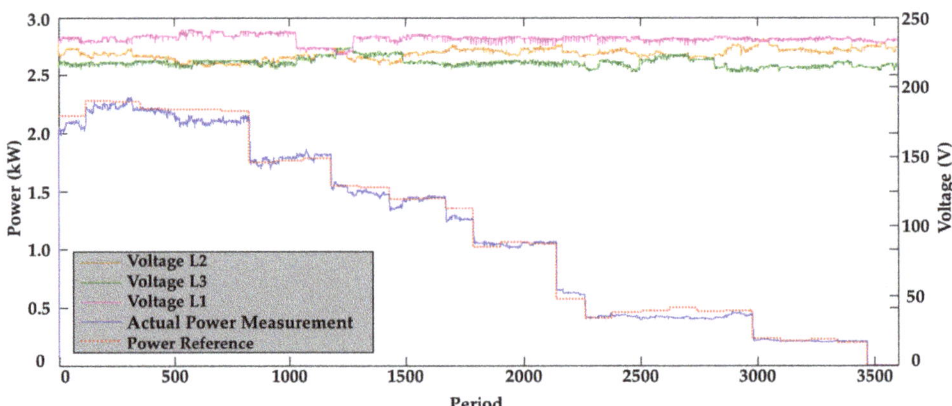

Figure 11. Experimental results adapted from OP5600 and Resistive Loads Bench.

Indeed, employing real-time simulation (OP5600) and laboratory equipment as HIL for emulating consumption profiles have several advantages. One of them is that we validate the actual demand reduction under the technical parameters of the grid, namely voltage variations (as shown in Figure 11). This leads to having a gap between the experimental and simulation results. This gap is clearly visible in Figure 11 between the red dashed line as Power Reference and the blue line as actual power measurement. Consequently, it is interesting to calculate and compare the remuneration costs of aggregator using both experimental and simulation results. Figure 12 shows the accumulated remuneration costs during the ramp period and the event using simulation profiles.

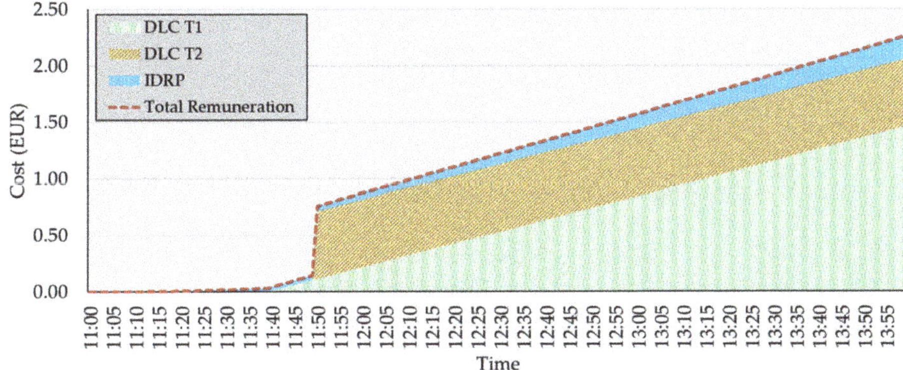

Figure 12. Accumulated remuneration costs of aggregator during the ramp period and the event.

As Figure 12 shows, there are a few remuneration costs for the IDRP program as the remuneration rate and the available capacity were not significant. Also, the costs of DLC T1 has a linear ascending gradient since the available capacity of this program was constant during the event. Finally, as DLC T2 has a fixed remuneration rate per event, it has a constant ratio in the aggregator's remuneration expenses. Table 3 demonstrates the detailed cost calculation for each program. In Table 3, the main focus is given to the first 30 min of the IDRP program as the real-time simulation and HIL methodology have been implemented for this specific program. The actual and simulation profiles are respectively the blue (actual power measurement) and red dashed line (power reference) in Figure 11.

Table 3. Remuneration costs for each program paid by the aggregator to DR participants.

Program Type	DLC T1	DLC T2	IDRP		
			11:00 to 11:29		11:30 to 14:00
Applied Period	11:00 to 14:00	11:00 to 14:00	Actual Profile	Simulation Profile	
Cost (EUR)	1.4583	0.6	0.0083	0.0081	0.1907

Total Cost = 2.2573 EUR (using actual profile); 2.2571 EUR (using simulation profile).

As Table 3 shows, the calculated remuneration cost between 11:00 to 11:29 in IDRP has a difference between the actual and simulation profiles. This cost difference is not significant because in this specific model it is only for six fan heater devices as a part of the IDRP program, which has a little reduction capacity for 30 min. Suppose that the aggregator has a huge number of DR participants, namely 1 million customers with a longer DR event. Therefore, this little difference becomes remarkable in this case as it would mean a huge amount of cost variation between what it is expected and what occurs in actual cases.

6. Conclusions

Using renewable energy resources and distributed generation has an important role to reduce the peak of greenhouse gas emissions. Innovative management strategies, such as integrating demand response programs, are required. This paper presented a precise vision of the demand response timeline in an aggregator model. The proposed aggregator has been considered as a third party between the upstream and downstream sides of the network, to aggregate small scale demand response resources.

The time needed in the short and real-time demand response programs to notify all participants, to wait for their response, and evaluate the available resources is addressed.

For real-time simulation, a set of resistive loads to emulate the actual demand reduction of some demand response participants have been used. The numerical results highlight that the costs related to the periods prior to the event, such as ramp period, should be taken into account as in the most of model, demand response costs are only related to the period between the starting and ending point of the event. It should always be considered that normally the aggregator does not reach the desired reduction level immediately, and it requires some time to reach the desired reduction level. Regarding the remuneration, while the consumption is being reduced, consumers expect to receive remunerations for the related consumption reduction, even if the reduction has occurred prior to the starting of the event.

The experimental results obtained through emulation of loads indicate that there is a gap between the expected and actual results. In this way, laboratory tests play an important role to reveal technical issues of any model under practical challenges, namely voltage variations, frequency instabilities, and other electrical grid conditions.

Author Contributions: Conceptualization, P.F.; investigation, O.A.; methodology, P.F.; project administration, Z.V.; resources, O.A.; writing—original draft, O.A.; writing—review and editing, P.F. and Z.V. All authors have read and agreed to the published version of the manuscript.

Acknowledgments: The present work was done and funded in the scope of the following projects: UIDB/00760/2020, CEECIND/02887/2017, and COLORS PTDC/EEI-EEE/28967/2017, funded by FEDER Funds through the COMPETE program. This work has also received funding from the European Union's Horizon 2020 research and innovation programme under project DOMINOES (grant agreement No 771066).

Conflicts of Interest: The authors declare no conflict of interest.

Nomenclature

T_{DR}	Required time to achieve demand response baseline
α_{DR}	Notification time specified for each demand response program
β_{DR}	Actual response time of demand response program
ψ	Maximum paid reduction
ϕ	Forecasted reduction baseline
σ	Reduction baseline
δ	Announcement deadline
ω	Deployment deadline
θ	Reduction deadline
λ	The finishing point of the demand response event
ΔE	The forecast error margin of reduction baseline
C	Number of consumers
I_{DR}	Incentive paid for each demand response program
$P_{DR\,N}$	The non-schedulable demand response program
$P_{DR\,S}$	The schedulable demand response program

References

1. Wang, S.; Tarroja, B.; Schell, L.S.; Shaffer, B.; Samuelsen, S. Prioritizing among the end uses of excess renewable energy for cost-effective greenhouse gas emission reductions. *Appl. Energy* **2019**, *235*, 284–298. [CrossRef]
2. Leonard, M.D.; Michaelides, E.E.; Michaelides, D.N. Energy storage needs for the substitution of fossil fuel power plants with renewables. *Renew. Energy* **2020**, *145*, 951–962. [CrossRef]
3. Ghorbani, N.; Aghahosseini, A.; Breyer, C. Assessment of a cost-optimal power system fully based on renewable energy for Iran by 2050—Achieving zero greenhouse gas emissions and overcoming the water crisis. *Renew. Energy* **2020**, *146*, 125–148. [CrossRef]
4. Abrishambaf, O.; Lezama, F.; Faria, P.; Vale, Z. Towards transactive energy systems: An analysis on current trends. *Energy Strategy Rev.* **2019**, *26*, 100418. [CrossRef]

5. Sikorski, T.; Jasiński, M.; Ropuszyńska-Surma, E.; Węglarz, M.; Kaczorowska, D.; Kostyła, P.; Leonowicz, Z.; Lis, R.; Rezmer, J.; Rojewski, W.; et al. A Case Study on Distributed Energy Resources and Energy-Storage Systems in a Virtual Power Plant Concept: Economic Aspects. *Energies* **2019**, *12*, 4447. [CrossRef]

6. Abrishambaf, O.; Faria, P.; Gomes, L.; Spínola, J.; Vale, Z.; Corchado, J. Implementation of a Real-Time Microgrid Simulation Platform Based on Centralized and Distributed Management. *Energies* **2017**, *10*, 806. [CrossRef]

7. Morales González, R.; Shariat Torbaghan, S.; Gibescu, M.; Cobben, S. Harnessing the Flexibility of Thermostatic Loads in Microgrids with Solar Power Generation. *Energies* **2016**, *9*, 547. [CrossRef]

8. Yi, W.; Zhang, Y.; Zhao, Z.; Huang, Y. Multiobjective Robust Scheduling for Smart Distribution Grids: Considering Renewable Energy and Demand Response Uncertainty. *IEEE Access* **2018**, *6*, 45715–45724. [CrossRef]

9. Althaher, S.; Mancarella, P.; Mutale, J. Automated Demand Response From Home Energy Management System Under Dynamic Pricing and Power and Comfort Constraints. *IEEE Trans. Smart Grid* **2015**, *6*, 1874–1883. [CrossRef]

10. Federal Energy Regulatory Commission. *Assessment of Demand Response & Advanced Metering*; Federal Energy Regulatory Commission: Washington, DC, USA, 2011.

11. Behboodi, S.; Chassin, D.P.; Djilali, N.; Crawford, C. Transactive control of fast-acting demand response based on thermostatic loads in real-time retail electricity markets. *Appl. Energy* **2018**, *210*, 1310–1320. [CrossRef]

12. Abrishambaf, O.; Faria, P.; Vale, Z. Application of an optimization-based curtailment service provider in real-time simulation. *Energy Inform.* **2018**, *1*, 3. [CrossRef]

13. Alasseri, R.; Rao, T.J.; Sreekanth, K.J. Conceptual framework for introducing incentive-based demand response programs for retail electricity markets. *Energy Strategy Rev.* **2018**, *19*, 44–62. [CrossRef]

14. Du, P.; Lu, N.; Zhong, H. *Demand Response in Smart Grids*; Springer: Berlin/Heidelberg, Germany, 2019; ISBN 978-3-030-19768-1.

15. Bakr, S.; Cranefield, S. Using the Shapley Value for Fair Consumer Compensation in Energy Demand Response Programs: Comparing Algorithms. In Proceedings of the 2015 IEEE International Conference on Data Science and Data Intensive Systems, Sydney, Australia, 11–13 December 2015; pp. 440–447.

16. Martínez Ceseña, E.A.; Good, N.; Mancarella, P. Electrical network capacity support from demand side response: Techno-economic assessment of potential business cases for small commercial and residential end-users. *Energy Policy* **2015**, *82*, 222–232. [CrossRef]

17. Abrishambaf, O.; Faria, P.; Vale, Z.; Corchado, J.M. Real-Time Simulation of a Curtailment Service Provider for Demand Response Participation. In Proceedings of the 2018 IEEE/PES Transmission and Distribution Conference and Exposition (T&D), Colorado, CO, USA, 16–19 April 2018; pp. 1–9.

18. Kwon, S.; Ntaimo, L.; Gautam, N. Optimal Day-Ahead Power Procurement with Renewable Energy and Demand Response. *IEEE Trans. Power Syst.* **2017**, *32*, 3924–3933. [CrossRef]

19. Della Giustina, D.; Ponci, F.; Repo, S. Automation for smart grids in Europe. In *Application of Smart Grid Technologies*; Elsevier: Amsterdam, The Netherlands, 2018; pp. 231–274.

20. Abrishambaf, O.; Faria, P.; Vale, Z. SCADA Office Building Implementation in the Context of an Aggregator. In Proceedings of the 2018 IEEE 16th International Conference on Industrial Informatics (INDIN), Porto, Portugal, 18–20 July 2018; pp. 984–989.

21. Khorram, M.; Abrishambaf, O.; Faria, P.; Vale, Z. Office building participation in demand response programs supported by intelligent lighting management. *Energy Inform.* **2018**, *1*, 1–14. [CrossRef]

22. Spinola, J.; Abrishambaf, O.; Faria, P.; Vale, Z. Identified Short and Real-Time Demand Response Opportunities and the Corresponding Requirements and Concise Systematization of the Conceived and Developed DR Programs—Final Release. Available online: http://dream-go.ipp.pt/PDF/DREAM-GO_Deliverable2-3_v3.0.pdf (accessed on 15 January 2020).

23. Zhong, H.; Xie, L.; Xia, Q. Coupon Incentive-Based Demand Response: Theory and Case Study. *IEEE Trans. Power Syst.* **2013**, *28*, 1266–1276. [CrossRef]

24. Mahmoudi, N.; Heydarian-Forushani, E.; Shafie-khah, M.; Saha, T.K.; Golshan, M.E.H.; Siano, P. A bottom-up approach for demand response aggregators' participation in electricity markets. *Electr. Power Syst. Res.* **2017**, *143*, 121–129. [CrossRef]

25. Do Prado, J.C.; Qiao, W. A Stochastic Decision-Making Model for an Electricity Retailer with Intermittent Renewable Energy and Short-Term Demand Response. *IEEE Trans. Smart Grid* **2019**, *10*, 2581–2592. [CrossRef]

26. Vahid-Ghavidel, M.; Mahmoudi, N.; Mohammadi-Ivatloo, B. Self-Scheduling of Demand Response Aggregators in Short-Term Markets Based on Information Gap Decision Theory. *IEEE Trans. Smart Grid* **2019**, *10*, 2115–2126. [CrossRef]
27. Wu, J.; Cheng, B.; Wang, M.; Chen, J. Energy-Efficient Bandwidth Aggregation for Delay-Constrained Video over Heterogeneous Wireless Networks. *IEEE J. Sel. Areas Commun.* **2016**, *35*, 30–49. [CrossRef]
28. Wu, J.; Cheng, B.; Wang, M.; Chen, J. Quality-Aware Energy Optimization in Wireless Video Communication With Multipath TCP. *IEEE/ACM Trans. Netw.* **2017**, *25*, 2701–2718. [CrossRef]
29. Abrishambaf, O.; Faria, P.; Vale, Z. Participation of a Smart Community of Consumers in Demand Response Programs. In Proceedings of the 2018 IEEE Clemson University Power Systems Conference (PSC), Clemson, SC, USA, 8–11 March 2018; pp. 1–5.
30. Faria, P.; Barreto, R.; Vale, Z. Demand Response in Energy Communities Considering the Share of Photovoltaic Generation from Public Buildings. In Proceedings of the 2019 IEEE International Conference on Smart Energy Systems and Technologies (SEST), Porto, Portugal, 9–11 September 2019; pp. 1–6.

Article

Impact of Social Welfare Metrics on Energy Allocation in Multi-Objective Optimization

Anders Clausen [1,*], Aisha Umair [1], Yves Demazeau [2] and Bo Nørregaard Jørgensen [1]

[1] Center for Energy Informatics, University of Southern Denmark, Campusvej 55, 5230 Odense M, Denmark; aisha039@hotmail.com (A.U.); bnj@mmmi.sdu.dk (B.N.J.)

[2] Laboratoire d'Informatique de Grenoble, CNRS, Batiment IMAG—700 avenue Centrale, Domaine Universitaire—CS 40700, F-38058 Grenoble cx 9, France; Yves.Demazeau@imag.fr

* Correspondence: ancla@mmmi.sdu.dk

Received: 24 April 2020; Accepted: 31 May 2020; Published: 9 June 2020

Abstract: Resource allocation problems are at the core of the smart grid where energy supply and demand must match. Multi-objective optimization can be applied in such cases to find the optimal allocation of energy resources among consumers considering energy domain factors such as variable and intermittent production, market prices, or demand response events. In this regard, this paper considers consumer energy demand and system-wide energy constraints to be individual objectives and optimization variables to be the allocation of energy over time to each of the consumers. This paper considers a case in which multi-objective optimization is used to generate Pareto sets of solutions containing possible allocations for multiple energy intensive consumers constituted by commercial greenhouse growers. We consider the problem of selecting a final solution from these Pareto sets, one of maximizing the social welfare between objectives. Social welfare is a set of metrics often applied to multi-agent systems to evaluate the overall system performance. We introduce and apply social welfare ordering using different social welfare metrics to select solutions from these sets to investigate the impact of the type of social welfare metric on the optimization outcome. The results of our experiments indicate how different social welfare metrics affect the optimization outcome and how that translates to general resource allocation strategies.

Keywords: multi-objective optimization; social welfare metrics; energy allocation

---

1. Introduction

Multi-objective optimization techniques have commonly been used in the context of resource allocation problems. These problems can be modeled as agent systems [1,2], with agents that express rational behavior, but where objectives between agents may be conflicting and evaluation function outcomes are incomparable. Often, such problems are modeled in a way where the immediate outcome of the optimization is a Pareto set of viable solutions [3,4]. In cases with conflicting objectives among agents, the Pareto set contains multiple solutions that represent different trade-offs in conflicts over resources between objectives. The goal then is to select the best trade-off for the particular domain in which the optimization is applied.

Selecting the best trade-off solution from a Pareto set is a problem on its own. In multi-objective optimization, each objective has an individual objective function, which means that the cost values returned by these objective functions are not necessarily directly comparable. Even in cases where the cost values are comparable, selecting a best trade-off is not possible without a metric to determine which solution is best [5]. To this end, the concept of social welfare [6] has been used to rank solutions in Pareto sets depending on their social properties [5]. The authors in [7,8] used such a notion of social welfare to achieve a fair compromise direction when exploring the negotiation space and obtain a solution that can be considered fair. Another notion of social fairness proposed by [9] is to find a solution from a Pareto set, which ensures that total utility is fairly divided among negotiating agents. The authors in [10] used multi-objective optimization to achieve a fair balance in requirement fulfillment for users of hand-held communication devices. Here, the authors minimized the standard deviation in the number of fulfilled requirements per customer to achieve a notion of fairness in the final solution. Other approaches have focused on the wellbeing of the society as a whole, where they consider only the combined utility. The authors in [11] used a genetic algorithm on a climate control optimization problem to create a set of Pareto-optimal solutions. They then applied a utilitarian social metric to select a final solution from the generated Pareto set. The authors in [12] maximized social welfare in terms of the Nash product on scores given by agents to distributions of indivisible goods in order to select the best distribution.

The authors in [5] found that the state-of-the-art in many cases considered derivable and quasi-concave utility functions, thus failing to consider cases with non-linear utility functions. Further approaches in the state-of-the-art often apply a social welfare criterion during agent negotiation and use these criteria on agent utility values that are not normalized. This means that while the approaches do offer a metric to compare and rank solutions, they assume homogenous utility functions with comparable values. Finally, the work in [5] found that work on comparing the performance of different social welfare metrics was limited in existing literature. To this end, the work in [5] made an effort to compare and evaluate different notions of social welfare on a common problem to get an indication of their properties. Here, Umair et al. introduced the concept of a relative importance graph and post-optimization normalization of objective function values to ensure compatibility between heterogeneous objectives while avoiding the use of weighted functions that do not translate to application domains. This also addresses another challenge in applying multi-objective optimization to a particular problem: the practical application of multi-objective optimization often associates objectives with an implied order of importance. As an example, consider a case where a multi-objective algorithm is to optimize the control of actuators in a building. While minimizing energy consumption is certainly an objective, it is often more important to maintain the comfort of the individuals in the building. This implies a certain order between objectives, where a certain preference exists from the perspective of the building owner/operator. Such a preference has often been modeled using a weighted sum method [4,13] where a set of weights or a preference function is defined, in order to select a solution from the Pareto set, which has some desired characteristics. However, using this approach is non-trivial: while the concept of weights is easy to understand, determining weights for a particular problem instance is a challenge on its own.

This paper provides a complete revision of and an extension to the work of [5]. This paper contains the following contributions in this regard:

- A formal definition and a posteriori normalization method used to normalize cost value vectors in the presented social welfare metrics.
- An overview of social welfare metrics and the uses identified in the literature in table form.
- A revised version of the social welfare metric definitions, the concepts of relative importance, the relative importance graph, and social welfare ordering, as well as a revised presentation of experimental results.
- An extended discussion of the implication of social welfare metrics and the use of relative importance in multi-objective optimization by generalizing on the results of the experiments made with commercial greenhouse growers.

The rest of the paper is organized as follows. We start in Section 2 by introducing an a posteriori normalization method used to make social welfare metrics applicable to problems with heterogeneous objective functions. We then introduce 10 social welfare metrics along with their mathematical definition in Section 3. In Section 4, we present the concepts of relative importance and social welfare ordering, and then, in Section 5, we present the case and formulate it as a multi-objective optimization problem. We also introduce the optimization framework that we utilize along with its configuration.

In Section 6, we present the results of the experiments, and in Sections 7 and 8, we discuss the experimental results and conclude the paper.

2. A Posteriori Normalization

This paper considers the problem of selecting a final solution, $C_f \in P_f$, from the Pareto set P_f generated by a multi-objective optimization algorithm. Each solution C contains values for each of the variables being optimized and is associated with a cost value vector, $q_C = q_{1,C}, q_{2,C}, \cdots, q_{n,C}$, with n elements, where n is the total number of objectives in the optimization. Element $q_{i,C}$ then constitutes the cost values for objective i. The cost value vector q_C is used to determine if solution C belongs to the Pareto set using the Pareto criterion. The Pareto criterion states that a solution C is said to dominate a solution C' if it is at least as good for all objectives and better for at least one objective. A solution belongs to the Pareto set if it is not dominated by any other solution in the set.

The objective functions may be heterogeneous, meaning that the costs of objective i and objective $i+1$ may not be directly comparable. This does not constitute a problem towards determining Pareto optimality, but it does pose a problem when selecting a final solution from the Pareto set. Here, the use of the weighted sum method [13] can in theory normalize and prioritize cost values returned by heterogeneous objective functions. The trade-off is the inherent drawbacks of using weights in a combined objective function, as these are non-trivial to determine, are only stable within a known range, and do not map to any properties in the problem domain. Another approach is to perform an a priori normalization where each objective function is made to return a value in the range of 0–1. Obviously, this normalizes the values yielded by the objective functions, but it is only possible to do if the entire non-normalized range, over which the objective function spans, is known a priori, before the optimization. That is, we must know the best possible values for the variables over which we optimize for each objective before we begin optimizing.

To cope with these challenges, Umair et al. [5] suggested an a posteriori-based normalization. As the objective functions may operate on different scales, a normalized cost value vector $q'_C = \{q'_{t_1,C}, q'_{t_2,C} \cdots, q'_{t_n,C}\} \in R^n$ with elements containing the normalized cost values for solution C is created for each solution C. Here, each element $q'_{t_i,C}$ is normalized on a 0–1 scale defined by the minimum and maximum value for objective i from any $C \in P_f$ [14] for a given instance of P_f. Thus, the minimum and maximum values could vary between optimization instances, yielding a dynamic normalization mechanism that always takes into account the current solution space. The method is formally defined in Equation (1).

$$q'_{t_i,C} = \frac{q_{t_n,C} - qmin_{t_n,C}}{qmax_{t_n,C} - qmin_{t_n,C}} \tag{1}$$

3. Social Welfare Metrics

One approach to rank solutions in the Pareto set P_f is to translate the cost value vector into a scalar to compare the magnitude of the scalar between different solutions. To this end, Umair et al. [5] applied social welfare metrics.

In the context of resource allocation, the purpose of social welfare metrics is to allocate resources in a way that maximizes the social welfare among society members. When resource allocation is performed by means of multi-objective optimization, objectives constitute individuals, and all objectives in a given optimization form the society. The goal is to find values for the optimization variables that represent a desired resource allocation according to the applied social welfare metric.

The work of Umair et al. [5] surveyed the literature to identify different social welfare metrics and categorize them based on their behavior on a specific resource allocation problem with energy distribution among commercial greenhouse growers. Table 1 presents the information in aggregated form along with sources in the literature where these methods have found application. From Table 1, it can be noticed that each social welfare metric has a type associated with it. In total, three types of social welfare metrics were identified by Umair et al. [5]. These include the inequality based metrics, the fairness based metrics (presented as equality based metrics by Umair et al), and the overall utility based metrics. The social welfare types and the individual metrics are formally defined in the context of multi-objective optimization in the following subsections. We assumed that costs are normalized according to the a posteriori normalization method described in Section 2. This means that each solution C is associated with an a posteriori normalized cost for each objective $i \in n$.

Table 1. Table of social welfare metrics.

Name	Type	Source(s)
Elitist Social Welfare	Inequality Based	[15]
Egalitarian Social Welfare	Fairness Based	[15,16]
Lexi-min Ordering	Fairness Based	[15]
Approximated Fairness	Fairness Based	[9]
Fairness Analysis	Fairness Based	[10]
Quantitative Fairness	Fairness Based	[17]
Entropy	Fairness Based	[18]
Utilitarian Social Welfare	Overall Utility Based	[15,16]
Nash Product	Overall Utility Based	[15,16]
Median Rank Dictators	Overall Utility Based	[15]

3.1. Inequality Based Metrics

Inequality based metrics comprise metrics that favor the best off individual with the purpose of securing at least one objective in the society to be as good off as possible. This could happen at the expense of all other individuals in the society.

Elitist Social Welfare

The elitist social welfare metric strives to ensure that one individual is favored to the extent possible. The elitist social welfare metric is defined in Equation (2).

$$f_{EL}(C) = \min\{q'_{t_i,C} \mid i \in n\}$$

(2)

Equation (2) shows how the elitist social welfare metric ranks solutions based on the normalized cost of the best off objective. Hence, this metric prefers solution C over solution C' if and only if $\overset{\uparrow}{q'}_{t_1,C} < \overset{\uparrow}{q'}_{t_1,C'}$, where $\overset{\uparrow}{q'}_C$ represents the reordering of the normalized cost value vector $q'_C = \{q'_{t_1,C}, q'_{t_2,C}, \ldots, q'_{t_n,C}\}$ in ascending order.

3.2. Fairness Based Metrics

Fairness based metrics are social welfare metrics that include a notion of fairness. Fairness in the context of optimization social welfare is ensuring, to the extent possible, fair adherence to the goals of all objectives in the optimization. That is, all objectives should be considered equal when selecting a final solution.

3.2.1. Egalitarian Social Welfare

The egalitarian social welfare metric defines fairness as ensuring that the worst off objective is as good off as possible. The egalitarian social welfare metric is defined in Equation (3).

$$f_E(C) = \max\{q'_{t_i,C} \mid i \in n\}$$

(3)

This means that the egalitarian social welfare metric prefers a solution C over solution C' if and only if $\overset{\downarrow}{q}'_{t_1,C} < \overset{\downarrow}{q}'_{t_1,C'}$ where $\overset{\downarrow}{q}'_C$ contains the values of the normalized cost vector $q'_C = \{q'_{t_1,C}, q'_{t_2,C}, \ldots, q'_{t_n,C}\}$ rearranged in descending order.

The egalitarian social welfare metric has a weakness in that it only takes into account the normalized cost of the worst off objective while defining the ordering of solutions. Consider three normalized cost value vectors (1,1,0), (1,1,1), and (1,0,0) for solutions C_1, C_2, and C_3, respectively. Each of the vectors is comprised of normalized cost values in decreasing order. The highest cost value in all the cost value vectors is one. In this case, the egalitarian social welfare metric fails to distinguish between solutions that have the same highest (worst) cost and assigns the same order to all three solutions, irrespective of the fact that C_3 is better objectively than C_1 and C_2.

3.2.2. Lexi-Min Ordering

The lexi-min ordering social welfare metric prefers a solution C over solution C' if and only if an integer $r \in \{1, \ldots, n\}$ exists such that the following two conditions are satisfied:

1. $(\overset{\downarrow}{q}'_C)_i = (\overset{\downarrow}{q}'_{C'})_i$ for all $i < r$

2. $(\overset{\downarrow}{q}'_C)_r < (\overset{\downarrow}{q}'_{C'})_r$

Here again, $\overset{\downarrow}{q}'_C$ is the normalized cost vector q'_C arranged in descending order.

This means that the lexi-min ordering social welfare metric addresses the weakness of the egalitarian social welfare metric by considering the next worst off cost value in compared solutions until their values no longer coincide. The lexi-min ordering social welfare metric will successfully distinguish among three solutions used in the example of the egalitarian social welfare metric and assign different rank to them, $C_3 = 1$, $C_1 = 2$ and $C_2 = 3$, respectively.

While the lexi-min ordering social welfare metric offers an improvement over the egalitarian social welfare metric, it deviates from the other social welfare metrics in that it offers no collective cost value. Rather, it presents a method for direct comparison between solutions.

3.2.3. Approximated Fairness

The approximated fairness social welfare metric [9] has adopted the concept of simple fair division, explained in social choice and game theory [19], in order to implement the concept of fairness. The metric is defined in Equation (4).

$$f_{AF}(C) = \sum_{i=1}^{n} \frac{(q'_{t_i,C} - q'_{avg,C})^2}{n} \tag{4}$$

$$\text{where } q'_{avg,C} = \frac{\sum_{i=1}^{n}(q'_{t_i,C})}{n}$$

Equation (4) shows how the approximated fairness social welfare metric ranks solutions based on the sum of the squared difference between the individual, normalized cost values of the objectives, and the average normalized cost value across all objectives. This means that the approximated fairness social welfare metric prefers a solution C over solution C' if and only if $f_{AF}(C) < f_{AF}(C')$.

3.2.4. Fairness Analysis

The concept behind the fairness analysis social welfare metric is very much similar to the concept behind the approximated fairness social welfare metric. This metric is used to balance the requirement fulfillment between customers of Motorola Company for hand-held communication devices [10]. The aim is to minimize the standard deviation of the number of fulfilled requirements for each

3.3.3. Median Rank Dictators

The median rank dictators [15] social welfare metric ranks solutions based on their median cost value as shown in Equation (13). Here, $(\overset{\downarrow}{q}'_C)_r$ is the rth cost value in $\overset{\downarrow}{q}'_C$, which is the normalized cost value vector rearranged in descending order. Here, r is calculated as $r = n/2$, when the number of individuals n is even and $r = (n+1)/2$ when the number of individuals n is odd.

$$f_{MRD}(C) = (\overset{\downarrow}{q}'_C)_r \tag{13}$$

The median rank dictator social welfare metric prefers a solution C over a solution C' if and only if $(\overset{\downarrow}{q}'_C)_r < (\overset{\downarrow}{q}'_{C'})_r$.

4. Relative Importance and Social Welfare Ordering

While social welfare metrics do offer a way of prioritizing solutions in a Pareto set, they pose a limitation. The social welfare metrics are not able to rank objectives according to importance. To this end, Umair et al. [5] introduced the notion of relative importance and the Relative Importance Graph (RIG). By combining social welfare metrics and relative importance, we get a social welfare ordering method. Both relative importance and social welfare ordering are explained below.

4.1. Relative Importance and the Relative Importance Graph

The relative importance between objectives c_1 and c_2 can be expressed through the integer values -1, 1, and 0. Here, a value of -1 means that objective c_1 is relatively more important than c_2. Likewise, a value of 1 reflects that objective c_1 is relatively less important than c_2. Finally, a value of zero reflects that objective c_1 and objective c_2 are equally important. Under this assumption we define an abstraction, relation, which allows us to specify the relative importance of an objective towards another objective. When all possible relation instances have been created, a list of sorted objectives can be created by adding all objectives that are not part of a relation in which they are relatively less important than any other objective to the top of the list. This (or these in case of multiple) objective(s) are then followed by the objectives that are only relatively less important than the objectives already added to the list. This continues until all objectives have been added to the list.

An RIG can be constructed as a directed graph, $RIG = <N, E>$, based on the list of sorted objectives. Here, $N = \{N_1, N_2, \ldots, N_n\}$ is a set of nodes where each node N_i comprises an objective or a group of objectives that are equally important, and $E = \{E_1, E_2, \ldots, E_m\}$ is a set of edges that contains edges between all nodes N, where $E_{i,j}$ represents a direction between node N_i and node N_j based on the relative importance of objectives residing in nodes N_i and N_j. The graph is constructed by creating a root node containing the first objective c_1 from the sorted list of objectives. The next objective in the sorted list, c_2, is then added to the RIG through comparison with objective c_1: If c_2 is relatively equally important to objective c_1, then it is added to the same node as shown in Figure 1a. If objective c_2 is relatively less important than objective c_1, then it is added to a new empty node N_2 as illustrated in Figure 1b. In this case, a directed edge is created from node N_1 to node N_2 to represent the hierarchy between the objectives contained in the nodes. This process is repeated until all nodes in the sorted list have been added to the RIG, at which point, the RIG constitutes a graph in which nodes (and consequently objectives) can be traversed in order of relative importance. While this creates an absolute hierarchy of objectives, this voids the task of assigning an absolute importance to the individual objectives.

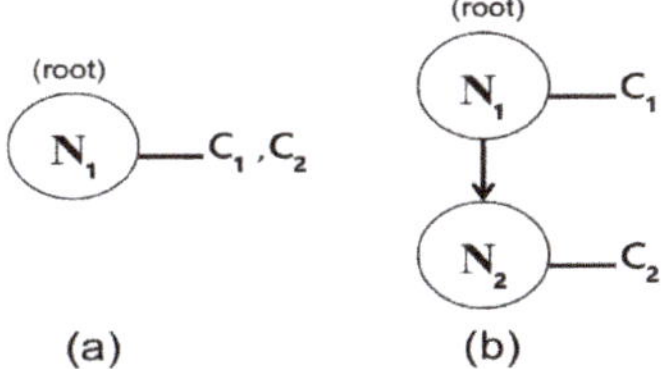

Figure 1. Examples of relative importance graphs. (**a**) Objectives c_1 and c_2 are relatively equally important. (**b**) Objective c_1 is relatively more important than objective c_2.

4.2. Social Welfare Ordering

The social welfare ordering is constructed by assigning a social welfare metric to each node N_i in the RIG. This then applies that specific notion of social welfare to the group of objectives within that node, which enables selecting a final solution from the Pareto set through a simple algorithm, which will start by visiting node N_1 in the RIG and apply the social welfare criterion associated with that particular node to the part of the normalized cost vector that contains costs from objectives in node N_1. This will yield a subset of one or more solutions that are equal (in terms of the value returned by the applied social welfare metric) from the perspective of objectives in node N_1. If this subset contains only one solution, the algorithm will terminate and return the solution. Otherwise, the algorithm will continue to traverse the RIG until (a) a subset containing exactly one solution is found or (b) the entire RIG has been traversed. In the latter case, the algorithm will return an arbitrary solution from the final subset, as every solution in this subset is considered equally good from the perspective of every objective in the optimization instance. The pseudo code for the social welfare ordering algorithm can be found in Algorithm 1. The concept of social welfare ordering ensures that higher ranking objectives are prioritized over lower ranking ones during the selection of the final solution. Further, the application of a social welfare metric to each node in the RIG ensures that any conflicts are resolved based on the social welfare metric applied to the particular node level.

Algorithm 1 Pseudo code of the social welfare ordering algorithm.

```
1:  set ← P_F                                    ▷ set initially contains the entire Pareto set.
2:
3:  while set.Size() > 1 and rig.HasNextNode() do
4:
5:      root ← rig.NextNode()
6:
7:      set ← root.SocialWelfareMetric(set)
8:
9:  end while
10:
11: finalSolution ← set.GetSolution()
```

---

5. Experimental Setup

To test the influence of the social welfare ordering and different social welfare metrics, we considered a set of experiments built around a case with commercial greenhouse growers. The case and the associated formal problem definition are described in this section along with the optimization framework and its associated configuration used to solve the problem.

5.1. Case

In northern countries, Commercial Greenhouse Growers (CGGs) are heavy consumers of energy used to power supplementary lighting, which is needed to facilitate plant growth [14]. We considered a case in which a Resource Domain (RD) was responsible for allocating energy to three CGGs. In this scenario, the task of the RD was to ensure that energy allocation never exceeded system-wide energy limits while adhering to the best of its ability to the energy needs of the CGGs. CGGs use energy for supplemental light to increase plant yield. However, light is not provided to the plants arbitrarily [20]. Long-day photo-periodic plants require long, uninterrupted periods of light to bloom. Here, supplemental light is often used to extend the natural day length. Short-day plants on the other hand require shorter days and are sensitive to long, uninterrupted periods of darkness. Here, supplemental light is used only when natural daylight is scarce, and only for a limited period of time. Day-neutral plants are insensitive to the length of the day or night and will always be able to utilize the provided supplemental light. To simplify the complexity of the optimization problem, we assumed that the CGGs had only two consumption levels, namely on, at which stage it would consume 1 MW of electricity, or off, where consumption would be 0 MW. However, each CGG can be on or off independent of the other CGGs. This type of problem translates to any domain in which energy allocation is necessary to run a facility.

The case was modeled as a multi-objective multi-variable optimization problem with 7 objectives and 3 variable vectors in total. We optimized to find a 24 h energy allocation schedule for each of the three CGGs.

We used 3 variable vectors to represent the hourly planned electricity allocation for each of the CGGs. To represent 24 h, we set $n = 24$. Here, each element could take on a value of either 0 or 1 (MW). Their definition is found in Equation (14).

$$\vec{x} = [x_1, ..., x_n]$$
$$\vec{y} = [y_1, ..., y_n]$$
$$\vec{z} = [z_1, ..., z_n] \tag{14}$$
$$\text{where}$$
$$x_i, y_i, z_i \in [0; 1]$$

The 7 objectives in the optimization problem include 1 system-wide energy constraint objective and 6 objectives that map to the 3 CGGs, 2 objectives for each CGG.

Equation (15) formulates the objective function for the system-wide energy limit objective. It considers the combined energy allocation for each of the CGGs and compares it to its preference vector $\vec{p_{ec}}$ by summarizing the absolute distance between elements of the combined allocation vector and the preference vector.

$$f_1\left(\vec{x}, \vec{y}, \vec{z}\right) = \sum_{i=1}^{n} |(x_i + y_i + z_i) - p_{ec,i}| \tag{15}$$

Each of the CGGs has 2 objectives representing it. One objective is focused on the amount of energy allocated and the time at which it is allocated for the CGG, whereas the other objective is concerned only with the total amount of energy allocated to the CGG, disregarding the time at which it is allocated. This may seem redundant, but in the context of this optimization problem, this means that shifting energy is perceived as better for the consumers than shedding or removing the energy, in cases where insufficient energy is available in one or more time slots.

Equations (16) and (17) show the definitions for the objective that is concerned with the allocated amount and time of allocation and the objective that is concerned only with the combined allocation across the entire schedule for CGG_1. Equations (18)–(21) are similar objective definitions for CGG_2 and CGG_3.

$$f_2\left(\vec{x}\right) = \sum_{i=1}^{n} |x_i - p_{cgg1,i}| \tag{16}$$

$$f_3\left(\vec{x}\right) = \left| \sum_{i=1}^{n} x_i - \sum_{i=1}^{n} p_{cgg1,i} \right| \tag{17}$$

$$f_4\left(\vec{y}\right) = \sum_{i=1}^{n} \left| y_i - p_{cgg2,i} \right| \tag{18}$$

$$f_5\left(\vec{y}\right) = \left| \sum_{i=1}^{n} y_i - \sum_{i=1}^{n} p_{cgg2,i} \right| \tag{19}$$

$$f_6\left(\vec{z}\right) = \sum_{i=1}^{n} \left| z_i - p_{cgg3,i} \right| \tag{20}$$

$$f_7\left(\vec{z}\right) = \left| \sum_{i=1}^{n} z_i - \sum_{i=1}^{n} p_{cgg3,i} \right| \tag{21}$$

The combined optimization problem is defined in Equation (22).

$$min\left(f_1\left(\vec{x},\vec{y},\vec{z}\right), f_2\left(\vec{x}\right), f_3\left(\vec{x}\right), f_4\left(\vec{y}\right), f_5\left(\vec{y}\right), f_6\left(\vec{z}\right), f_7\left(\vec{z}\right)\right)$$
$$s.t. \tag{22}$$
$$x_i, y_i, z_i \in [0; 1]$$

5.2. Optimization Framework

The multi-objective optimization problem described in Section 5.1 was solved using Controleum [21,22]. Controleum is an object-oriented genetic algorithm framework. Controleum is one framework to employ a Pareto set to store outcomes. The Pareto set is used to store populations of solutions between evolutions, and once optimization has terminated, a final solution is drafted from the resulting Pareto set. To this end, Controleum uses objective priority and a non-normalized utilitarian social welfare to select a final solution from the Pareto frontier.

Controleum uses the concern abstraction to represent objectives and the issue abstraction to represent optimization variables as vectors. An optimization context in Controleum is defined by N concerns that negotiate over M issues. Each concern c_n, $n \in N$ defines an evaluation function that takes into account values of each of the issues, e_m, $m \in M$, which are part of the objective function. Further, the evaluation function takes into account a concern-specific preference vector, which describes a target that the concern is aiming for. An issue, $e_m = \{e_{1,m}, e_{2,m}, \ldots e_{t,m}\}$ defines a number of vector elements t, as well as the range from which these elements can draw values. In the context of Controleum, a solution, C, is a set of values for each of the M issues defined. The solution is defined as $C = s_{C,1}, \ldots, s_{C,m}$, where $s_{C,i}$ is the value for issue e_i in solution C.

The initiation of the algorithm and the optimization process itself was explained in depth in [5], in which Controleum was also extended by (1) introducing the notion of the relative importance of objectives to determine their ordering, (2) utilizing normalized costs of objectives in the selection of a final outcome, and (3) implementing the notions of social welfare presented in Section 3. This enables the configuration of Controleum to use a specific social welfare ordering by deciding on a social welfare metric and by assigning relative importance to objectives.

5.2.1. Issue Configuration

The variable vectors described in Section 5.1 were translated to three corresponding issues, $M = 3$, in Controleum. Each issue was configured with $t = 24$ to cover 24 1 h slots. This meant that each solution comprised 3 vectors, each with 24 values of either 0 or 1. This constituted a complete allocation for each of the CGGs presented as part of the case.

5.2.2. Concern Configuration

Each of the objectives described in Section 5.1 was translated 1:1 to concerns. This meant that we had 7 concerns in Controleum representing the 7 objectives in the optimization problem.

These 7 concerns were distributed across 3 types of concerns; we had 1 system-wide energy concern (*SEC*), 3 CGG resource concerns (*CRC*), which consider the amount of energy allocated and the time at which it is allocated for each of the CGGs, and finally, 3 CGG resource sum concerns (*CRC − sum*) that consider the total amount of energy allocated and disregard the temporal dimension. Algorithms 2–4 show a pseudo code representation of each of these types of concerns.

Algorithm 2 Pseudo code representation of the evaluation method in the system-wide energy concern.

Precondition: n is the number of hours being allocated.

1:
Precondition: $pref_{SEC}$ is the system-wide energy limit vector with n elements.

2:
3: **function** EVALUATE(solution)

4:
5:     $alloc_{sum} \leftarrow Int[n]$

6:
7:     **while** $solution.HasNextIssueValue()$ **do**

8:
9:         $alloc_{sum} \leftarrow alloc_{sum} + solution.NextIssueValue()$

10:
11:     **end while**

12:
13:     $result \leftarrow calcManhattenDistance(pref_{SEC}, alloc_{sum})$

14:
15:     **return** $result$

16:
17: **end function**

Algorithm 3 Pseudo code representation of evaluation method of the CGG resource concern (*CRC*).

Precondition: n is the number of hours being allocated, and x is a CGG ID from 1-3.

1:
Precondition: $pref_x$ is the demand for energy in CGG x.

2:
3: **function** EVALUATE(solution)

4:
5:     $alloc_x \leftarrow solution.getIssueValue(x)$

6:
7:     $result \leftarrow calcManhattenDistance(pref_x, alloc_x)$

8:
9:     **return** $result$

10:
11: **end function**

Algorithm 4 Pseudo code representation of evaluation method of the CGG resource sum concern (*CRC*).

Precondition: n is the number of hours being allocated, and x is a CGG ID from 1–3.
1:
Precondition: $pref_x$ is the demand for energy in CGG x.
2:
3: **function** EVALUATE(solution)
4:
5:     $alloc_x \leftarrow solution.getIssueValue(x)$
6:
7:     $allocSum_x \leftarrow 0, prefSum_x \leftarrow 0$
8:
9:     **for** $i = 0, n$ **do**   ▷ Sum all elements in both vectors
10:
11:         $allocSum_x \leftarrow allocSum_x + alloc_x[i]$
12:
13:         $prefSum_x \leftarrow prefSum_x + pref_x[i]$
14:
15:     **end for**
16:
17:     $result = abs(allocSum_x - prefSum_x)$ ▷ Absolutes sum difference
18:
19:     **return** $result$
20:
21: **end function**

The variable $pref_{SEC}$ denotes the system-wide energy constraint. As seen from the pseudo code in Algorithm 2, the absolute differences between elements in this vector and the sum of elements in the allocations for each of the CGGs are considered. This means that the system will not tolerate allocations above or below the system-wide energy constraint. We ran two sets of experiments where we varied the value of $pref_{SEC}$. In one case, the value of the $pref_{SEC}$ vector was equal to the sum of elements in each of the demand vectors of the CGGs. This is illustrated in Figure 2, where the columns are the demands presented by each of the CGGs and the dotted line is a plot of $pref_{SEC}$. This translates into a resource allocation problem where sufficient resources are available and where the problem is one of allocating these resources to each of the CGGs in a way where every CGGs gets what it demands.

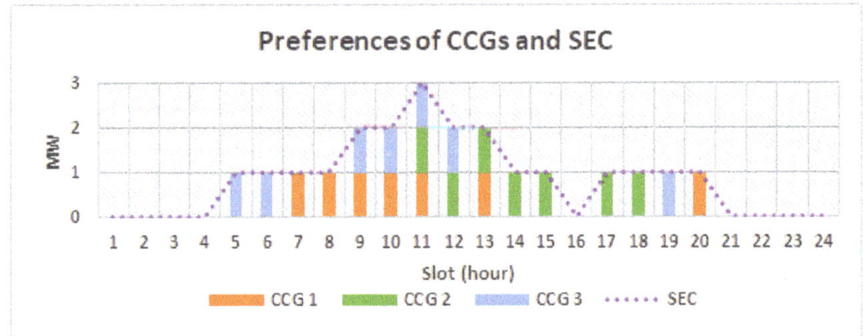

Figure 2. Demand profiles of Commercial Greenhouse Growers (CGGs) and the system-wide energy constraint (line) with sufficient energy. SEC, system-wide energy concern.

In the other case, we introduced a constraint on the amount of resources available to the system, by altering $pref_{SEC}$ as shown in Figure 3. Here, we see that $pref_{SEC}$ did not match the need for energy in Slots 11, 12, and 13. In a resource allocation context, the task of the system in this case was to gracefully degrade. Here, the definition of graceful degradation depended on the resource allocation strategy: some systems would like to see the consequences distributed across as many entities as possible. Here, a strategy that distributed resources across entities in a fair manner was needed. In other cases, the number of entities affected by the system state should be reduced. Here, the resource allocation strategy should ensure that the least possible entities were influenced, at the expense of fewer entities, which would be influenced more severely.

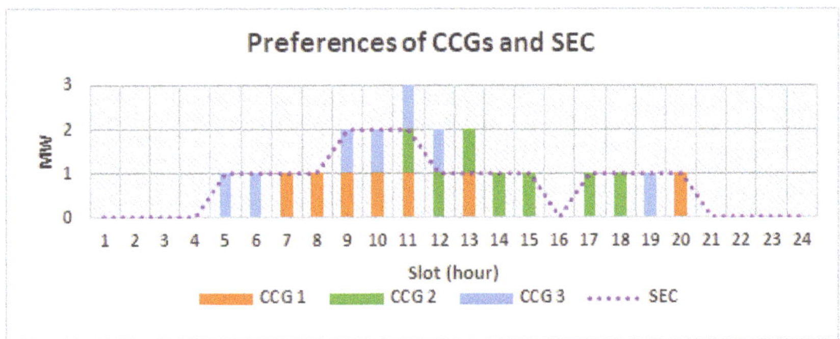

Figure 3. Demand profiles of CGGs and the system-wide energy constraint (line) with insufficient energy.

5.3. Social Welfare Ordering

The goal of the system as presented in the case of this section was to adhere to system energy limits while allocating energy to each of the CGGs to the best of its ability under this constraint. Thus, the SEC was more important than the six CGG concerns in terms of relative importance. As the case did not describe any priority between each of the CGGs, the remaining six concerns were relatively equally important. The resulting relative importance graph is shown in Figure 4, where the SEC is placed in the node immediately above the CRCand the CRC-sum concerns. This configuration had several properties. First, the SEC was always prioritized over any of the CRC concerns, meaning that solutions that were ideal for the SEC were selected from the Pareto set first. The 10 social welfare metrics described in Section 3 were then applied to node N_1 in separate experiments to apply different methods for selecting a final solution from the Pareto set. This enabled us to analyze the impact of each of the social welfare metrics on the resulting resource allocation strategy in a society of 3 identical production entities.

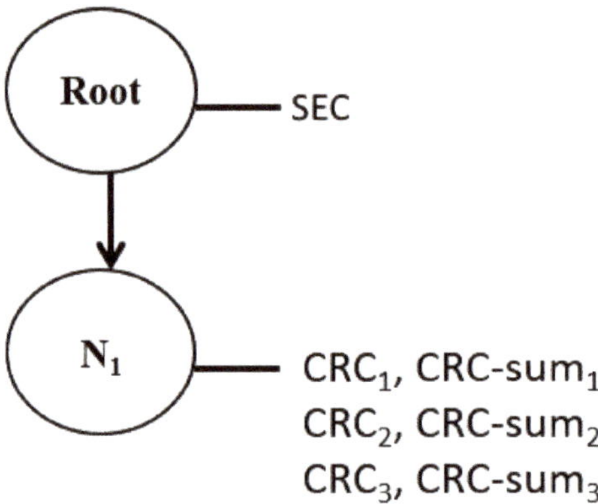

Figure 4. Relative importance graph of the Resource Domain (RD).

6. Experiments

Based on the experimental setup described in Section 5, two sets of experiments were conducted. The experiments were conducted in [5]. Here, we present the results, and in Section 7, we extrapolate the results to resource allocation strategies. The two sets of experiments describe two scenarios that are well known within resource allocation, namely one with sufficient resources and one with insufficient resources. From the experiments, we observed how different notions of social welfare (described in Section 3) translated to resource allocation strategies and how this manifested as system consequences with respect to graceful degradation in the horticulture domain. Each set of experiments was executed 20 times to induce resilience towards the potential random behavior sparked by the use of an evolutionary algorithm.

6.1. Sufficient Resources

This experiment was conducted for each of the social welfare metrics described in Section 3. However, here, the social welfare metric did not influence the outcome, which is displayed in the graph of Figure 5 for all 10 metrics. We see the demand profile of each of the CGGs as stronger colored bars and allocations as weaker colored bars. When a weaker and a stronger colored bar of same color is placed in a time slot, this means that allocation and demand matched. We see from the figure that allocation matched demand in the case of sufficient resources irrespective of the employed resource allocation strategy.

6.2. Insufficient Resources

In these experiments, insufficient resources, in terms of available electricity power, were made available in Slots 11, 12 and 13. We conducted one series of experiments for each of the 10 metrics described in Section 3. From the experiments, we notice that the resulting resource allocation strategy depended on the type of social welfare metric rather than the social welfare metric itself: similar types of social welfare metrics yielded the same results. Thus, Figures 6–8 show the results for the inequality based, the 6 fairness based, and the 3 overall utility based social welfare metrics, respectively. As a result, we present the results for each type of social welfare metric in the subsections below.

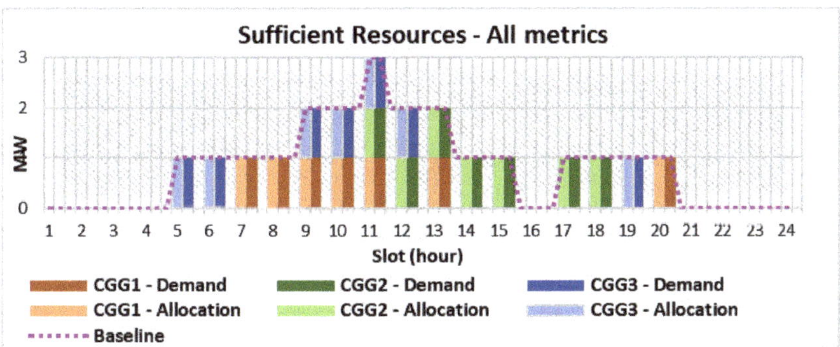

Figure 5. Results of the baseline experiment.

6.2.1. Results: Inequality Based Metrics

With an inequality based metric, we expected that one entity is best off, and indeed, we observed that CGG_2 was allocated sufficient energy to match its demand at the expense of CGG_1 and CGG_3, which were compromised of 1 and 2 MWh, respectively, as seen from Figure 6.

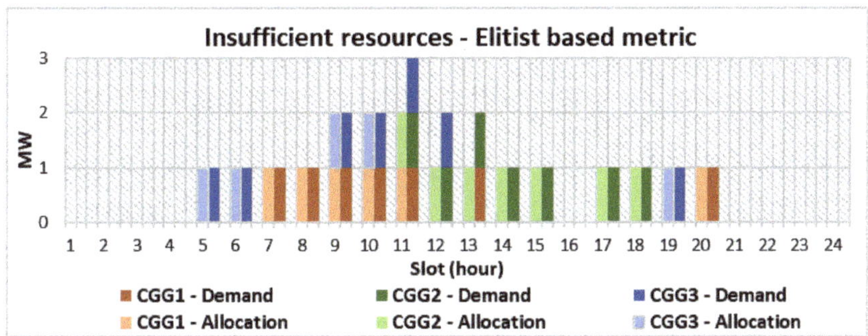

Figure 6. Results: insufficient resources-elitist based metric. Insufficient allocation for CGG_1 in Slot 13 and CGG_3 in Slots 11 and 12.

6.2.2. Results: Fairness Based Metrics

With fairness based metrics, we would expect that the negative impact of insufficient resources was equally distributed across all entities. From Figure 7, we see that each of the entities lacked one 1 MWh of electricity allocation to meet their demand. As the entities in our model valued energy equally in all of their demand hours, this meant that this result fell within our expectations.

6.2.3. Results: Overall Utility Based Metrics

The results of applying either of the overall utility based metrics are seen in Figure 8. Here, we see that a single entity (CGG_2) carried the entire consequence of having insufficient resources available in the system, leaving the two other entities unaffected.

6.3. Summary

The experiments are summarized in Table 2 where we disregard the temporal dimension. The baseline demand of each CGG was 7 MWh, with the first set of experiments allocating this amount to each CGG entity regardless of the social welfare metric. The remaining three experiments showed how different types of social welfare metrics performed different tradeoffs between the entities.

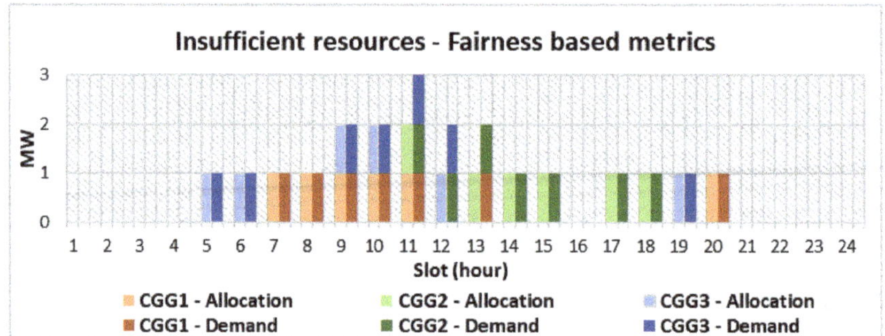

Figure 7. Results: insufficient resources-fairness based metrics. Insufficient allocation for CGG_3, CGG_2, and CGG_1 in Slots 11, 12, and 13, respectively.

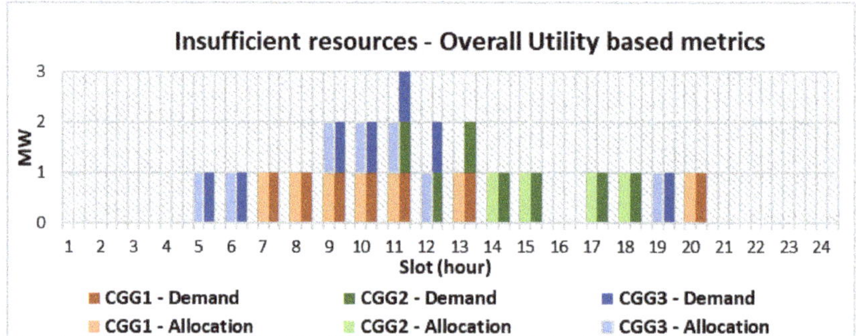

Figure 8. Results: insufficient resources-overall utility based metrics. Insufficient allocation for CGG_2 in Slots 11, 12, and 13.

Table 2. Summary of the experimental results.

Experiment	Allocation (MWh)		
	CGG_1	CGG_2	CGG_3
Sufficient-all metrics	7	7	7
Insufficient-elitist	6	7	5
Insufficient-fairness	6	6	6
Insufficient-overall utility	7	4	7

7. Discussion

The experiments presented in this paper and by Umair et al. [5] showed that the choice of social welfare metric was arbitrary in cases with sufficient resources. In the case with insufficient resources, the type of social welfare metric had an impact on the resource allocation strategy, while we observed no variance between social welfare metrics of a similar type. However, the variance between social welfare metric types manifested itself in vastly different resource allocation strategies and resulting domain consequences.

In the experiment with insufficient resources, we observed that one CGG was better off than the two others when using the inequality based social welfare metric. Here, two entities shared the consequences of the resource insufficiency. In the context of resource allocation, this type of resource allocation strategy could be employed to ensure sufficient allocation of resources to mission-critical processes, which requires an exact match of its demand to run without interruptions. This then happens at the expense of the non-mission critical processes, which are compromised, although in our case, one more than the other. In a grower domain, this resource allocation strategy could be relevant in the production of long-day photoperiodic plants, such as Rudbeckia and California poppy, which are sensitive to the duration of day-night. This strategy ensures that at least one of the CGGs gets the required amount of light, thereby preserving the production yield in that entity.

With the fairness based social welfare metrics, we saw that consequences were evenly distributed across all CGGs, with all entities compromising on their operation to an extent. This sort of resource allocation strategy could appeal to cases where processes can be interrupted without severe consequences and where no process should compromise more than any other. With long-day photo-periodic plants, this would be an inefficient strategy, as plants in all entities could be destroyed during such an event. However, for short-day plants like chrysanthemum, Christmas cactus, and poinsettia, such a strategy would not have any severe impact on production, as these plants would be able to compromise between the entities to reach their relatively sparse requirements for day light [23].

Finally, with the overall utility based social welfare metrics, the consequences of the resource insufficiency are carried entirely by a single entity. This type of resource allocation strategy seems immediately appealing in that two entities are in fact entirely unaffected by the lack of resources. However, the mathematical properties of these metrics do nothing to guarantee this. Rather, the target is to maximize the overall utility of the society regardless of the consequences. This could be ideal for plant productions with day-neutral plants, which do not require any specific day length, such as rose and tomatoes. Here, the overall yield is more important than the wellbeing of any one entity.

The experiments presented in this paper and in Umair et al. [5] failed to shed light on the variance between metrics of similar types. While this implies that these metrics are interchangeable, we believe that this maps back to the specific problem instance employed throughout the experiments. The problem presented a fixed constraint on resources, which, when combined with the nature of the objective functions, meant that no entity would ever attempt to shift its consumption to another time slot as it would have an adverse effect on the overall system state (and not be Pareto optimal). We believe that experiments where the RD suggests the CGGs to shift their consumption, by increasing allocation in alternative time slots, could help to shed light on inter-type differences between metrics.

8. Conclusions

This paper discussed the properties and impact of different notions of social welfare on a resource allocation problem solved through multi-objective optimization that generated a Pareto based solution set.

We presented 10 different notions of social welfare and categorized them based on their perceived behavior. To show their properties in the context of resource allocation, we considered a case with three commercial greenhouse growers with independent energy demands in a setting with a resource domain responsible for allocating energy to each of these entities. We formulated this as a multi-objective optimization problem and solved the problem with Controleum [21,22], a multi-objective optimization framework.

We carried out two sets of experiments: one set with sufficient resources and one set with insufficient resources. An experiment was conducted for each metric in each of the sets of experiments. Each experiment was repeated 20 times to mitigate the potential impact of randomness due to the nature of the genetic algorithm employed by Controleum. The results showed that different notions of social welfare led to different resource allocation strategies. These then translated to different impacts in the grower domain. This in turn meant that the choice of social welfare metric was not arbitrary and depended on the behavior that was expected of the system as a whole.

Author Contributions: Conceptualization, A.U., B.N.J., A.C. and Y.D.; methodology, A.C. and A.U.; software, A.C. and A.U.; validation, A.C., A.U. and B.N.J.; formal analysis, A.U. and A.C.; investigation, A.U. and A.C.; resources, B.N.J.; data curation, A.U.; writing–original draft preparation, A.C.; writing–review and editing, Y.D., B.N.J., A.U.; visualization, A.U. and A.C.; supervision, B.N.J. and Y.D.; project administration: B.N.J. All authors have read and agreed to the published version of the manuscript.

Funding: This research received no external funding.

Conflicts of Interest: The authors declare no conflict of interest.

References

1. Hosseini, H. Dynamic Multiagent Resource Allocation: Integrating Auctions and MDPs for Real-Time Decisions. In Proceedings of the Seventeenth AAAI/SIGART Doctoral Consortium, ON, Canada, 22–23 June 2012.

2. Briola, D.; Mascardi, V. Multi Agent Resource Allocation: A Comparison of Five Negotiation Protocols. In Proceedings of the 12th Workshop on Objects and Agents (WOA 2011), Rende (CS), Italy, 4–6 July 2011; pp. 95–104.

3. Clausen, A.; Umair, A.; Demazeau, Y.; Jørgensen, B.N. Agent-Based Integration of Complex and Heterogeneous Distributed Energy Resources in Virtual Power Plants. In *Advances in Practical Applications of Cyber-Physical Multi-Agent Systems: The PAAMS Collection*; PAAMS 2017; Lecture Notes in Computer Science; Demazeau, Y., Davidsson, P., Bajo, J., Vale, Z., Eds; Springer: Cham, Switzerland, 2017; Volume 10349, pp. 43–55.

4. Chaharsooghi, S.K.; Kermani, A.H.M. An effective ant colony optimization algorithm (ACO) for multi-objective resource allocation problem (MORAP). *Appl. Math. Comput.* **2008**, *200*, 167–177. [CrossRef]

5. Umair., A.; Clausen., A.; Demazeau., Y.; Jørgensen., B.N. Impact of Social Welfare Methods on Multi-objective Resource Allocation in Energy Systems. In Proceedings of the 8th International Conference on Smart Cities and Green ICT Systems-Volume 1: SMARTGREENS, Heraklion, Greece, 3–5 May 2019; pp. 179–186. [CrossRef]

6. Feldman, A.M.; Serrano, R. *Welfare Economics and Social Choice Theory*; Springer Science & Business Media: New York, NY, USA, 2006.

7. Li, M.; Vo, Q.B.; Kowalczyk, R. Searching for fair joint gains in agent-based negotiation. In Proceedings of The 8th International Conference on Autonomous Agents and Multiagent Systems-Volume 2, Budapest, Hungary, 10–15 May 2009; pp. 1049–1056.

8. Heiskanen, P.; Ehtamo, H.; Hämäläinen, R.P. Constraint proposal method for computing Pareto solutions in multi-party negotiations. *Eur. J. Op. Res.* **2001**, *133*, 44–61. [CrossRef]

9. Fujita, K.; Ito, T.; Klein, M. A secure and fair protocol that addresses weaknesses of the Nash bargaining solution in nonlinear negotiation. *Group Decis. Negotiat.* **2012**, *21*, 29–47. [CrossRef]

10. Finkelstein, A.; Harman, M.; Mansouri, S.A.; Ren, J.; Zhang, Y. A search based approach to fairness analysis in requirement assignments to aid negotiation, mediation and decision making. *Requir. Eng.* **2009**, *14*, 231–245. [CrossRef]

11. Sørensen, J.C.; Jørgensen, B.N. An extensible component-based multi-objective evolutionary algorithm framework. In Proceedings of the 6th International Conference on Software and Computer Applications, Bangkok, Thailand, 26–28 February 2017; pp. 191–197.

12. Darmann, A.; Schauer, J. Maximizing Nash product social welfare in allocating indivisible goods. *Eur. J. Op. Res.* **2015**, *247*, 548–559. [CrossRef]

13. Marler, R.T.; Arora, J.S. The weighted sum method for multi-objective optimization: New insights. *Struct. Multidiscip. Optim.* **2010**, *41*, 853–862. [CrossRef]

14. Umair, A. An Agent Based Approach to Coordination of Resource Allocation and Process Performance. Ph.D. Thesis, University of Southern Denmark, Odense, Denmark, 2018.

15. Chevaleyre, Y.; Dunne, P.E.; Endriss, U.; Lang, J.; Lemaitre, M.; Maudet, N.; Padget, J.; Phelps, S.; Rodriguez-Aguilar, J.A.; Sousa, P. Issues in multiagent resource allocation. *Informatica* **2006**. Available online: https://staff.fnwi.uva.nl/u.endriss/MARA/mara-survey.pdf (accessed on 5 June 2020).

16. Nguyen, N.T.; Nguyen, T.T.; Roos, M.; Rothe, J. Computational complexity and approximability of social welfare optimization in multiagent resource allocation. *Auton. Agents Multi-Agent Syst.* **2014**, *28*, 256–289. [CrossRef]

17. Huaizhou, S.; Prasad, R.V.; Onur, E.; Niemegeers, I. Fairness in wireless networks: Issues, measures and challenges. *IEEE Commun. Surv. Tutor.* **2014**, *16*, 5–24. [CrossRef]

18. Shannon, C.E. A mathematical theory of communication. *ACM SIGMOBILE Mobile Comput. Commun. Rev.* **2001**, *5*, 3–55. [CrossRef]

19. Robertson, J.; Webb, W. *Cake-Cutting Algorithms: Be Fair if You Can*; CRC Press: Boca Raton, FL, USA, 1998.

20. Clausen, A.; Maersk-Moeller, H.; Soerensen, J.C.; Joergensen, B.; Kjaer, K.; Ottosen, C. Integrating commercial greenhouses in the smart grid with demand response based control of supplemental lighting. In Proceedings of the 2015 International Conference on Industrial Technology and Management Science, Tianjin, China, 27–28 March 2015.

21. Sørensen, J.C.; Jørgensen, B.N.; Klein, M.; Demazeau, Y. An agent-based extensible climate control system for sustainable greenhouse production. In *Agents in Principle, Agents in Practice*; PRIMA 2011; Lecture Notes in Computer Science; Kinny, D., Hsu, J.Y., Governatori, G., Ghose, A.K., Eds; Springer: Heidelberg, Berlin, 2011; Volume 7047, pp. 218–233.

22. Clausen, A.; Demazeau, Y.; Jørgensen, B.N. An agent-based framework for aggregation of manageable distributed energy resources. In *Highlights of Practical Applications of Heterogeneous Multi-Agent Systems*; The PAAMS Collection; PAAMS 2014; Communications in Computer and Information Science; Corchado, J.M., Eds; Springer: Cham, Switzerland, 2014; Volume 430, pp. 214–225.

23. Kumpf, J. Lighting Systems. 2019. Available online: https://courses.cit.cornell.edu/hort494/greenhouse/lighting/lightlft.html (accessed on 4 January 2019).

Article

Quantifying the Flexibility of Electric Vehicles in Germany and California—A Case Study

Michel Zade *, Zhengjie You, Babu Kumaran Nalini, Peter Tzscheutschler and Ulrich Wagner

Chair of Energy Economy and Application Technology, TUM Department of Electrical and
Computer Engineering, Technical University of Munich, 80333 Munich, Germany; zhengjie.you@tum.de (Z.Y.);
babu.kumaran-nalini@tum.de (B.K.N.); ptzscheu@tum.de (P.T.); uwagner@tum.de (U.W.)
* Correspondence: michel.zade@tum.de; Tel.: +49-89-23972

Received: 9 September 2020; Accepted: 21 October 2020; Published: 27 October 2020

Abstract: The adoption of electric vehicles is incentivized by governments around the world to decarbonize the mobility sector. Simultaneously, the continuously increasing amount of renewable energy sources and electric devices such as heat pumps and electric vehicles leads to congested grids. To meet this challenge, several forms of flexibility markets are currently being researched. So far, no analysis has calculated the actual flexibility potential of electric vehicles with different operating strategies, electricity tariffs and charging power levels while taking into account realistic user behavior. Therefore, this paper presents a detailed case study of the flexibility potential of electric vehicles for fixed and dynamic prices, for three charging power levels in consideration of Californian and German user behavior. The model developed uses vehicle and mobility data that is publicly available from field trials in the USA and Germany, cost-optimizes the charging process of the vehicles, and then calculates the flexibility of each electric vehicle for every 15 min. The results show that positive flexibility is mostly available during either the evening or early morning hours. Negative flexibility follows the periodic vehicle availability at home if the user chooses to charge the vehicle as late as possible. Increased charging power levels lead to increased amounts of flexibility. Future research will focus on the integration of stochastic forecasts for vehicle availability and electricity tariffs.

Keywords: charging strategy; optimization; electricity pricing; electric vehicle; flexibility; flexibility market; home energy management system

---

1. Introduction

Scarcity of fossil fuels, oil price fluctuations, and increased awareness of the negative impacts caused by anthropogenic climate change have led to an increasing use of variable renewable energy (VRE) sources. With the agreed goal of limiting anthropogenic global warming to well below 2 degrees Celsius, this trend is expected to continue and even accelerate. While hydropower and biomass are, in their operational behavior comparable to conventional power plants, the power generation of photovoltaic and wind systems is variable, and generation prediction challenging and subject to uncertainty. Introducing flexibility products to the power system is one measure to cope with this variability and uncertainty.

Ma et al. define flexibility as the "the ability of a power system to cope with variability and uncertainty in both generation and demand, while maintaining a satisfactory level of reliability at a reasonable cost, over different time horizons" [1]. While this definition describes the general characteristic of flexibility, the Union of the Electricity Industry—Eurelectric—defines flexibility in a more application-oriented way, as "the modification of generation injection and/or consumption patterns in reaction to an external signal (price signal or activation) in order to provide a service within the energy system. The parameters used to characterize flexibility include the amount of power modulation, the duration, the rate of change, the response time, the location, etc." [2].

Nowadays, large-scale flexibility products with a capacity greater than 1 MW are widely used to stabilize grid frequency. On the supply side, system operators (SO) use measures such as redispatch and feed-in management. On the demand side, they use sheddable loads and industrial demand-side-management as grid ancillary services [3–5]. While regulation and various market designs for energy trading already exist, academic and industrial research now focus on introducing unique flexibility platforms [6]. Such new platforms will allow residential consumers and prosumers to participate with their distributed energy resources (DER)—such as combined-heat-and-power units (CHP), electric vehicles (EV), residential heat pumps (HP), photovoltaic systems (PV), and battery storage units—as well as large industrial parties to offer flexibility [7–9]. In the future, SO will be able to manage grid congestions in a less resource-intensive manner and potentially avoid costly grid expansions and the curtailment of VRE [10,11]. Such flexibility platforms differ from existing energy market mechanisms in that they trade power instead of energy. SOs place their flexibility demand on the platform and are matched with residential and industrial flexibility providers.

Flexibility can be both negative and positive. Negative flexibility refers to the delay of grid feed-in or the consumption of non-scheduled energy. Positive flexibility is the delay of grid energy consumption or the non-scheduled grid feed-in.

Home energy management systems (HEMS) can quantify, price, and offer flexibility from private DER to such platforms and re-schedule devices based on the platform response. Beaudin et al. conclude that an HEMS is a demand response tool with the goal of optimizing consumption and production profiles in a house that communicates with household devices, utilities, and forecasting service provider [12]. The most important components of such a system required for calculating flexibility offers are visualized in Figure 1.

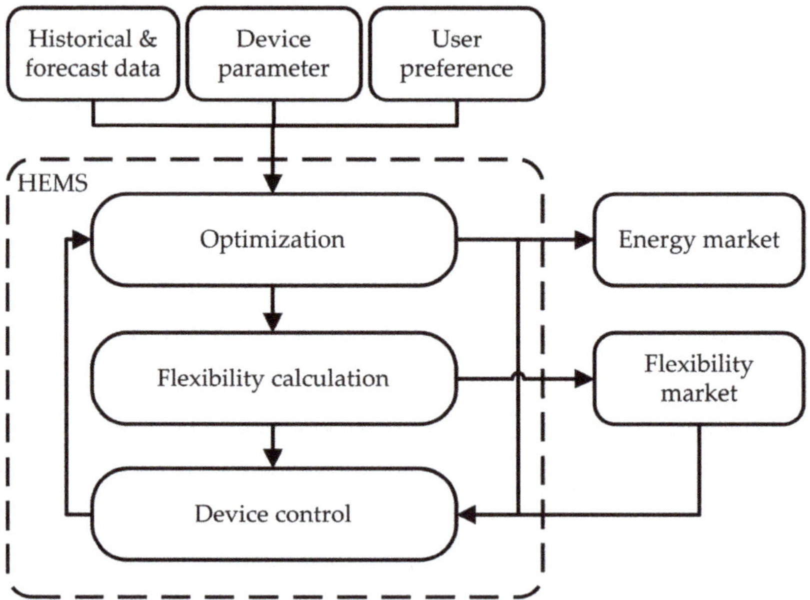

Figure 1. Generalized structure of an HEMS. Historical and forecast data refers to weather data, historical consumption and production data and expected energy prices.

A review of HEMS concluded that cost optimization is the most frequently implemented objective function [12]. Yan et al. state that price-driven demand response is an important demand response measure [13]. Therefore, the type and structure of the electricity price signal is of crucial importance for the optimization problems of HEMSs.

Eurelectric differentiates fixed-priced offers and various types of dynamic pricing [14]. Nowadays, the majority of residents in the US, for example, have a fixed-priced electricity tariff [15]. Besides fixed-priced offers, utilities offer different types of dynamic pricing: time-of-use (ToU), real-time pricing (RTP), and others, such as critical peak pricing (CPP). ToU tariffs offer static pricing schemes with pre-defined prices for specified periods and seasons. As such, ToU tariffs are easy to follow for any customer, however, for the SO they run the risk of creating demand peaks of higher magnitude than the ones caused by fixed-priced offers [13]. In RTP, prices vary over short periods and are communicated to customers one day or less in advance. California was one of the first states to introduce RTP, in 1985 [16]. Nowadays, only a few RTP programs, such as ComEd's Hourly Pricing, exist because they are technically difficult to implement and hard for customers to understand. A lot of studies have investigated the impact of different electricity tariffs on the peak demand of a distribution grid and concluded that simple ToU strategies can lead to increased peak demand [17,18]. However, the literature rarely discusses the impact of different electricity tariffs on flexibility.

Zade et al. published an HEMS model that optimizes the charging process of an electric vehicle (EV), and calculates the flexibility based on synthetic electricity prices, vehicle availabilities, and energy demands [9]. In order to analyze the realistic flexibility potential of EVs in a distribution grid, this paper describes a detailed case study conducted with vehicle field trial data from California, USA and Germany, three electricity tariffs, two controller strategies, and three charging power levels.

2. Materials and Methods

Figure 1 provides a functional overview of the generalized structure of an HEMS. As defined above, the primary objective of the HEMS is to fulfill the electricity, heat, and mobility demand of the household. For this purpose, the HEMS retrieves historical load profiles from an internal database and various other input data, e.g., user preferences, weather, and price forecasts from external sources. Then, an optimizer inside the HEMS calculates cost-optimal operating strategies for all controllable devices. Based on those operating schedules, the HEMS buys and sells energy on the energy market. Afterwards, the HEMS can offer deviations from the cost-optimal operating strategy as flexibility to SOs via a flexibility platform.

2.1. Input Data, Generation, and Consumption Forecast

The HEMS receives data from various parties, e.g., household inhabitants, forecast providers and weather stations. In an initial configuration step, household inhabitants insert device parameters like the charging station's maximal charging power, or EV's battery capacity, etc. More frequently, inhabitants update operational constraints, such as the daytime when an EV needs to be fully charged or the room temperature they find comfortable. Besides those user inputs, the HEMS is fed with different forecasts such as the upcoming weather conditions and expected energy prices. Finally, the optimization is triggered whenever new input data arrives or a certain amount of time has passed.

2.2. Optimization Approach

The calculation of the cost-optimal charging schedule is based on [19] but has been modified in order to incorporate constraints such as EV availabilities over time. This section describes the mixed-integer linear programming (MILP) model that has been used to calculate the cost-optimal charging schedule. In this work, it is assumed that the prosumer prefers a cost-optimized solution to the scheduling problem. Therefore, the target function in Equation (1) is formulated as a cost minimization.

$$\min\left(\sum_{t=1}^{T} (p_t^{im} \cdot c_t^{im} - p_t^{ex} \cdot c_t^{ex} + p_t^{gas} \cdot c_t^{gas}) \cdot \Delta t + K \cdot \Delta SoC_t\right) \tag{1}$$

Here T is the set of time steps (t) considered throughout the scheduling horizon. Δt is the duration of each time step, p^{im} and p^{ex} are the electrical import and export power, while c^{im} and c^{ex} are the corresponding electricity costs or revenues. p^{gas} and c^{gas} denote the volume of natural gas used and its specific cost, respectively. K represents a penalty coefficient that is multiplied with the variable ΔSoC_t which is the difference between the desired final state of charge (SoC) and the actual SoC at the end of charging. This penalty term allows the optimizer to create a feasible problem even though the available time is not sufficient to charge a vehicle fully. Thereby, infeasible problems are avoided.

In addition, the energy balance and constraints for each appliance are critical to reflect a correct and realistic optimization. The constraints are as follows. The energy balance for electricity is represented by Equation (2) and for heat by Equation (3).

$$p_t^{im} + \sum_{\delta \in Flex_{el}} p_t^{\delta} - p_t^{ex} - p_t^{load} = 0 \qquad \forall t \in [1, T] \tag{2}$$

$$\sum_{\delta \in Flex_{th}} p_t^{\delta} - q_t^{load} = 0 \qquad \forall t \in [1, T] \tag{3}$$

p^{load} and q^{load} represent the electrical and thermal load of the household. p^{δ} is the power of one specific device, which belongs to one of the flexible appliance groups $Flex_{el} \in \{EV, CHP, HP, PV, Bat\}$ or $Flex_{th} \in \{CHP, HP, TES\}$. The group $Flex_{el}$ includes electric vehicles (EV), combined heat and power (CHP), heat pumps (HP), photovoltaic (PV), and battery systems (Bat). The group $Flex_{th}$ covers CHP, HP and thermal energy storages (TES). CHPs and HPs are classified into both groups because CHP produce electricity and heat at the same time and heat pump can convert power to heat.

For each flexible appliance and storage system, multiple constraints exist concerning their operation. Because this paper focuses on the quantification of flexibility of EVs, Equations (4)–(10) only list the constraints for EVs.

$$SoC_t = SoC_0 + \frac{1}{C_{Bat}} \sum_{s=1}^{t} \left(p_t^{EV} \cdot \eta - p_t^{cons} \right) \cdot \Delta t \qquad \forall t \in [1, T] \tag{4}$$

$$SoC_{min} \leq SoC_t \leq SoC_{max} \qquad \forall t \in [1, T] \tag{5}$$

$$SoC_t \geq SoC_t^e - \Delta SoC_t \qquad \forall t \in \left\{ t_1^e, t_2^e, \ldots \right\} \tag{6}$$

$$SoC_t \leq SoC_t^s \qquad \forall t \in \left\{ t_1^s, t_2^s, \ldots \right\} \tag{7}$$

$$0 \leq p_t^{EV} \leq A_t \cdot P_{max}^{EV} \qquad \forall t \in [1, T] \tag{8}$$

$$A_t = \begin{cases} 1 & \forall t \in \left[t_1^s, t_1^e \right] \cup \left[t_2^s, t_2^e \right] \cup \ldots \\ 0 & \forall t \in \left[0, t_1^s \right) \cup \left(t_1^e, t_2^s \right) \cup \ldots \end{cases} \tag{9}$$

$$p_t^{cons} \geq 0 \qquad \forall t \in [1, T] \tag{10}$$

In Equations (4)–(7), SoC represents the SoC of the EV battery. SoC_0 denotes the initial SoC of the EV, SoC_{min} and SoC_{max} are minimal and maximal SoC. SoC_t^s refers to the SoC at time step zero and SoC_t^e represents the SoC at the last available time step. SoC_t^e may be reduced by ΔSoC_t if the time the vehicle is parked at home is not sufficient to charge the battery from start SoC to the desired end SoC. t^s and t^e refer to the start and end time of each availability period of the EV. C_{Bat} is the battery capacity, p_t^{EV} is the charging power of the EV, and η is the charging efficiency. p_t^{cons} is a variable that covers the power demand if the EV is available for multiple periods and energy has been discharged from the battery in between.

Equations (8)–(10) describe the constraints for p_t^{EV}. A_t describes the vehicle availability for charging, which is 1 between t^s and t^e, and otherwise zero. Integrated into Equation (8), the availability does not allow the charging power to be greater than 0 if the vehicle is not available for charging.

If the vehicle is available for charging, the charging power can be as high as the maximal charging power P^{EV}_{max}.

The formulated MILP model can be solved using commercial and open-source solvers, such as GLPK or Gurobi. Depending on the problem complexity, a conventional computer (e.g., Intel i7, 4 Cores, 24 GB RAM) presents a solution within a few seconds.

After optimizing the device's operating strategy, a market agent in the HEMS trades its excess and required energy on the energy market and a controller schedules the device's operation accordingly. After successful interaction with the energy market, the HEMS can start the flexibility calculation.

2.3. Flexibility Offers

Based on the optimal operating strategy of the devices described in the previous subsection, the HEMS calculates flexibility offers. Such a flexibility offer consists of the flexible power they can offer, the duration they can offer it, at what time, at which position in the grid, and at what price. The following subsections describe the calculation of these parameters in detail.

2.3.1. Location, Negative, and Positive Flexibility

Since in the setting considered here all flexible devices are stationary, the location of a flexible device is considered to be constant and is described by a unique identifier. In Germany, the Bundesnetzagentur introduced a 11-digit identifier (MaLo-ID: market location identifier) to simplify the market communication. Therefore, each HEMS that offers flexibility must attach the MaLo-ID to their bids.

Generally, all flexible devices are able to offer positive and negative flexibility, some even at the same time. For example, an EV charging station can offer positive flexibility by stopping an ongoing charging process or by reducing the charging power. Negative flexibility can be offered by charging a vehicle even though it has not been scheduled or by increasing the charging power while charging.

2.3.2. Power, Duration, and Energy

The DERs have different operating types. For both operating types, flexibility can be determined using Equations (11) and (12).

$$p^{\delta}_{pos} = \begin{cases} P^{\delta}_{max} - p^{\delta} & \forall \delta \in G \\ p^{\delta} & \forall \delta \in C \end{cases} \tag{11}$$

$$p^{\delta}_{neg} = \begin{cases} -p^{\delta} & \forall \delta \in G \\ p^{\delta} - P^{\delta}_{max} & \forall \delta \in C \end{cases} \tag{12}$$

As mentioned in Section 2.2, p^{δ} denotes the power of the electricity consumer (C) and generator (G). P^{δ}_{max} is the maximal power of each flexible device. p^{δ}_{pos} and p^{δ}_{neg} represent the resulting positive and negative flexibility power for each device. Note that the positive and negative flexibility power is always positive and negative, respectively, or zero. As described above, one device can offer positive and negative flexibility at the same time.

Equations (13)–(15) describe the duration that flexibility is available. d^{δ}_{flex} is the maximal duration that flexibility can be offered by a flexible device.

$$\max_{t^e_{flex} \in [t^s_{flex},\, T]} d^{\delta}_{flex} = \left(t^e_{flex} - t^s_{flex} \right) \Delta t \tag{13}$$

s.t.

$$f_t\left(p^{\delta'},\, p^{\delta}_{flex} \right) = 0 \qquad \begin{aligned} &\forall \delta' \in A_f - \delta \\ &\forall t \in \left[t^s_{flex},\, t^e_{flex} \right] \end{aligned} \tag{14}$$

$$g_t\left(\hat{p}^{\delta'}, p_{flex}^{\delta}\right) \leq 0 \qquad \begin{array}{l} \forall \delta' \in A_f - \delta \\ \forall t \in \left[t_{flex}^s, t_{flex}^e\right] \end{array} \tag{15}$$

This optimization problem is solved for each time step. The start time t_{flex}^s of each flexibility is exactly the time step chosen by each iteration. t_{flex}^e is variable in this problem and should be maximized without violating constraints f_t and g_t, which are abstracted from the equality and inequality constraints discussed in Section 2.2, respectively. Subscript t for f and g indicates that the constraints shall be satisfied in the whole domain of t. p_{flex}^{δ} is the positive or negative power of one specific flexible appliance, whereas $\hat{p}^{\delta'}$ refers to all other flexible appliances that still follow the cost-optimal schedules.

Once the flexibility duration has been acquired, the flexible energy is calculated by Equation (16).

$$e_{flex,t}^{\delta} = \sum_{s=t}^{t+d_{flex}^{\delta}-1} p_{flex,t}^{\delta} \cdot \Delta t \tag{16}$$

Hence, the duration for which flexibility can be offered depends on the device's current state. In the case of an EV, the flexibility that can be offered depends on the battery's SoC, maximal charging power and availability.

In a final step, the flexibility would need a price tag in order to be offerable on a flexibility platform. However, this paper focuses on the quantification of flexibility of EVs and therefore the pricing is excluded from this analysis. Nevertheless, one possible pricing mechanism for flexibility of EV is described in [9].

Finally, the HEMS transfers the calculated flexibility parameters to a flexibility platform and waits for flexibility calls. Once a provider is called for flexibility, user preferences change, or new forecasts are available, the HEMS reinitiates the entire procedure from optimization to flexibility calculation and updates the offers on the flexibility platform.

The model is open-source and accessible via the link in the Supplementary Material.

3. Case Study

Figure 2 visualizes the general design of the case study. In a first step, we computed vehicle availabilities based on field trial data, collected by the California Department of Transportation and the Karlsruhe Institute of Technology. After gathering and pre-processing the vehicle availabilities and electricity tariffs, the cost-optimal charging schedules and the flexibility for each vehicle availability is calculated using the model described in Section 2. In order to analyze the aggregated flexibility potential of more than 4000 Californian and more than 11,000 German vehicle availabilities, the final results are aggregated. The following paragraphs describe the case study setup in detail. The link in the Supplementary Material contains an open-source script for the case study.

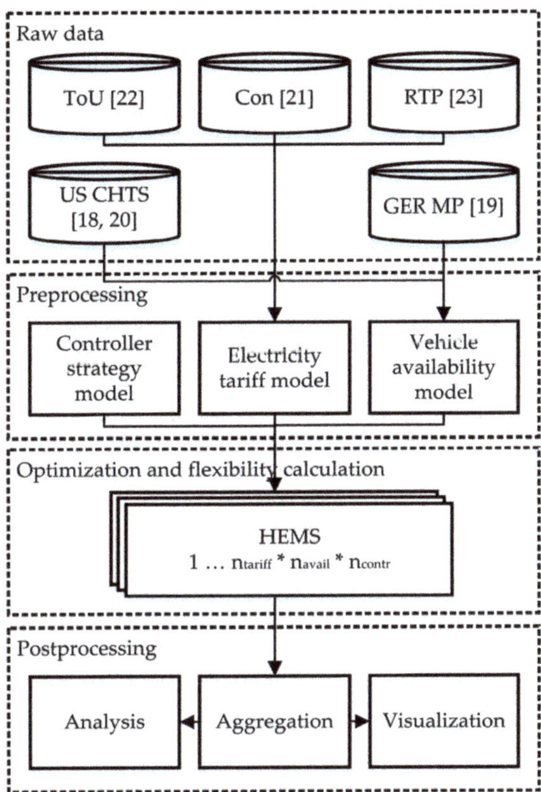

Figure 2. Case study design, US CHTS, and GER MP refers to the Californian and German field trial data used to calculate the vehicle availabilities [18–20]. Con refers to the constant electricity rate in California [21], ToU refers to the 'ToU-D-Prime' electricity tariff of Southern California Edison [22], and RTP refers to the Hourly Real-Time prices of ComEd in Illinois [23], US. n_{tariff} refers to the number of electricity tariffs, n_{avail} refers to the number of vehicle availabilities, and n_{contr} refers to the number of controller strategies investigated in this case study.

3.1. Data Input and Preprocessing

For the case study, vehicle availabilities at home are computed based on the data sets of the Californian Household Travel Survey (US CHTS) and the German Mobility Panel (GER MP) [20,21]. Table 1 summarizes the parameters that characterize a vehicle availability.

Table 1. Vehicle availability characterizing parameters.

Parameter	Unit	Description
ID_{veh}	-	Vehicle identifier
$t_{arrival}$	Timestamp	Arrival time at home
$t_{departure}$	Timestamp	Departure time from home
$d_{travelled}$	km	Distance traveled since last departure from home
$\Delta t_{available}$	s	Available time at home

The following subsections provide a brief summary of the most important characteristics and differences of the data sets used, as well as a short analysis of the computed vehicle availabilities.

3.1.1. Californian Household Travel Survey

Between February 2012 and March 2013, 677 Californian vehicles were equipped with in-vehicle GPS tracking devices. Every trip made by each vehicle was tracked for one week. The publicly available data set contains information about start and end times, start and end location, average speed and miles driven for every trip [22]. In the data set, 19,075 trips by 662 unique vehicles are recorded. The distance traveled varies from less than 0.5 km up to 1289 km. The average distance traveled is 35 km for all conducted trips (see Table 2). Start and end locations are categorized in four categories: HOME, WORK, SCHOOL, and OTHER.

Table 2. Mean, maximum, minimum and 95%-ile of distance traveled since last departure from home.

Parameter	Unit	US CHTS	GER MP
$d_{\text{travelled,avg}}$	km	35	38
$d_{\text{travelled,max}}$	km	1289	1992
$d_{\text{travelled,min}}$	km	0.5	0.2
$d_{\text{travelled,95\%}}$	km	131	130

Based on these parameters, the availability of vehicles at 'HOME' can be extracted, with arrival and departure time. Furthermore, the distance traveled is used to calculate the energy used by an average vehicle from its last departure from home. This procedure results in 4062 vehicle availabilities of 592 unique vehicles that contain a vehicle identifier, the distance traveled since the vehicle's last departure from home, its arrival time at and departure time from home. Inconsistencies in the GPS data set, e.g., a vehicle arrives at home but departs for the next trip from another location, are neglected in this analysis.

3.1.2. German Mobility Panel

Between September and November 2017, 3867 persons from 1881 households logged their daily mobility behavior in a travel diary. After plausibility checks conducted by the Karlsruhe Institute of Technology (KIT), a total of 70,252 trips were gathered from 1850 persons [21]. The final trip data set includes information about the date, the trip's start and end time, purpose, mode of transport used, duration, distance, and household. In 33,250 of the 70,252 trips logged, the person recorded having driven a vehicle as a driver either as a first, second or third "mode of transport used". In 13,550 of the 33,250 trips, the purpose of the trip was to return home.

Based on the 33,250 trips, a total of 11,458 vehicle availabilities at home are computed by considering household and person identifier, trip purpose, and the trips' chronology.

3.1.3. Vehicle Availabilities

In this subsection, the calculated vehicle availabilities are visualized and analyzed. In order to analyze the number of available vehicles at home during an average week, all vehicle availabilities are summed up for each time step of the week. Thereafter, the sums are averaged over all weeks of the field trials. Figure 3 shows the results for the Californian and German data sets. The average number of vehicles available in the Californian data set is 4.5 vehicles and for the German data set 21.3 vehicles. This can be explained by the compressed German field trial period of three months and the higher number of field trial participants (see [20,21]). Despite the differences in quantity, the vehicle availabilities indicate the same trends. At night, the number of vehicles increases until 12 a.m. and then decreases until 12 p.m. This behavior is repeated every day of the week. On weekends, however, the magnitude of the oscillation decreases approximately by a factor of three.

Figure 4 visualizes a histogram of the total number of available vehicles over the distance traveled since their last departure from home. The results of both data sets show an exponential decay with only a few outliers. Ninety-five percent of the American and German vehicles arrive at home with less than 130 km driven since their last departure from home (see Table 2).

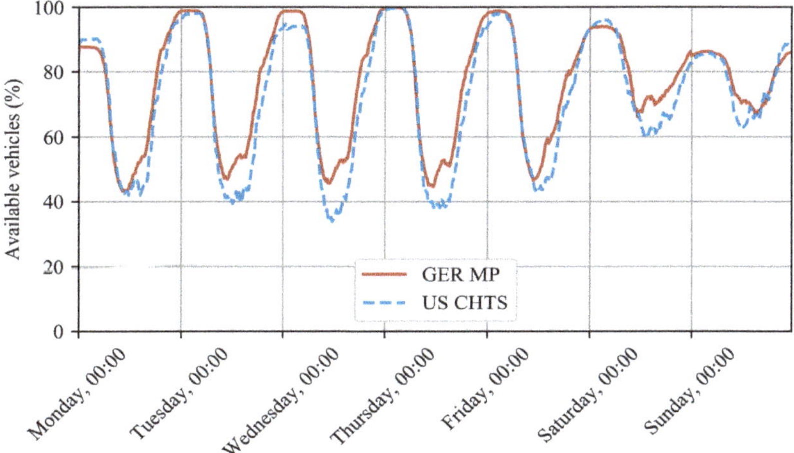

Figure 3. Percentage of vehicles available at home during an average week. The maximum number of vehicles available at home is 6 for the US CHTS data and 120 for the GER MP.

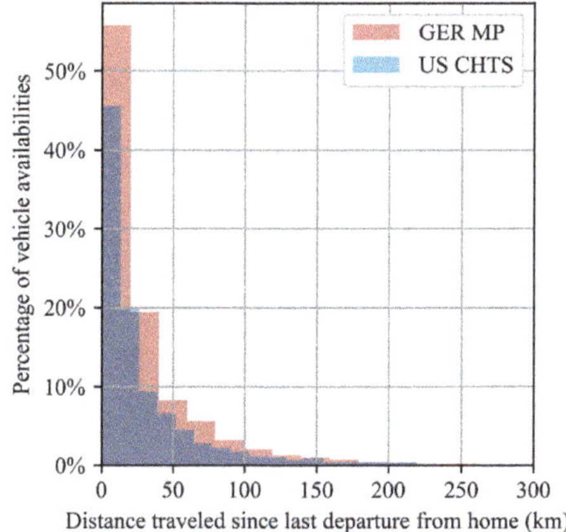

Figure 4. Distribution of the distance traveled since the vehicle's last departure from home—sampled from all modeled vehicle availabilities.

Figure 5 visualizes a relative frequency histogram of the total number of vehicles over the period the vehicles are available at home. Both data sets (US CHTS and GER MP) indicate a periodic behavior with a decreasing amplitude for an increasing time of availability. Most vehicles are available either for less than 1 to 3 h or for 7 to 25 h. Far fewer vehicles are available for 4 to 6 h or for more than 25 h. Table 3 summarizes the mean, maximum, minimum and 95%-ile of the available time for both data sets.

Figure 6 visualizes the distribution of the number of vehicles arriving at home over the hour of the day and the day of the week. Most vehicles arrive at home during the afternoon hours from 3 to 6 p.m. and depart from home between 6 and 9 a.m. Vehicles arrive at and depart from home more frequently during the week than on the weekend. Neither result is surprising considering conventional 9 to 5 working hours.

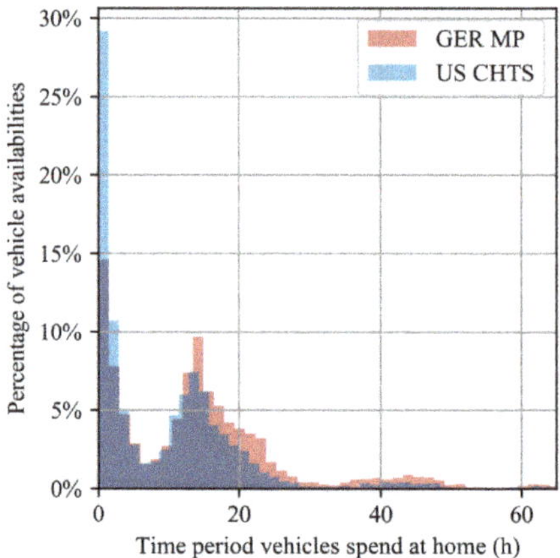

Figure 5. Distribution of the time period vehicles spend at home—sampled from all modeled vehicle availabilities.

Table 3. Mean, maximum, minimum, and 95%-ile of time of availability at home.

Parameter	Unit	US CHTS	GER MP
$\Delta t_{\text{available, avg}}$	h	10.43	16.16
$\Delta t_{\text{available, max}}$	h	142.38	148.67
$\Delta t_{\text{available, min}}$	h	0	0.02
$\Delta t_{\text{available, 95\%}}$	h	34	47.5

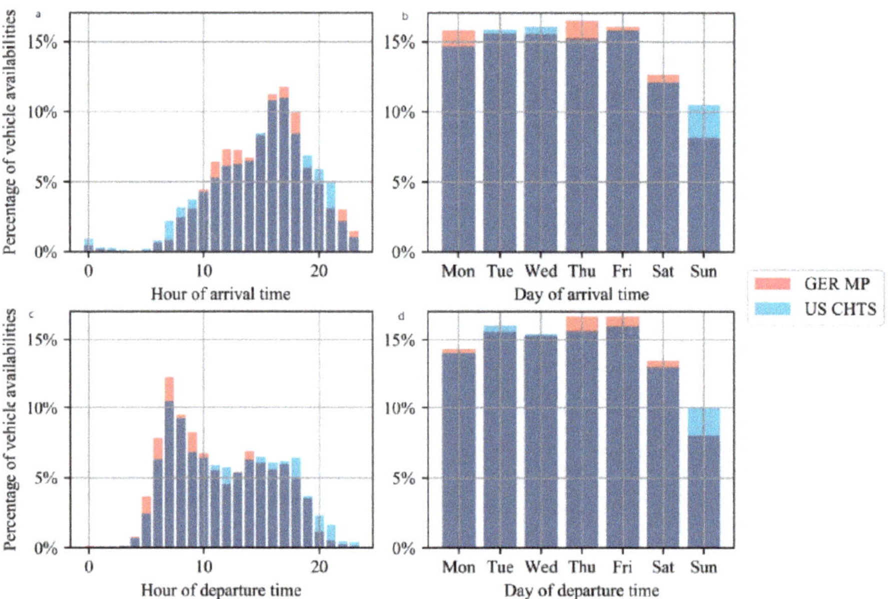

Figure 6. Percentage of vehicle availabilities over (**a**) hour, (**b**) day of arrival time, (**c**) hour, and (**d**) day of departure time.

3.1.4. Electricity Tariffs

In order to quantify the impact of different electricity tariffs on the flexibility potential of electric vehicles, three tariffs are used in this case study.

The first tariff is a constant tariff '*Con*' in which the electricity price does not vary. The price is set to 0.19 \$/kWh, which was the average electricity price in California in 2018 [23].

'*ToU*' tariffs are offered throughout the United States and in other countries to motivate the reduction of electricity consumption in peak demand periods. Southern California Edison offers multiple *ToU* tariffs for residential customers on their website. For the simulations, the 'ToU-D-Prime' tariff as published on the website in the beginning of 2020 has been used. The tariff differentiates between winter and summer, weekday and weekend, and hour of day. In the winter, weekdays and weekends are priced equally. Between 4 and 9 p.m. the mid-peak tariff is active during the entire winter (0.36 \$/kWh) and on weekends in the summer (0.27 \$/kWh). During the summer, on weekdays from 4 to 9 p.m., the on-peak tariff for 0.39 \$/kWh is active. From 9 to 4 p.m. the off or super-off-peak tariff at 0.14 and 0.13 \$/kWh is active. This tariff motivates customer to reduce their electricity consumption in the late afternoon and early evening.

The third tariff integrated in this analysis is *RTP*. California already implemented two *RTP* programs in 1985 and 1987. However, both *RTP* programs had been canceled by 2003 [16]. Nowadays, California only offers *Con* and *ToU* tariffs. Therefore, a publicly available *RTP* tariff from ComEd, an energy supplier in Illinois, US is chosen [24]. In order to equalize the electricity prices for all tariffs, $p_{\text{RTP,offset}}$ is added to the real-time prices. $p_{\text{RTP,offset}}$ is the equal to the constant electricity price of 0.19 \$/kWh minus the mean of all RTP. Since an analysis of the forecasting error of *RTP* is beyond the scope of this publication, the *RTP* tariff is assumed to be a perfect forecast of the electricity prices.

$$p_{\text{RTP,offset}} = p_{\text{Con}} - \frac{1}{N} \sum_{t=1}^{T} p_{\text{RTP,i}} \tag{17}$$

3.1.5. Controller Strategies

In order to quantify the impact of different controller strategies or user preferences on the flexibility potential of electric vehicles, we implemented two controller strategies.

The first controller strategy is to charge the vehicle at minimal costs but as soon as possible. Such behavior can be simulated by adding a minimal price increment onto the electricity prices (see Equation (18)).

$$c_t^{\text{im/ex}} = c_t^{\text{grid, im/ex}} + c_t^{\text{contr}} \qquad \forall t \in [1, T] \tag{18}$$

$c_t^{\text{grid, im/ex}}$ denotes the actual electricity prices/revenues and c_t^{contr} the term that is added in accordance with the controller strategy. In the case of the first controller strategy, minimal price increments are added in the range of 0.00001 to 0.00002 \$/kWh and therefore do not affect the actual price of electricity for the user. In the case of constant electricity prices, the optimizer would choose the first possible time steps in order to charge at minimal costs. For the rest of this publication, this operating strategy is denoted as "+*MI*".

In order to conserve battery life, a second controller strategy is to charge the vehicle as late as possible and therefore to keep the SoC of the EV battery as low as possible as long as possible. This controller strategy can be implemented either by the addition of a minimal price decrement in Equation (18) or in the optimizer by default. In our case, this behavior was implemented by default in the solver. Therefore, this controller strategy is not separately labeled.

Table 4 lists the five simulated operating strategies that represent the combination of the three electricity tariffs and the two controller strategies.

Table 4. Simulated operating strategies that represent the combination of electricity tariffs and controller strategies.

Operating Strategy	Electricity Tariff	Controller Strategy
Con	Con	Minimal decrements
ToU	ToU	Minimal decrements
Con + MI	Con	Minimal increments
ToU + MI	ToU	Minimal increments
RTP	RTP	-

3.2. Flexibility Calculation

In order to use the vehicle availabilities described in Section 3.1 as EV input parameters for the model described in Section 2, the energy demand is calculated based on the distance traveled. The energy required is the product of the specific energy consumption of the EV $e_{kWh/km}$ and the distance traveled $d_{travelled}$.

$$E_{EV,i} = e_{kWh/km} * d_{travelled} \tag{19}$$

For this case study, a specific energy consumption of 0.2 kWh/km is used for all vehicle availabilities [25,26]. Furthermore, the user preference for the desired SoC of the vehicle at the time of departure $t_{departure}$ was set to 100 %. The charging efficiency is set to 98 %.

In order to investigate the impact of the maximal charging power, the maximal charging power is varied in three steps: $P_{charge,max} \in \{3.7\ kW,\ 11\ kW, 22\ kW\}$. This variation allows all current and possible future residential charging station configurations to be analyzed.

While the HEMS is capable of calculating the flexibility of HP, CHP, PV, and batteries, all other possible inputs, such as additional electrical or thermal loads or generation, are set to zero.

For every one of the five operating strategies listed in Table 4 and every $P_{charge,max}$, the model calculates the optimal charging schedule and flexibility potential as a time series. This procedure resulted in a total of 165,870 for GER MP and 60,930 for US CHTS executions of the model.

3.3. Data Aggregation

Once optimal charging schedules and flexibility have been calculated for more than 15,000 vehicle availabilities for 5 operating strategies and 3 maximal charging powers, the results are aggregated.

First, all available vehicles, charging schedules, flexible power and energies are summed up for every time step of the field trial periods. The result is a data set that shows the total number of available vehicles at home, charging powers, flexible power and energy for every time step of the field trial.

In a final step, the summed data is clustered into weekly time steps (e.g., "Monday, 09:00"), and weekdays and weekends. The clusters are then averaged over the field trial duration.

4. Results

This chapter visualizes and describes the results of the case study in detail. The cost-optimal charging schedules are shown in the top two rows of plots, whereas the flexibility potential is shown in the bottom two rows of plots (Figure 7). The first section describes the cost-optimal charging schedules, and the second section the flexibility potentials of the vehicle availabilities from both data sets for the five operating strategies.

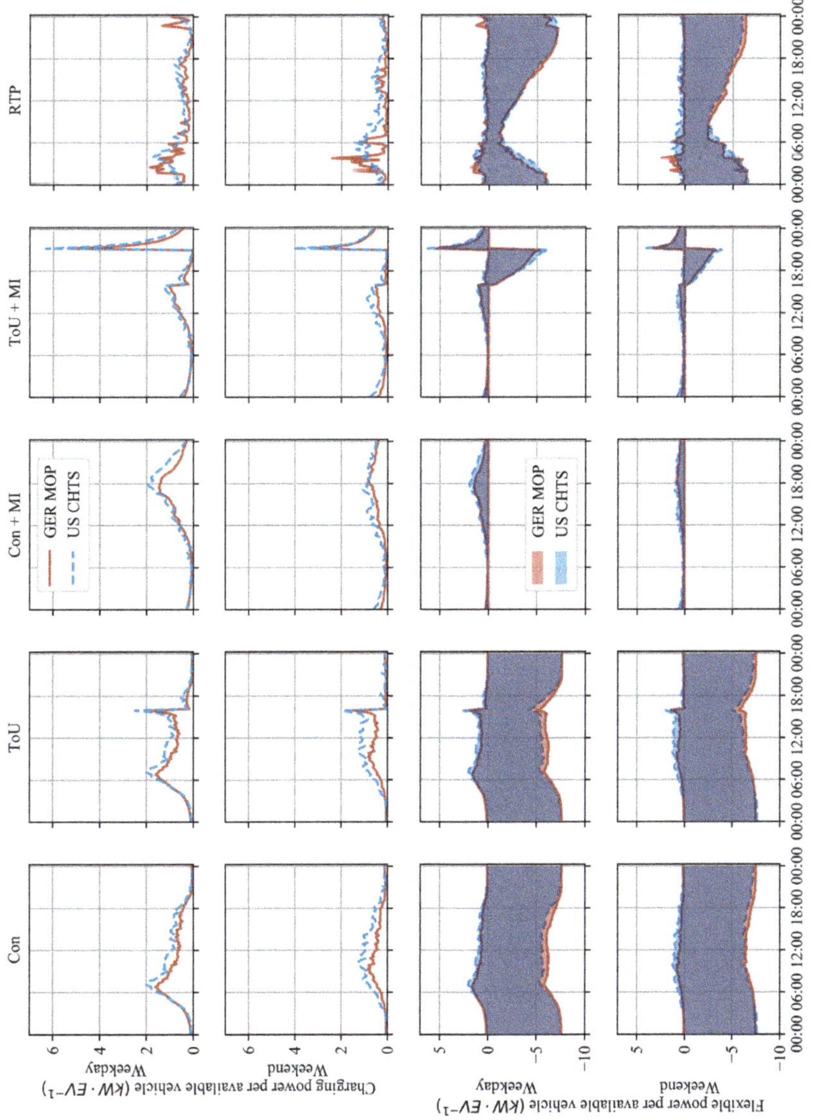

Figure 7. The plot series visualizes the charging and flexible power per available EV for the five operating strategies on weekdays and weekends. The first two rows of plots show the charging power per available EV for the five operating strategies on weekdays and weekends. The third and fourth rows of plots show the flexible power per available EV for the five operating strategies on weekdays and weekends.

4.1. Cost-Optimal Charging Schedules

In the top two rows of plots, Figure 7 shows the cost-optimal power demand per vehicle for weekdays in the first row and for weekends in the second row. The curves of the *Con* and the *ToU* operating strategy are almost identical and can only be distinguished by their behavior between 3 p.m. and 9 p.m. At 4 p.m., the *ToU* curves show a smaller second peak compared to the early morning hours. This behavior can be explained by the optimizer logic and mid/on-peak tariffs. The optimizer implemented charges the vehicles as late as possible and as cheaply as possible. Considering the mid/on-peak tariffs starting at 4 p.m., the optimizer schedules all vehicles that depart between 4 and 9 PM to charge right before 4 p.m. Therefore, this trend is consistent with the implemented optimizer logic. Besides the difference mentioned in the early afternoon, the power curves for the *ToU* and *Con* operating strategy show the same trend as visualized by the histogram of the departure times in Figure 6. The amplitude ranges from 0 to 2.5 kW/EV for both data sets. On weekends, the power ranges from 0 to 1.9 kW/EV and is more spread out throughout the day. Generally, the results indicate that the Californian vehicles require greater power per vehicle compared to the German vehicles.

The *RTP* operating strategy causes charging peaks that are spread out from 11 p.m. to 8 a.m. The peaks are more irregular than the ones for the *Con* and *ToU* operating strategies. While the power curves for the *Con* and *ToU* operating strategy indicate similar trends for the US CHTS and the GER MP data set, the cost-optimal charging power differ significantly between the German and the Californian data set in the *RTP* operating strategy. The charging power for the Californian data set looks rather smooth, whereas the results of the German vehicles look much spikier. Since the German data set was collected over a period of three months, a single drop in the real-time prices and the corresponding peak of charging power have a greater impact on the average charging power than those that occurred during the 12-month Californian field trial with only a few vehicles. However, the amplitude ranges also from 0 to 2.5 kW/EV for both data sets. Since real-time prices are much more difficult to forecast and exhibit erratic short-term changes, the demand peaks are most probably overestimated in these results.

The cost-optimal charging power for the operating strategy *Con + MI* indicates a shifted charging behavior. Whereas the *Con* operating strategy schedules vehicle charging right before their departure in the morning hours, the minimal price increments force the optimizer to charge the vehicles right after they arrive home. Therefore, the charging power curve for the *Con + MI* follows the almost Gaussian distribution of the arrival times shown in Figure 6. The amplitude ranges from 0 to 2 kW/EV, which is comparable to the curves of the *Con* and *ToU* operating strategies.

Nevertheless, *ToU + MI* cause the greatest charging power peaks (see Figure 7). Every day at 9 p.m., the optimizer schedules the vehicles that arrived between 4 and 9 p.m. to start charging at the same time. This leads to power peaks of more than 6 kW/EV for both data sets.

Overall, the US CHTS and the GER MP results show similar trends for the cost-optimal charging power for the five operating strategies simulated.

4.2. Flexibility

4.2.1. Operating Strategies

In the bottom two plots of Figure 7 the ranges of flexibility for the five operating strategies simulated are visualized.

For EVs, positive flexibility is equivalent to a pause or postponement of the charging process. Therefore, the upper boundary of the flexibility is equal to the optimal charging power.

According to the definition in Section 1, negative flexibility is the ability to consume electricity ahead of its schedule. Considering the operating strategy *Con + MI* and a cost optimization, no negative flexibility can be offered. Therefore, the lower boundary of the simulation results is congruent with the zero line (see Figure 7).

Similar to the aforementioned operating strategy, *ToU + MI* result in no negative flexibility between 9 p.m. and 4 p.m. From 4 p.m. to 9 p.m., the negative flexibility increases linearly as vehicles arrive home, and their charging process is scheduled from 9 p.m. onwards owing to lower electricity prices. At 9 p.m., negative flexibility drops back to zero.

The operating strategy *Con* and *ToU* result in almost identical negative flexibility results. Furthermore, the negative flexibility that can be offered follows the vehicle availability curves discussed in Section 3.1.3). Periodically, at night time, negative flexibility increases and reaches its maximum around 1 to 3 a.m. During the morning hours before 12 a.m., the flexibility decreases. Negative flexibility ranges from −5 kW/EV to −7.5 kW/EV with the *Con* and *ToU* operating strategies. On weekends, the ranges are smaller since vehicle fluctuations also decrease. At 4 p.m., the *ToU* operating strategy causes a minor drop in negative flexibility due to the charging of vehicles that depart between 4 and 9 p.m.

The *RTP* operating strategy also follows the vehicle availability described in Section 3.1.3). However, in contrast to the results of the *ToU* and *Con* operating strategies, the maximum negative flexibility is available right before midnight. After midnight, when electricity prices are the lowest, the vehicles are charged and the available negative flexibility decreases. On weekends, the range of negative flexibility that can be offered decreases slightly as the fluctuations in vehicle availabilities also decrease. The negative flexibility that can be offered ranges between −2 kW/EV and −7 kW/EV for both data sets. Therefore, *RTP* prices lead to less offerable negative flexibility than a *Con* or *ToU* operating strategy.

Having described the impact of the five operating strategies, the next subsections describe the impact of the maximal charging power on the offerable flexibility of EVs.

4.2.2. Maximal Charging Power

To analyze the impact of the maximum charging power level, the optimization and flexibility calculation for all vehicle availabilities are repeated for three maximum charging power levels: 3.7 kW, 11 kW, and 22 kW. Figure 8 shows the positive and negative flexible power that can be offered for the GER MP data set. Each operating strategy corresponds to a row and each maximum charging power level to a column of heat maps. Tables 5 and 6 summarize the maximal flexible power and average flexible power for all five operating strategies. Generally, the positive flexibility is also representative of the cost-optimal charging power.

Table 5. Maximum and average positive flexible power for five operating strategies and three maximum charging powers.

Operating Strategy	$P_{\text{pos,max}}$ [kW/EV]			$P_{\text{pos,avg}}$ [kW/EV]		
	3.7 kW	11 kW	22 kW	3.7 kW	11 kW	22 kW
Con	1.1	1.8	2.1	0.4	0.5	0.5
ToU	1.1	2.1	2.8	0.4	0.5	0.5
Con + MI	1.1	1.6	2.0	0.4	0.5	0.5
ToU + MI	2.4	5.7	9.1	0.4	0.4	0.5
RTP	1.9	4.0	6.2	0.4	0.4	0.4

Table 6. Maximum and average negative flexible power for five operating strategies and three maximum charging powers.

Operating Strategy	$P_{\text{neg,max}}$ [kW/EV]			$P_{\text{neg,avg}}$ [kW/EV]		
	3.7 kW	11 kW	22 kW	3.7 kW	11 kW	22 kW
Con	−2.9	−7.8	−12.8	−1.9	−6.9	−11.3
ToU	−2.9	−7.8	−12.8	−1.9	−6.8	−11.1
Con + MI	0.0	0.0	0.0	0.0	0.0	0.0
ToU + MI	−2.3	−5.6	−9.0	0.0	−0.6	−0.9
RTP	−2.9	−7.6	−12.2	−0.2	−4.3	−6.7

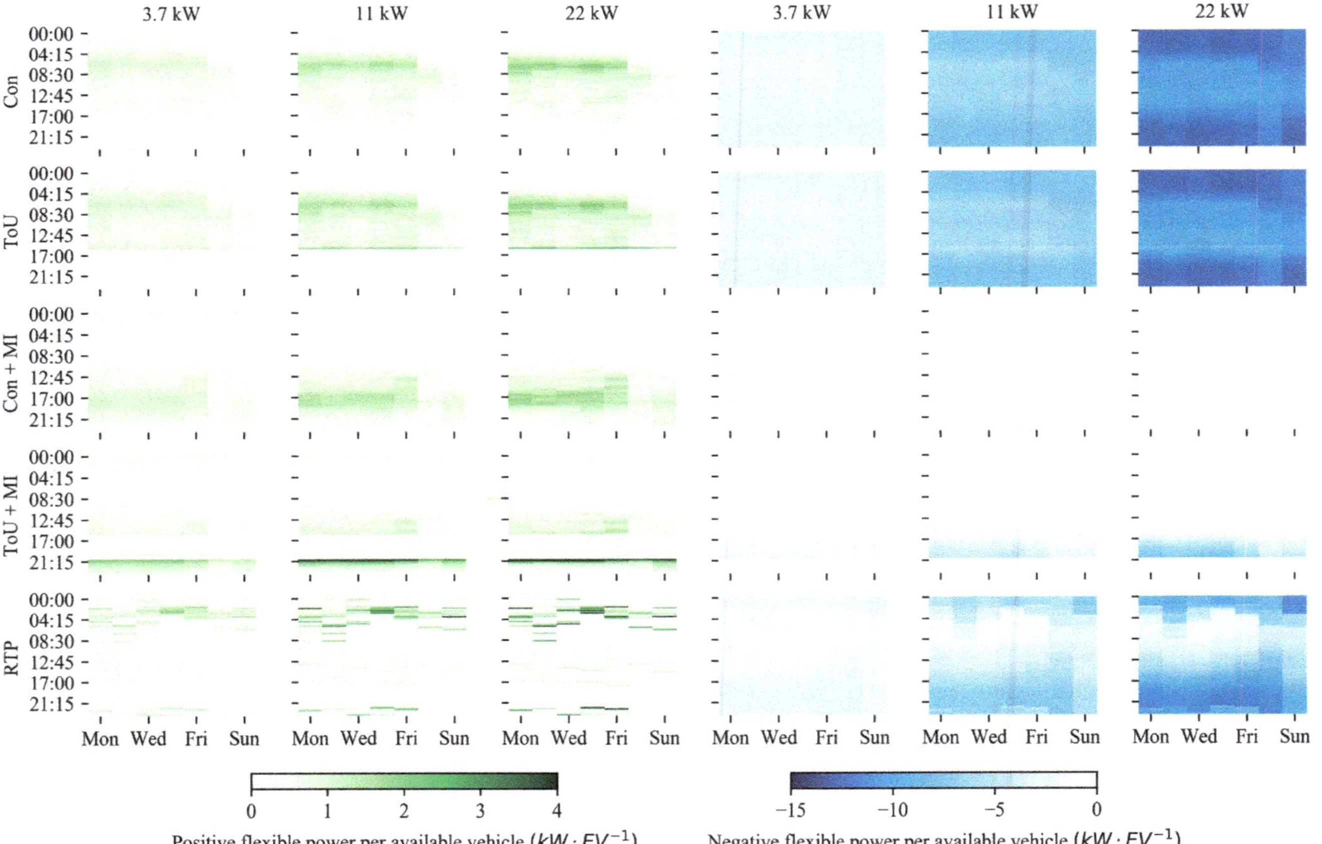

Figure 8. Resulting positive and negative flexible power per available EV. These results are based on the GER MP vehicle availabilities, the five operating strategies {*Con, ToU, Con + MI, ToU + MI, RTP*}, and three maximal charging power levels {3.7, 11, 22 kW}.

The results of the *ToU* and *Con* operating strategy show similar behavior. As described in Section 4.1, the optimizer schedules EV chargings at the latest possible time. This leads to higher charging powers and a high level of positive flexibility from 3 a.m. to 8 a.m. on weekdays. A higher maximum charging power increases the maximal and average positive flexible power that can be offered (see Figure 8). However, the duration of positive flexibility seems to decrease with increasing maximum charging power in Figure 8. With a higher charging power level, the energy required is charged over a shorter time period, and therefore leads to a compressed availability of positive flexibility. The increase of maximal and average positive flexible power with an increasing maximum charging power can be explained by assuming that with a low charging power, e.g., 3.7 kW, not all vehicles are completely charged. In this case, the vehicle cannot offer any flexibility. With a higher maximum charging power, the charging station charges the EV over a shorter period and can therefore offer more flexibility. However, this relation is not linear, since the average positive flexible power seems to only change insignificantly from 11 to 22 kW. Therefore, the two observations described complement each other.

Considering the cost optimization described in IV.A, the results for positive flexible power for the *ToU + MI*, *Con + MI*, and *RTP* operating strategies indicate a similar behavior. With an increasing charging power, the positive flexible power that can be offered is compressed in time whereas the maximal power increases. Furthermore, the average quantity of positive flexibility increases (see Table 5). Both effects are explained in the previous paragraph.

Nevertheless, the *ToU + MI* and *RTP* operating strategies cause such high charging peaks at 9 p.m. (*ToU + MI*) and overnight (*RTP*) that the average maximal positive flexible power is three and two times higher than in the remaining operating strategies. Whereas the impact of the *RTP* might be overestimated since in a real-world scenario prices cannot be predicted as easily, the *ToU + MI* operating strategy can pose a major threat to grid stability.

Considering *ToU* and *Con* operating strategies, most negative flexibility can be offered at night and on weekends, when most vehicles are at home. Operating strategies with minimal price increments result in no flexibility (*Con + MI*) or only for short durations from 4 to 9 p.m. (*ToU + MI*). The causes have been discussed in the previous section. An *RTP* operating strategy shows similar trends as the *ToU* and *Con* operating strategies for weekdays. Most negative flexibility is offered at nighttime, from 5 p.m. to 3 p.m. On weekends, negative flexibility is at a high level and homogenously distributed for the *Con* and *ToU* operating strategy, whereas *RTP* results indicate a similar behavior as during the week. Such behavior can be explained by the time-varying electricity prices that are lower at nighttime throughout the entire week.

As discussed in the previous subsection the *ToU*, *Con*, and *RTP* operating strategies show similar trends in offerable negative flexibility. Table 6 displays the absolute differences between the operating strategies and the maximal charging power.

A variation in charging power results in an increase in maximal and average negative flexibility for all five simulated pricing scenarios. In order to identify a mathematical relationship between the maximum charging power and the amount of negative flexibility further simulations are required.

With all results summarize, the next chapter discusses the validity and limitations of the applied method.

5. Discussion

This paper presents a thorough analysis of cost-optimal charging schedules and flexibility potential of more than 15,000 vehicle availabilities at home for five operating strategies, and three maximal charging power levels. While the calculation of cost-optimal charging schedules is state of the art, the quantification and analysis of the available flexibility of EV complements and enhances existing literature.

In this analysis, perfect price forecasts have been used to analyze the flexibility of EVs. For the first four operating strategies, which were based on *Con* and *ToU* tariffs, the consideration of perfect price forecasts would not have led to any other results. However, in the case of *RTP* the effect of the perfect price forecast is not negligible. Since *RTP* cannot be forecasted precisely and multiple methods lead to a range of results, the absolute impact of *RTP* is expected to be smaller in reality. Therefore, future research will investigate the impact of the uncertainty of price forecasts on the flexibility that can be offered.

Overall, the *ToU + MI* operating strategy leads to the least favorable charging behavior and flexibility offers. The average charging power indicates major peaks at 9 p.m. and a smaller peak at 3:45 p.m. Both peaks are caused by the mid- and on-peak prices between 4 and 9 p.m. These peaks occur every weekday with similar power levels and therefore represent a significant stress for grid operation. The original assumption that the network could be relieved by time-varying discrete tariffs will become obsolete in the near future, when charging processes will be optimized and automated. This conclusion is in line with the existing literature [13,17]. Nevertheless, *ToU* operating strategies lead to the overall minimum charging costs compared to the other operating strategies (see Figure 9). Despite the seemingly cheaper *ToU* tariffs, regulators should omit operating strategies that offer pre-known price differences in the future for the sake of grid stability and security of supply.

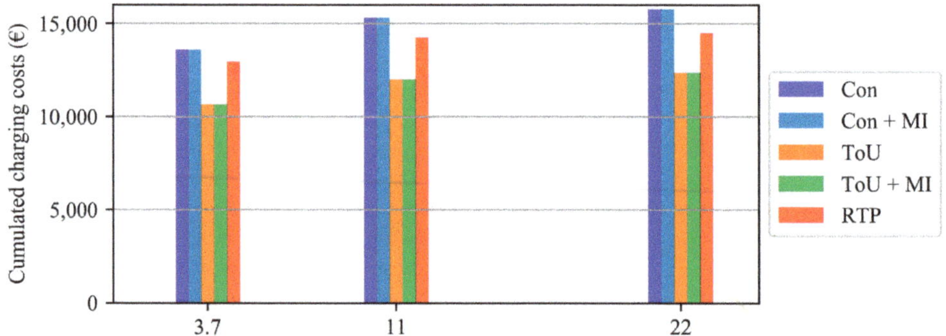

Figure 9. Cumulated charging costs in € for five operating strategies and three maximum charging power levels. For this analysis, the 11,103 vehicle availabilities of the GER MP field trial were used.

In order to achieve grid-friendly user behavior and not to create further grid congestions, we will investigate the integration of local energy markets (LEM). LEM enable participants to trade and exchange their electricity locally. Market agents within HEMS predict the vehicle availability, post bids on the LEM, and adjust their bids automatically based on market results. With this approach, different prices are calculated locally, and users are motivated to consume electricity in times of high generation and to generate electricity in times of high demand.

For this case study, the most recent publicly available data sets with all required parameters were chosen. Since the field trial data was collected from a wide variety of households with different types of vehicles and only contain information about the distances traveled, departure and arrival times, the results can only be representative for realistic user behavior but not for specific types of vehicles. The energy demands of the vehicles were calculated based on the distances traveled. Even though the two data sets are not from the same year (2012/2013 and 2017), the results do not indicate any major differences. Furthermore, the report on the GER MP state that the trends in transportation and individual mobility have remained almost constant over the last 10 years [21]. Therefore, the effect of the different survey periods is considered insignificant. Nevertheless, the continuation of the coronavirus pandemic may mean that employees will be able to work from home to a greater extent, and that vehicle availability may therefore change in the long term. This effect has not yet been taken into account in this study but would be an interesting new aspect.

The gathered flexibility results of this case study are based on availabilities of vehicles at home. However, the method described is neither limited to those two regions nor to quantify flexibility based on EVs at home. This method is applicable to any region/data set that contains information about trip start and end times, purpose or start and end location of the trip, means of transport and distance travelled. Further investigations will investigate differences from other world regions and the quantification of flexibility at other locations, such as workplaces.

6. Conclusions

This paper describes in detail a model that calculates cost-optimal charging schedules and quantifies the flexibility of EV. A case study with more than 15,000 vehicle availabilities from Germany and the USA was conducted and the results visualized for weekdays, weekends, and an average week. Furthermore, the impact of five operating strategies and three charging power levels on the offerable flexibility were analyzed.

Based on these results, the following key findings can be drawn:

1. *ToU* tariffs in combination with the user preference to charge the vehicle as soon as possible (*ToU + MI*) leads to significant increased grid congestions.
2. Positive flexibility is mostly available during either the evening hours or early morning hours depending on the user's preferred charging time (*MI*).
3. No negative flexibility is available if the user is charged a constant electricity rate and chooses to charge as soon as possible (*Con + MI*).
4. Negative flexibility follows the periodic availability of vehicle availabilities at home if the user chooses to charge the vehicle as late as possible (*Con*).
5. Increased charging power levels lead to higher absolute positive and negative flexibility power levels and also increase the total offerable flexibility of EVs.

In conclusion, the model presented in Section 2 is able to quantify EV flexibility. Regulators, researchers, and system operators can use this model to investigate various influences such as tariff structures, user preferences, charging power levels etc. on the flexibility of EVs. Furthermore, the presented HEMS model can calculate the flexibility of heat pumps, combined heat and power, photovoltaic and battery systems. Once completed, this model will be a new helpful tool for tasks such as flexibility calculation, grid expansion planning, and the design and implementation of future electricity regulations.

Supplementary Materials: The model and the script to perform the ev case study is open-source and accessible via the following link: https://zenodo.org/badge/latestdoi/212816117.

Author Contributions: Conceptualization, M.Z.; data curation, M.Z.; formal analysis, M.Z. and P.T. funding acquisition, P.T. and U.W.; investigation, M.Z. and Z.Y.; methodology, M.Z. and Z.Y.; project administration, P.T. and U.W.; software, M.Z., Z.Y., and B.K.N.; supervision, P.T. and U.W.; validation, M.Z.; visualization, M.Z.; writing—original draft, M.Z. and Z.Y.; writing—review and editing, M.Z., Z.Y., B.K.N., P.T. and U.W. All authors have read and agreed to the published version of the manuscript.

Funding: The German Federal Ministry for Economic Affairs and Energy, Bundesministerium für Wirtschaft und Energie, funded this research under the grant number 03SIN109.

Conflicts of Interest: The authors declare no conflict of interest. The funders had no role in the design of the study; in the collection, analyses, or interpretation of data; in the writing of the manuscript, or in the decision to publish the results.

References

1. Ma, J.; Silva, V.; Belhomme, R.; Kirschen, D.S.; Ochoa, L.F. Evaluating and Planning Flexibility in Sustainable Power Systems. *IEEE Trans. Sustain. Energy* **2012**, *4*, 200–209. [CrossRef]
2. Eurelectric (2014)—Flexibility-and-Aggregation. 2014. Available online: https://www.usef.energy/app/uploads/2016/12/EURELECTRIC-Flexibility-and-Aggregation-jan-2014.pdf (accessed on 26 October 2020).

3. Paulus, M.; Borggrefe, F. The potential of demand-side management in energy-intensive industries for electricity markets in Germany. *Appl. Energy* **2011**, *88*, 432–441. [CrossRef]

4. Strbac, G. Demand side management: Benefits and challenges. *Energy Policy* **2008**, *36*, 4419–4426. [CrossRef]

5. Zhou, K.; Yang, S. Demand side management in China: The context of China's power industry reform. *Renew. Sustain. Energy Rev.* **2015**, *47*, 954–965. [CrossRef]

6. Radecke, J.; Hefele, J.; Hirth, L. Markets for Local Flexibility in Distribution Networks. Kiel, Hamburg. 2019. Available online: https://www.econstor.eu/bitstream/10419/204559/1/Radecke%2C%20Hefele%20%26%20Hirth%202019%20%20Markets%20for%20Local%20Flexibility%20in%20Distribution%20Networks.pdf (accessed on 1 June 2020).

7. Nalini, B.K.; Eldakadosi, M.; You, Z.; Zade, M.; Tzscheutschler, P.; Wagner, U. Towards Prosumer Flexibility Markets: A Photovoltaic and Battery Storage Model. In Proceedings of the 2019 IEEE PES Innovative Smart Grid Technologies Europe (ISGT-Europe), Bucharest, Romania, 29 September 2019–2 October 2019; pp. 1–5.

8. You, Z.; Nalini, B.K.; Zade, M.; Tzscheutschler, P.; Wagner, U. Flexibility quantification and pricing of household heat pump and combined heat and power unit. In Proceedings of the 2019 IEEE PES Innovative Smart Grid Technologies Europe (ISGT-Europe), Bucharest, Romania, 29 September 2019–2 October 2019; pp. 1–5.

9. Zade, M.; Incedag, Y.; El-Baz, W.; Tzscheutschler, P.; Wagner, U. Prosumer Integration in Flexibility Markets: A Bid Development and Pricing Model. In Proceedings of the 2018 2nd IEEE Conference on Energy Internet and Energy System Integration (EI2), Beijing, China, 20–22 October 2018; pp. 1–9.

10. Kumar, A.; Srivastava, S.; Singh, S. Congestion management in competitive power market: A bibliographical survey. *Electr. Power Syst. Res.* **2005**, *76*, 153–164. [CrossRef]

11. Yusoff, N.I.; Zin, A.A.M.; Bin Khairuddin, A. Congestion management in power system: A review. In Proceedings of the 2017 3rd International Conference on Power Generation Systems and Renewable Energy Technologies (PGSRET), Johor Bahru, Malaysia, 4–6 April 2017; pp. 22–27.

12. Beaudin, M.; Zareipour, H. Home energy management systems: A review of modelling and complexity. *Renew. Sustain. Energy Rev.* **2015**, *45*, 318–335. [CrossRef]

13. Yan, X.; Ozturk, Y.; Hu, Z.; Song, Y. A review on price-driven residential demand response. *Renew. Sustain. Energy Rev.* **2018**, *96*, 411–419. [CrossRef]

14. Eurelectric—Union of the Electricity Industry. Dynamic Pricing in Electricity Supply. February 2017. Available online: http://www.eemg-mediators.eu/downloads/dynamic_pricing_in_electricity_supply-2017-2520-0003-01-e.pdf (accessed on 18 June 2020).

15. Limmer, S. Dynamic Pricing for Electric Vehicle Charging—A Literature Review. *Energies* **2019**, *12*, 3574. [CrossRef]

16. Barbose, G.; Goldman, C.; Neenan, B. *A Survey of Utility Experience with Real Time Pricing*; LBNL: Berkeley, CA, USA, 2004.

17. Muratori, M.; Rizzoni, G. Residential Demand Response: Dynamic Energy Management and Time-Varying Electricity Pricing. *IEEE Trans. Power Syst.* **2015**, *31*, 1108–1117. [CrossRef]

18. Veldman, E.; Verzijlbergh, R.A. Distribution Grid Impacts of Smart Electric Vehicle Charging From Different Perspectives. *IEEE Trans. Smart Grid* **2014**, *6*, 333–342. [CrossRef]

19. Atabay, D. An open-source model for optimal design and operation of industrial energy systems. *Energy* **2017**, *121*, 803–821. [CrossRef]

20. NuStats, L.L.C. 2010–2012 California Household Travel Survey: Final Report. June 2013. Available online: https://www.nrel.gov/transportation/secure-transportation-data/assets/pdfs/calif_household_travel_survey.pdf (accessed on 23 July 2020).

21. Ecke, L.; Chlond, B.; Magdolen, M.; Hilgert, T.; Vortisch, P. Deutsches Mobilitätspanel (MOP): Wissenschaftliche Begleitung und Auswertung Bericht 2018/2019: Alltagsmobilität und Fahrleistung. *Ger. Mobil. Panel* **2020**.

22. California Department of Transportation. 2010–2012 California Household Travel Survey. National Renewable Energy Laboratory. Available online: https://www.nrel.gov/transportation/secure-transportation-data/tsdc-california-travel-survey.html (accessed on 23 July 2020).

23. U.S. Energy Information Administration. 2018 Average Monthly Bill—Residential: (Data from Forms EIA-861- Schedules 4A-D, EIA-861S and EIA-861U). 2018. Available online: https://www.eia.gov/electricity/sales_revenue_price/pdf/table5_a.pdf (accessed on 23 July 2020).

24. Commonwealth Edison Company. Comed's Hourly Pricing Program. Available online: https://hourlypricing.comed.com/ (accessed on 23 July 2020).
25. Electric Vehicle Database: Energy Consumption of Full Electric Vehicles. Available online: https://ev-database.org/cheatsheet/energy-consumption-electric-car (accessed on 23 July 2020).
26. Hao, X.; Wang, H.; Lin, Z.; Ouyang, M. Seasonal effects on electric vehicle energy consumption and driving range: A case study on personal, taxi, and ridesharing vehicles. *J. Clean. Prod.* **2020**, *249*, 119403. [CrossRef]

Publisher's Note: MDPI stays neutral with regard to jurisdictional claims in published maps and institutional affiliations.

Article

A Methodology to Systematically Identify and Characterize Energy Flexibility Measures in Industrial Systems

Alejandro Tristán [1,2,*], Flurina Heuberger [1] and Alexander Sauer [1,2]

[1] Institute for Energy Efficiency in Production (EEP), University of Stuttgart, Nobelstr. 12, 70569 Stuttgart, Germany; hflurina@outlook.com (F.H.); alexander.sauer@ipa.fraunhofer.de (A.S.)

[2] Fraunhofer Institute for Manufacturing Engineering and Automation (IPA), Nobelstr. 12, 70569 Stuttgart, Germany

* Correspondence: alejandro.tristan.jimenez@ipa.fraunhofer.de

Received: 13 September 2020; Accepted: 7 November 2020; Published: 11 November 2020

Abstract: Industrial energy flexibility enables companies to optimize their energy-associated production costs and support the energy transition towards renewable energy sources. The first step towards achieving energy flexible operation in a production facility is to identify and characterize the energy flexibility measures available in the industrial systems that comprise it. These industrial systems are both the manufacturing systems that directly execute the production tasks and the systems performing supporting tasks or tasks necessary for the operation of these manufacturing systems. Energy flexibility measures are conscious and quantifiable actions to carry out a defined change of operative state in an industrial system. This work proposes a methodology to identify and characterize the available energy flexibility measures in industrial systems regardless of the task they perform in the facility. This methodology is the basis of energy flexibility-oriented industrial energy audits, in juxtaposition with the current industrial energy audits that focus on energy efficiency. This audit will provide industrial enterprises with a qualitative and quantitative understanding of the capabilities of their industrial systems, and hence their production facilities, for energy flexible operation. The audit results facilitate a company's decision making towards the implementation, evaluation and management of these capabilities.

Keywords: energy flexibility; industrial energy management; demand response; demand side management

---

1. Introduction

Energy systems worldwide are undergoing a radical transition to low-carbon energy sources. This transition is necessary for countries to achieve their nationally determined contributions (NDCs) as per the Paris Agreement of 2015. The International Renewable Energy Association (IRENA) global roadmap for energy transformation, ReMap, has quantified that, for countries to achieve their NDCs, renewable energy sources should account for two-thirds of the total primary energy supply worldwide by 2050 [1].

ReMap also calls for large-scale electrification of the energy demand. Currently, electricity accounts for 20% of the final energy demand worldwide, according to the roadmap this ought to be 49% by 2050. Therefore, to meet the intended NDCs, considerable electrification of the final energy demand and a tripling of the installed capacity of renewable electricity sources, when compared to its current levels, should occur simultaneously around the globe. Additionally, due to their extended availability and continuously reducing costs, variable renewable energy sources (VRE), particularly wind and solar energy, are expected to be the primary sources of 61% of the total electricity generated worldwide [2].

Under this scenario, electrical grids worldwide will be subject to extreme stress on two fronts, a vertiginous growth in demand, while primary energy sources are substituted by considerably volatile replacements. Consequently, for the energy transition to be successful, grid operators have to be equipped with new grid management opportunities. Hitherto, these operators have relied on a combination of base and peak load power plants that adapt their output to balance changes in demand. Nonetheless, in an electrical grid that relies on a high share of VREs, new forms of grid flexibility are essential. Grid flexibility allows the balance of both end sides of the electrical grid, supply and demand, hence ensuring grid stability.

Grid flexibility options in a VRE-centered electrical grid include supply-side energy flexibility (SSEF), storage at grid level, grid expansion and demand-side energy flexibility (DSEF) [3]. SSEF consists of the diversification of primary energy sources, increasing the share of dispatchable sources, which can adapt their electrical output to offset any unexpected output deviation of VREs. Energy storage consists of a series of facilities connected directly to the grid with the sole purpose of storing different energy forms that act as a buffer between electrical supply and demand. Grid expansion goes in tandem with energy storage and involves the expansion of high-performance electricity grids, which can transport and distribute the electricity over wide territorial spaces, aggregating VREs with different generation profiles. All of these options involve a considerable additional investment in infrastructure that increases electricity costs and in some cases, might even involve the reliance on non-renewable energy sources [4].

The final option, DSEF, comprises the capacity of the demand sectors within the electrical grid to adapt (increase, reduce or shift) their electrical consumption over a specific duration to balance variations in the electrical supply [4,5].

Among the sectors that constitute the electrical demand, DSEF of the industrial sector or industrial energy flexibility (IEF) is of particular interest. From the perspective of the grid operators, the share size of the industrial sector in the electrical demand, which in the EU-28 represented 37.4% of the electrical consumption in 2017, and as mentioned is expected to grow, makes it a prime candidate to add flexibility to the electrical grid [6]. From the companies perspective, the high relative electricity costs when compared to the cost of other energy carriers, i.e., natural gas, added to the increased control in their energy consumption makes IEF an attractive optimization opportunity [7,8]. In contrast to SSEF, storage at grid-level and grid expansion, DSEF, in general, and IEF, in particular, allow the techno-economic optimization of energy consumption, potentially reducing instead of increasing the overall energy, predominantly electricity, costs for the industrial sector [9].

The tracking report of the International Energy Agency (IEA) in the topic of Demand Response (DR), one of the applications of DSEF, shows that, despite the above-mentioned benefits, the industrial sector still plays a minor role in the current and expected DR potential [10]. As the results of the industRE and SynErgie projects show, this in part due to a lack of knowledge from companies of the energy flexibility capabilities of their production facilities and of the prospective benefits they can obtain by exploiting these capabilities [8,11]. Hence, there is a need for third-party industrial energy audits that support the industrial sector towards systematically identifying and quantifying the energy flexibility capabilities of their facilities and estimating the associated benefits of exploiting such capabilities.

This article proposes an answer to this problem through a systematic, industrial system-focused methodology. The methodology consists of delimiting and classifying the different industrial systems that constitute a production facility, establishing which systems are suitable for flexible operation and, identifying and characterizing the available energy flexibility measures in those systems deemed suitable. The characterization includes a calculation of the potential economic benefit from the usage of these measures. The proposed methodology combines the existing practices of industrial energy auditing with the state of research in the topic of industrial energy flexibility through an innovative approach that complements the current widespread focus in industrial energy efficiency.

The article is structured as follows. The understanding of production facilities from an energy perspective is described in Section 2. Section 3 defines the key concepts of DSEF and IEF. In Section 4, the proposed methodology to systematically identify and characterize energy flexibility measures is explained in detail. Section 5 illustrates the application of the methodology by summarizing its application in an existing industrial system, a chilled-water air conditioning system. Section 6 discusses the proposed methodology and the results of its initial application. The paper concludes with several final insights and an outlook of the prospective applications of the proposed method.

2. Production Facilities from an Energy Perspective

A production facility, i.e., a factory, is defined as the representation of a local concentration of the primary factors of production: personnel, equipment, buildings and materials, and the derived factors knowledge, skills and capital [12]. Morphologically, a production facility consists of a series of industrial systems that work together to execute the intended production processes. These processes are a series of production tasks that yield specific products [13]. The industrial systems are in turn made-up by different components which are interconnected among them and to other systems by material, energy and information flows [14].

According to the task they perform on the production processes, the industrial systems are grouped in technical units to facilitate their analysis. Four technical units make up the core structure of modern factories, Manufacturing, Auxiliary Systems, Technical Building Services and Energy and Media. An additional technical unit called Energy and Manufacturing Control binds the different industrial systems together by managing and coordinating, through information flows, their operation. A brief overview of the technical units in a modern factory and a description of the industrial systems that constitute them are presented in Figure 1.

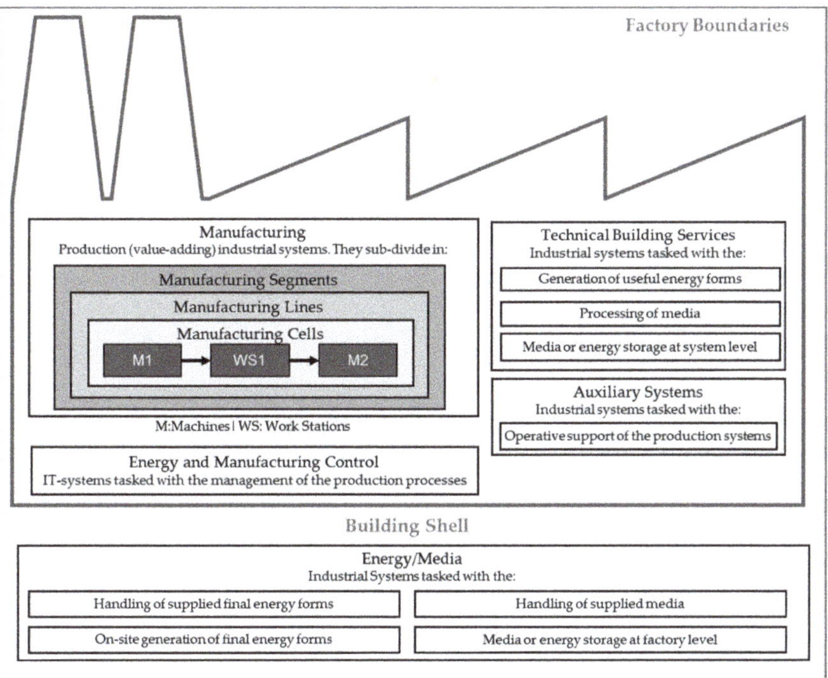

Figure 1. Technical units and sub-units of a factory (Adapted from References [12,14,15], own illustration).

Manufacturing (MA) is the central (value-adding) technical unit of the factory. It consists of all industrial systems that directly execute production tasks in the production processes, i.e., production equipment and human workforce that directly add value to the manufactured product. Although the industrial production systems that make up the MA technical unit differ in configuration, they all consist of an arrangement of two basic components, Production Machines (PM) and Workstations (WS). Groups of PMs and/or WSs that are commonly operated by a specific employee group are referred to as Manufacturing Cells. A series of Manufacturing Cells, that sequentially work together connected by a material-flow, constitute a Manufacturing Line. Based on their end goal, Manufacturing Lines can be in turn be grouped in Manufacturing Segments. Manufacturing Segments are self-contained groups of Manufacturing Lines limited by distinct boundaries from an organizational or management perspective. The segmentation of the MA technical unit is usually carried out by physically establishing separate areas, or even allocating different buildings to each segment [12].

The Auxiliary Systems (AS) technical unit consists of industrial systems that do not directly add value to the manufactured product but support the industrial production systems in the MA technical unit in the execution of their task. The industrial systems in the AS technical unit are further classified as centralized if they support a complete Manufacturing Segment or decentralized if they serve just a specific Manufacturing Line, Cell, PM or WS. Examples of industrial systems that belong to the AS technical unit include transport preparation systems like palletizing machines and logistic systems like conveyor belts and automated guided vehicles (AGVs).

The Technical Building Services (TBS) technical unit comprises industrial systems tasked with the generation, processing and/or storing of useful energy forms and media, demanded or emitted by the industrial systems within the MA and AS technical units. TBS include, for example, compressed air, process heating and cooling, and heating, ventilation and air-conditioning systems (HVAC). The industrial systems in the TBS technical unit can be further classified based on their operative function as generators, handlers or buffers [14].

The Energy/Media (EM) technical unit involves the industrial systems tasked with the buffering and conditioning of media and final energy forms supplied to, or any infrastructure intended for the generation of final energy forms directly at the factory. Final energy forms refer to all energy carriers that are in a form ready to be consumed. Examples include high-, medium-, and low-voltage electricity, natural gas, district heating, cogeneration and trigeneration systems and combustible fuels [15–17].

The boundary between the EM and the TBS technical units depends on the particularities of each facility, but generally, the EM technical unit will group industrial systems related to the generation, conditioning and storage of final energy forms and media at the factory-level. These final energy forms and media might be directly consumed or might then processed by the industrial systems in the TBS technical unit into useful energy forms or media for a specific application within the factory such as space and process heat, electricity, cooling media, mechanical energy (i.e., compressed air), light, etc.

Energy and media storage will take place throughout the MA, AS, TBS and, clearly, the EM technical units. If the storage serves a specific industrial system that belongs to the MA, AS or TBS technical units, the storage infrastructure belongs to this respective system. If, in turn, the storage supports multiple industrial systems across different technical units, the storage infrastructure is considered an industrial system by itself, which belongs to the EM technical unit.

The Energy and Manufacturing Control (EMC) technical unit englobes all the overarching data processing infrastructure that integrates the information flows to plan, monitor and control the operation of all the industrial systems across the other technical units and to coordinate the material and energy flows between them [12,15].

Finally, the factory boundaries delimit the factory's physical extension, determining its energy and media inputs and outputs. The building shell surrounds the factory's buildings, defining the impact of local climate on the factory and the emissions released by the factory into the surrounding environment.

The division of a production facility into its technical units and sub-units helps to delimit its constituting industrial systems. Nonetheless, the interdependences and interactions between these systems are explained by the material, energy and information flows connecting them. The material flows encompass the chain of production processes involving the handling, processing, storage and distribution of materials and goods within the factory. These flows usually start with raw materials and media entering the facility and end with products, by-products, emissions and waste, leaving it. In modern production facilities, the material flows are regulated via manufacturing orders, which are orders that stipulate the required manufacturing of a specific product on a specific volume and to a specific point in time [12].

Energy flows involve the energy transactions and conversions between the components of the industrial systems and between systems. As the factory space is essentially an open system, energy flows also include the interaction of the factory as an entity with its peripheries in the form of final energy forms entering and leaving its boundaries [14].

The information flows describe the information exchange relationships between the components in the different industrial systems in the factory and with actors in the periphery. The information flows are internal when they comprehend only the interaction among the industrial systems within the factory boundaries and external when they involve the communication of the factory as an entity with actors in the periphery [18].

In modern factories, where the different industrial systems are being progressively automated, information flows take place within a hierarchical automation infrastructure. The EMC technical unit is hence physically structured in the form of a communication and control pyramid, on which each level is defined by a specific set of hardware components, as presented in Figure 2 [19].

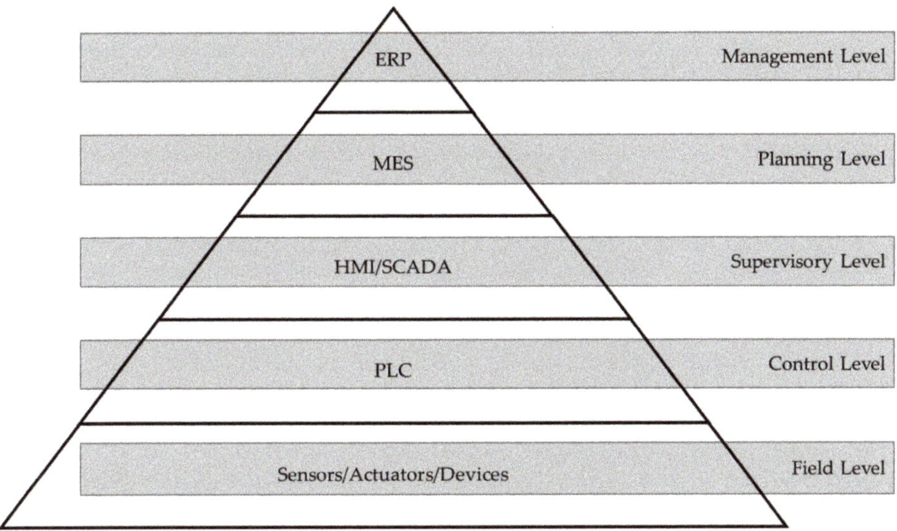

Figure 2. Hierarchical control pyramid (Adapted from References [15,19], own illustration).

The base level is the field level where sensors measure the necessary parameters and actuators execute the necessary actions to manage the operations of all industrial systems across the facility. The second level is the control level constituted by programmable logic controllers (PLCs) and embedded control systems. In this level, control systems react in real-time to discrete inputs from the field level that result in specific operative commands to an industrial system or its components. The third level is the supervisory level consisting of the human–machine interfaces (HMI) and the supervisory control and data acquisition system (SCADA), which essentially combines the previous levels (field and control) to access data and control industrial systems and their components from a single location. The supervisory level is in charge of the control and coordination of multiple industrial systems. The fourth level is the planning level entailing the manufacturing execution system (MES) which has a direct link to process automation and allows prompt monitoring and control of all the production processes. The top-level is the management level, involving the enterprise resource planning system (ERP), which concisely maps all business practices of a company. The ERP's main function is the strategic and tactical (long- and medium-term) planning and scheduling of the activities related to procurement, storage, production, accounting and finance across the factory [14,18,19].

3. Key Concepts of DSEF and IEF

IEF is understood as the ability of an industrial system to adapt quickly and cost-effectively to changes in the energy markets [20]. Usually, the concept of DSEF, and therefore IEF, and that of demand response (DR) are used as synonyms. Nonetheless, the Federal Energy Regulatory Commission of the United Stated defines DR as "Changes in electric usage by end-use customers from their normal consumption patterns in response to changes in the price of electricity over time, or to incentive payments designed to induce lower electricity use at times of high wholesale market prices or when system reliability is jeopardized" [21]. Meanwhile, the European Commission describes DR as "A series of programs sponsored by the electrical grid, the most common of which pays companies (commercial DR) or end-users (residential) to be on call to reduce electricity usage when the grid is stressed to capacity" [22]. As can be inferred from the definitions, DR describes the activity per se of adapting the electrical consumption to profit from financial incentives sponsored by grid operators. DSEF, on the other hand, describes the capability of an energy-consuming system, which in the case of IEF is an industrial system, to react to a triggering event and change its energy consumption. This capability enables the energy-consuming system to take part in DR schemes and programs; nonetheless, the possible applications of DSEF expand even further. In the case of IEF, potential outcomes or implementation objectives include [23]:

- An intelligent response to the volatility of energy prices: IEF, as mentioned before, have the capacity of optimizing the factory's energy costs—in its simplest form, this means reducing energy costs via the reactive adjustment of consumption to price fluctuations in the electrical markets.
- Proactive marketing of the energy flexibility potentials in the grid service markets: the combination of IEF and production planning, can allow its proactive offering in the ancillary service markets of the electrical grid, thus obtaining a financial incentive from the grid operators.
- Maximize the usage of local energy sources/maximize the use of the renewable energy portfolio: through IEF, the energy consumption of an industrial system can be adapted to match the production profiles of local (within the factory boundaries) or nearby electricity generation plants. Hence, achieving balanced or real energy self-sufficiency in the production facility [24]. In the specific case of renewable electricity sources, IEF can reduce the carbon footprint of the factory, thus reducing potential Green House Gases-emissions related costs.
- Peak shaving and load management: peak shaving and load management are both benefits of IEF, eliminating the need for over-capacity to supply the peaks of highly variable loads and reducing time-of-use-related costs and stress on the energy distribution infrastructure.
- Improvement of the resilience of the proprietary energy infrastructure: IEF also can assist the energy infrastructure to recover quickly from energy supply disruptions or support self-sufficient operation. Thus, avoiding the considerable costs of production disruption. IEF can also serve to avoid or delay energy infrastructure expansions and their investment cost, by adapting the consumption patterns of different industrial systems to the capacity of the existing infrastructure.

3.1. Energy Flexibility Measures and their Energy Flexibility Potential

IEF acquires a usable form by its formulation in an energy flexibility measure (EFM). An EFM is a conscious and quantifiable action to carry out a defined change of an operative state in an industrial system [20]. In this definition, an operative state refers to the energy demand rate of an industrial system at a specific point in time. Therefore, a change of operative state refers to the variation of this rate of energy demand for a definite period. The energy flexibility potential (EFP) is the quantification of the change in operative state that the EFM will induce on the industrial system. The EFP is, therefore, quantitatively described by a power component, the flexible power, and a temporal component, the active duration [25].

The quantification of the EFP is dependent on the characteristics of the industrial system and the features of its context, considered for its calculation. Therefore, a reference framework needs to be established to quantify the EFP.

This reference framework can be progressively developed to introduce additional system characteristics or context features, hence making its quantification more complex but attaining a more accurate EFP value. When the EFP is calculated only taking into consideration the physical characteristics of the industrial system as a reference framework, it will be theoretical. The theoretical EFP usually only takes the power rating of the industrial system and its operation time into consideration. The technical EFP, on the other hand, is calculated by adding the system's operative characteristics to the reference framework. The operative characteristics of the industrial system are attributes related to the patterns of operation that the industrial system follows to fulfil its task effectively. The practical EFP goes further and includes the relevant characteristics of the production facility of which the system is a part. These relevant characteristics relate to the production planning strategies prevailing in the factory. The economical EFP is the share of the practical EFP that is economically feasible, meaning when the revenues from making use of the EFM outperform its costs. These revenues are a function of pursued implementation objectives as defined in the last section. Finally, the viable EFP is the share of the economical EFP that also aligns with the company's investment approach, i.e., payback periods and risk policies, and, that outperforms other relevant investments, for example, energy efficiency measures. The different types of EFPs according to the different reference framework are presented in Figure 3.

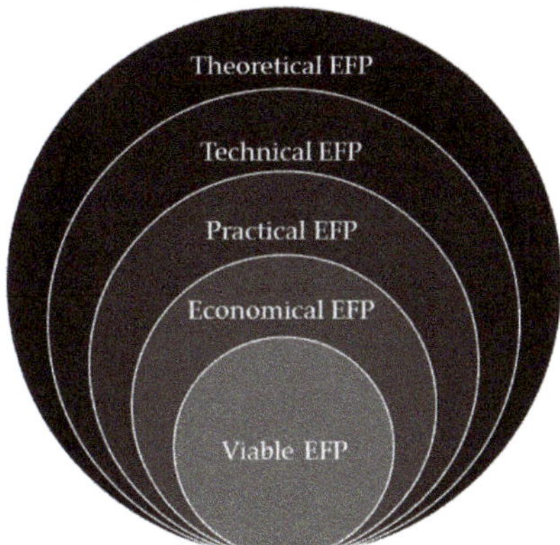

Figure 3. Types of energy flexibility potential based on different quantification reference frameworks. (Adapted from Reference [25], own illustration).

The different characteristics that constitute each reference framework, and hence each type of EFP, are described in Section 4. The division of the EFP in different types serves two purposes, first to be transparent on the scope of the quantification, and second, it allows estimating the influence each of the variations in the reference framework has on the EFP of the EFM.

3.2. Categorization of Energy Flexibility Measures

Based on their nature, EFMs can be classified as technical or organizational. Organizational EFMs involve actions that take advantage of the production strategy of the factory to modify the operative state of the industrial systems [26]. Usually, organizational EFMs do not alter the aggregated energy consumption of the respective industrial system. If this is the case, organizational EFMs will not influence the energy efficiency of the industrial system. Technical EFMs, on the other hand, influence the specific load profile of the industrial system by altering its operative pattern. They usually do alter the overall energy consumption and their influence must be carefully evaluated after the EFM has been characterized.

A list of general categories of EFMs in industrial systems was originally established in Reference [27] and further standardized in Reference [23]. Nonetheless, depending on the specific nature of the EFM, particularly if they are organizational or technical, specific EFMs only apply to industrial systems that belong to specific technical units. Table 1 lists and defines the established EFM general categories, classifies them as technical (T) or organizational (O), and pinpoints to which type of industrial system, as defined by the system's technical unit, the specific EFM category applies. This last point is referred to as applicability.

Table 1. Energy Flexibility Measure categories, classification and their applicability. Data from References [23,27].

Name	Classification [1]	Description	Applicability [2]
Adaptation of staff free time	O	Aligning the staff break times to fit different energy consumption profiles.	MA
Adaptation of working shifts	O	Aligning the work shift times to different energy consumption profiles.	MA
Adaptation of order execution sequence	O	Changing the chronological execution sequence of manufacturing orders to different energy demand patterns.	MA
Capacity planning adjustment	O	Changing the assignment of a product to a production resource (Production Machine/Work Station, Manufacturing Cell or Line) to alter the energy consumption profile.	MA
Defer of production start	O	Premature or delayed start of production (all manufacturing orders) within different periods to fit different energy consumption profiles.	MA
Manufacturing order interruption	O	Interruption of a manufacturing order and restart of the same order at a later time point. Duration might expand through minutes, hours or even full shifts.	MA
Adaptation of order production sequence	O	Changing the chronological production sequence of a specific manufacturing order to adjust to a different energy consumption profile.	MA
Adaptation of resource allocation	O	Targeted selection of specific components in an industrial system based on their energy consumption patterns.	MA TBS AS
Adaptation of operation parameters	T	Adaptation of the control variables of an industrial system to fit different energy consumption profiles.	MA TBS AS

Table 1. *Cont.*

Name	Classification [1]	Description	Applicability [2]
Operation interruption	T	Temporary suspension of the operation, and hence of the energy consumption, of an industrial system.	MA TBS AS
Adjustment of the operational sequence	T	Changing the operative sequence of an industrial system to adjust to different energy consumption profiles.	MA TBS AS
Inherent energy storage	T	Use the operative inertia of an industrial system as energy storage.	MA TBS AS EM
Bivalent operation	T	Switch between different energy carriers to supply the energy consumption of a specific industrial system.	MA TBS AS
Dedicated energy storage	T	Storage of energy in a suitable storage medium. The storage can take place within a system (system-level), or serve multiple industrial systems (factory-level).	MA TBS AS EM
Energy carrier exchange	T	Use of different energy carriers to supply the energetic demand of multiple industrial systems across the factory.	EM

[1] Refers to O: Organizational Energy Flexibility Measures, T: Technical Energy Flexibility Measures. [2] Refers to MA: Manufacturing technical unit, AS: Auxiliary Systems technical unit, TBS: Technical Building Services technical unit, EM: Energy and Media technical unit.

The execution of an EFM has two parts. The virtual part takes place on the data processing systems of the EMC technical unit and consists of a targeted response to a triggering event, i.e., change in electrical price, activation request from the electrical grid operators, peak consumption, etc. Once a response is defined, the physical part of the EFM occurs in the form of an actual change of the operative state of the industrial system. The EFM is hence operatively a proportional response to a triggering event. The nature of the triggering event is determined by the intended implementation objective of the EFM. As mentioned before, the virtual part of an EFM is restricted to the EMC technical unit. The physical part, on the other hand, takes place on industrial systems belonging to the MA, AS, TBS or EM technical unit.

The presented structured understanding of the factory and its industrial systems, the definition of the available EFM categories and the considerations to calculate their EFP constitute the theoretical foundation on which the proposed the methodology was developed. The next section explains the proposed methodology in detail.

4. Methodology to Systematically Identify and Characterize Energy Flexibility Measures

The development of the proposed methodology started by establishing specific requirements that must be fulfilled. These requirements are:

1. Systematicity: as is the case with current industrial energy audits [28,29], the methodology has to follow a structured procedure on which all industrial systems in a production facility and their characteristics are progressively analyzed and decisions regarding their energy flexibility capabilities respond to procedural considerations.
2. Focus in electrical flexibility: although different energy carriers are considered, the EFMs resulting from the application of the methodology should aim to optimize the electrical consumption of production facilities and its costs.

3. Applicable to a plethora of industrial systems and production facilities: the methodology has to apply to the heterogeneous nature of modern industrial systems and production facilities.

4. Agile: the methodology needs to be more agile, hence providing results in a shorter time-lapse, than a more exhaustive approach to identify EFMs, i.e., industrial system modelling [15].

5. Current operation-friendly: the methodology does not aim to redesign industrial systems for energy flexible operation but to identify EFMs based on their current operation patterns.

6. Outcome relevant for industrial stakeholders: the outcomes of the methodology should be qualitatively and quantitatively sufficient to inform the decision-making process of companies regarding the implementation and usage of the energy flexibility capabilities of their production facilities.

Based on the previously defined requirements, the steps presented in Figure 4 constitute the proposed methodology to identify and characterize EFMs in industrial systems. Each step is detailed in the following subsections.

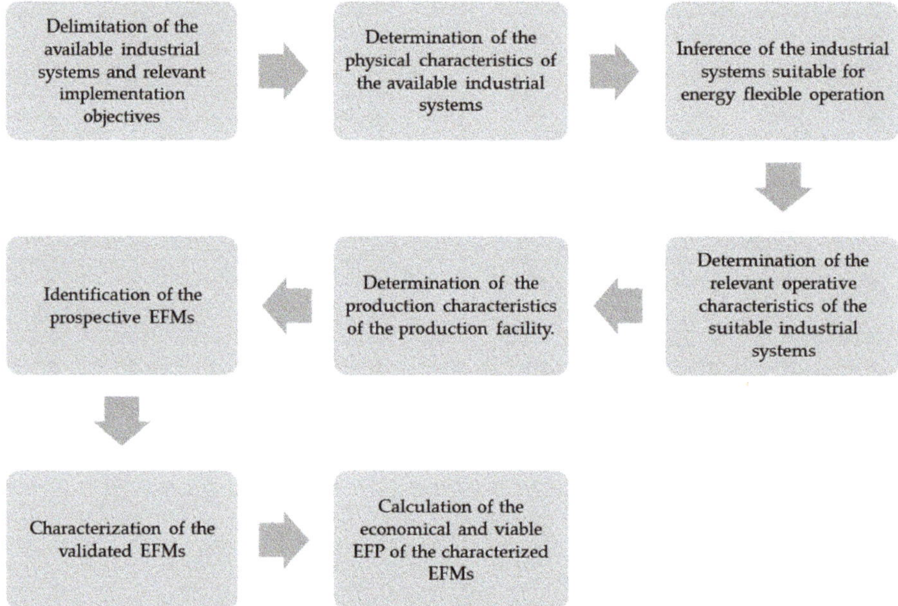

Figure 4. Steps to identify and characterize EFMs.

4.1. Delimitation of the Available Industrial Systems and Relevant Implementation Objectives

The starting point to identify and characterize energy flexibility measures is to delimit the available industrial systems in the analyzed production facility and the relevant implementation objectives for the analyzed production facility. For this purpose, the facility is conceptualized as the series of technical units described in Section 2. The different energy-consuming and -handling components in the facility are then assigned to these units according to the task they execute within the production processes. Components that work together towards completing a specific task are then grouped in systems. These groups of components will constitute the available industrial systems. The focus is therefore only on industrial systems, hence systems that collaborate directly or indirectly in the production processes of the facility. Other energy-consuming elements in the facility, for example, office spaces, although potentially relevant for energy flexibility, are not the purview of the presented methodology.

The relevant implementation objectives are the group of the implementation objectives described in Section 3, that when accomplished will techno-economically improve the analyzed facility's energy consumption. The relevance of each implementation objective depends on the energy context (available energy markets, quality of energy supply, energy costs, energy supply contracts, etc.) on which the production facility operates. These objectives are ought to be decided by the production facility stakeholders.

4.2. Determination of the Physical Characteristics of the Available Industrial Systems

Once the available industrial systems and the relevant implementation objectives in the production facility have been defined, the way each system consumes energy needs to be understood. Hence, the system needs to be energy transparent. An initial understanding of the energy consumption of the industrial system is achieved through its physical characteristics. The physical characteristics from an energy transparency perspective are [14]:

- Technical Unit: Already defined in the previous step, the technical unit to which the industrial system belongs provides relevant insights on the task the system performs on the production facility and hence its energy consumption patterns.
- Industrial system layout: The arrangement of all, but particularly the energy-consuming, components in the industrial system help to understand the energy consumption chains, or how energy is distributed and used, across the system.
- Power rating and maximum system output: The power rating is the maximum allowable power input, meaning the aggregated maximum rate of energy transfer, of the energy-consuming components in the system. The maximum output of the industrial system is the maximum material or energy production provided by a system on each operative cycle.
- Operative Time: Aggregated utilization time of the components in the system, also understood as the duration of the task or tasks the system performs.
- Control Concept: The course of action through which the behavior of the industrial system is managed.

For appropriately defined industrial systems, the physical characteristics can be easily inferred from surveying the technical specifications of the system's components. The physical characteristics provide an initial level of energy transparency and hence a very superficial overview of the available EFMs and, up to this point, theoretical EFP.

4.3. Inference of the Industrial Systems Suitable for Energy Flexible Operation

Once the available industrial systems and their physical characteristics have been settled, they need to be sorted based on their energy flexible operation suitability.

The suitability of an industrial system for energy flexible operation is assessed through three different criteria [26,30,31]:

1. Controllability: indicating how restrictive is the control concept of an industrial system in terms of additional variations in its operative state.
2. Criticality: specifying the grade on which a change of operative state in an industrial system might alter the quality of the manufactured product or the continuity of the production processes within the factory.
3. Input/output interdependence: defined by the level of decoupling between the energy input and the output of the industrial system along its operative cycle. The operative cycle of an industrial system is understood as the series of sequential tasks the system performs to achieve a unit of output.

These three criteria are gradual. Therefore, they can be divided into different cases that help to quantify the system's suitability for energy flexible operation.

Regarding the controllability of industrial systems, four different cases are discernible. These are referenced as levels with the abbreviation Co and progressive case numbers. The system can be process-controlled and hence have its operation fully defined in time and quantity, leaving virtually no possibility for energy flexible operation (Co0). The control concept of the industrial system might be dependent on state variables, i.e., temperature, hampering the system's flexible operation ability (Co1). The system might only be controllable over its operative time, switching operative state over fixed intervals, allowing for a considerable degree of energy flexible operation (Co2). Finally, the control concept might allow the industrial system to execute its tasks continuously and unrestricted in time and quantity (Co3), completely freeing the system to operate in an energy flexible manner. The controllability criterion is directly determined by analyzing the control concept of the system, defined as a physical characteristic of the system in the last step.

For the second criterion, criticality, four different cases are also distinguishable. Similarly, the criticality cases are considered as suitability levels with the abbreviation Cr and progressive case numbers. A change of operative state in the industrial system might reduce the product quality or induce a continuity failure of the respective production processes exceeding the acceptable risk for energy flexible operation (Cr0). The change of operative state in the system could have a limited, not failure-inducing, but a significantly negative, influence on the continuity production processes (Cr1). The influence that a change of operative state in the industrial system might have on the continuity of the production processes might be limited and only marginally negative (Cr2). What constitutes a significantly or marginally negative influence is usually associated with an increase in the system's operating costs. Therefore, it is case-specific and it needs to be discussed with relevant stakeholders in the production facility. Alternatively, the change of operative state might have a neutral or even positive influence in the production processes (Cr3). The influence that a change of operative state in the industrial system has on the production processes is usually extrapolated from the previously mentioned dialogue with the relevant stakeholders in the factory. Plus, an analysis of the system's tasks in the facility. This last point is a combined evaluation of the system's technical unit and its control concept.

The final criterion, the input/output interdependence, is also subdivided into four different cases. The interdependence cases are also referenced with the abbreviation In and the case number. The energy input might be completely coupled proportionally and instantly to the output of the industrial system without any type of decoupling capability and then leaving no tolerance for energy flexibility (In0). On the other hand, decoupling capabilities might exist between the energy input of the industrial system and the system's output, i.e., through energy or media storage. These capabilities might be inherent, owing to the operative characteristics of the system (In1). Alternatively, specific components in the system might exist that provide decoupling capabilities. These capabilities are limited if, when aggregated, they are smaller in capacity than the required input to complete an operative cycle (In2). Conversely, these capabilities might be comprehensive, if the decoupling capabilities along the system, when aggregated, are larger in capacity than the necessary input to complete an operative cycle (In3). The input and output interdependence assessment is a result of the analysis of the system layout and its control concept.

The suitability of industrial systems is assessed graphically in a radar graph where each axis signifies each of the criteria and the levels are used as a scale. This is presented in Figure 5. Industrial systems with a level zero (0) on any of the criteria are unsuitable for energy flexible operation and should not be further examined. On the other hand, those systems with a level three (3) on all the criteria are highly suitable for energy flexible operation. Industrial systems with combined levels, between one (1) and (3) present risks when operating in an energy flexible manner. These risks need to be factored and evaluated after the EFMs have been identified and characterized.

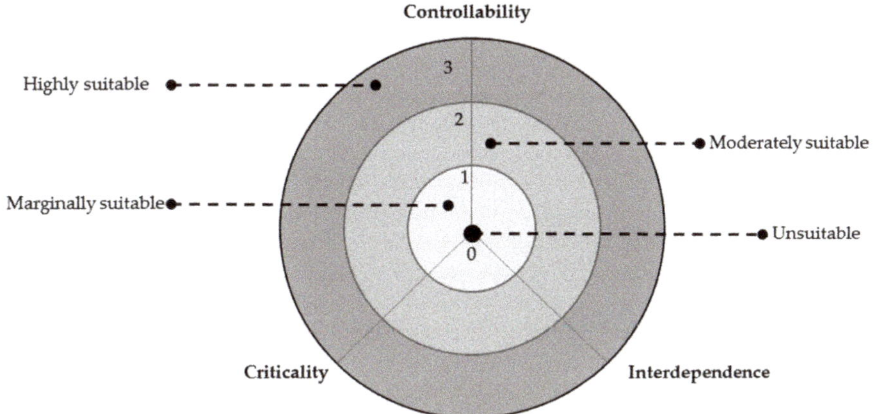

Level	Controllability (Co)	Criticality (Cr)	Input/output Interdependence (In)
3	Unrestricted control concept	Neutral or positive influence	Comprehensive decoupling capabilities (>Cycle)
2	Time-dependent control concept	Marginally negative influence	Limited decoupling capabilities (< Cycle)
1	State variable dependent control concept	Significantly negative influence	Inherent decoupling capabilities
0	Process-dependent control concept	Quality reducing or failure inducing influence	Direct input-output correspondence

Figure 5. Radar chart to assess the suitability of industrial systems for energy flexible operation.

The suitability can also be analyzed through the calculation of a suitability score by multiplying their level in each criterion. A score of zero (0) will render the system unsuitable for energy flexible operation. A score of one (1) will symbolize marginal suitability for energy flexible operation. A score between one (1) and eight (8) will denote moderate suitability for energy flexible operation, while a score over eight (8) will indicate high suitability. The suitability scores do not reflect the EFP or attractiveness of EFMs in the system. Therefore, they should not be used to prioritize the analysis of specific suitable systems. The suitability analysis is performed through a qualitative analysis of each available industrial system based on the known physical characteristics and in close cooperation with relevant facility stakeholders. Once the industrial systems suitable for energy flexible operation have been singled out, the next step is to determine their relevant operative characteristics.

4.4. Determination of the Relevant Operative Characteristics of the Suitable Industrial Systems

The physical characteristics serve to typify the industrial system and hence achieve an initial level of energy transparency. Nonetheless, they provide only reduced information about the dynamics of the industrial system's operation. Therefore, the operative characteristics of the industrial system serve to better understand the energy consumption patterns of an industrial system and the factors that influence them. The relevant operative characteristics of industrial systems from the energy transparency perspective are [14,23,32]:

- Typical load profile: Typical pattern of energy consumption of the industrial system. A load profile consists of the curve of energy input versus time in the industrial system for a specific period. The typical load profile is usually a synthesis of the energy consumption record for a longer period, i.e., a year. There are several techniques to obtain the typical load profile, or profiles, of a system. The state-of-the-art consists on performing K-means clustering to the raw energy-consumption record of the system resulting in different clusters, or profiles, and calculating the median of the data samples in the cluster to obtain the typical curve profile. The optimal number of clusters is determined by using silhouette analysis and selecting the number of clusters that provide the maximum average silhouette scores. In practice, the clusters respond to the modes of operation of the industrial system under different operating conditions. The selected approach follows the recommendations of several research works that have dealt with the optimal approach to obtain the typical load profiles of electrical loads using machine learning algorithms, exalting the usage of silhouette scores and the k-means algorithm as the most fitting approach [33–35].

- Controlled Variable: Independent parameter(s) that determine the operating state of an industrial system. Their variation will induce a change in the operative state.
- Control horizon and latency: The control horizon is the minimum time interval between the variation of a control variable and the occurrence of the change in the operative state of the system. The control latency is the amount of time it takes signals to traverse the system or systems in the EMC Technical Unit.
- Operative continuity: Consistency of the operative cycles of the system. Three types can be discerned:

  - Discontinuous, the operative cycle of the system, consists of multiple operative states that take place in irregular intervals throughout the operative time. The intervals are divided by irregular periods on which the system is idle.
  - Part continuous, the operative cycle of the system involves a single operative state that occurs in regular intervals throughout the operative time. The intervals are divided by regular periods on which the system is idle.
  - Continuous, the operative cycle of the system consists of a single operative state throughout the operative time. The system is never idle during its operative time.

- Operative Steps: Amount and type of successive steps that make up the operative cycle of the system.
- Output flexibility: The ability of a system to operate at a range of different output levels without incurring in major setup alterations.
- Bivalence or multivalence: The ability of a system to satisfy its energy demand with two or more energy carriers.
- Buffer Capability: Ability of an industrial system to store energy and/or media temporarily and locally. The storage capability might come from the system's operative inertia (i.e., thermal or mechanical inertia) or dedicated storage components.
- Redundancy: The ability of more than one component within a system (system level), or more than one system within a technical unit (technical unit level) to perform a specific task.
- Operative Shiftability: The ability of a system to shift the totality or a part of their operation cycle to an earlier or later time point.
- Interruptible: The ability of a system to stop its operation cycle and continue at a later time point.
- Task Flexibility: The ability of a system to execute a variety of tasks for a production process, i.e., perform a range of operations or produce a variety of products, without incurring in any major setup variation.
- Routing Flexibility: The ability of a system to execute its tasks via alternative operative sequences.

The operative characteristics of the system are determined through a detailed surveyal and analysis of the energy consumption data of the system, its material, energy and information flows and its design specifications. The operative characteristics provide a deeper level of energy transparency and hence a more realistic overview of the available EFMs and their EFP, which will be considered technical at this point.

4.5. Determination of the Production Characteristics of the Production Facility

Besides the physical and operative characteristics of the industrial system, it is necessary to understand the production approach of the production facility on which the system finds itself. The production approach is defined by a series of production characteristics common to the facility as a whole. These characteristics allow allocating the energy-to-cause and hence to understand the variation of energy consumption throughout time and within the system's modes of operation. The relevant production characteristics from an energy transparency perspective are [13,14,23,32]:

- Manufacturing principle: The manufacturing principle follows the expected volume and variety of the manufactured product by the market. Four different principles are discernible [36]:

  - Make-to-Stock (MTS), the product is made in their final form and stocked as finished goods.
  - Assemble-to-Order (ATO), the product is assembled to its final form based on the customer's purchase order.
  - Make-to-Order (MTO), the product is completely manufactured after a customer has issued a purchase order.
  - Engineer-to-Order (ETO), the product is designed and manufactured after the customer's purchase order.

- Production Method: The production method is the basic approach to production planning, they fall into four categories [37]:

  - Job processing, the production focuses on a single item at a time and usually requires a specific set of skills depending on the manufactured product.
  - Batch processing, the production takes place in specific groups of pieces or completed products in small pre-set batch sizes.
  - Flow processing, production involves passing of sub-assemblies or individual parts from one production station to the next until the final product is completed.
  - Continuous processing, similar to flow production but there is no possible stop between production stations.

- Working shift model: Amount and extension of the working shifts on which the factory conducts its production processes.
- Production planning horizon: Minimum time-lapse between the end of detailed production planning and the start of production.
- Change in manufacturing orders: The possibility to change manufacturing orders once they have been issued.
- Change horizon: Minimum time-lapse between a request for a change in a manufacturing order and the enactment of this change.
- Product-based divergences: The influence that the manufactured product has on the energy consumption of the involved industrial systems.
- Multiple energy carriers: The presence of different energy carriers on the facility that can supply the energy consumption of the industrial system.
- Labor flexibility: The ability of labor to execute a range of different tasks.
- Relevant costs: Energy, maintenance and labor costs in the facility.

The production characteristics can be determined through the surveyal of the production planning strategies established on the ERP. These characteristics give the final necessary perspective to achieve a level of energy transparency in the industrial system that allows the identification of the practically viable EFMs and their EFP which will be considered practical as it includes the physical and operative characteristics of the industrial system and the production characteristics of the production facility.

4.6. Identification of the Prospective Energy Flexibility Measures

The EFMs, present in an industrial system, are a function of the characteristics of this system and the production characteristics of the facility to which the system belongs. These characteristics have one of four different levels of influence on each specific EFM category. These levels of influence are:

- Crucial: A characteristic is crucial if it is decisive to the existence of an EFM belonging to a specific category. Meaning the way this characteristic manifests in the industrial system decides if the specific EFM-category is available in the industrial system.
- Influential: A characteristic is influential if it will delimit the EFP of the EFM belonging to the specific category.
- Relevant: A characteristic is relevant if it only serves to quantify additional characterization parameters, outside of the EFP, of the EFM.
- Irrelevant: the characteristic plays no role in the existence of a specific category of EFM or its characterization parameters.

Tables 2 and 3 present the level of influence each of the characteristics examined in the last steps has on the previously defined organizational and technical EFM categories respectively.

Table 2. Level of influence of the physical, operative and production characteristics in the existence of organizational EFMs [1].

Characteristic Name	Organizational EFM-Categories							
	Adaptation of Staff Free Time	Adaptation of Working Shifts	Adaptation of Order Execution Sequence	Capacity Planning Adjustment	Defer of Production Start	Manufacturing Order Interruption	Adaptation of Order Production Sequence	Adaptation of Resource Allocation
Technical unit	●●●	●●●	●●●	●●●	●●●	●●●	●●●	●●●
Industrial system layout	○○○	○○○	○○○	●●●	○○○	○○○	●●●	●●●
Power rating	●●○	●●○	●●○	●●○	●●○	●●○	●●○	●●○
Operative time	●●○	●●○	●●○	●●○	●●○	●●○	●●○	●●○
Control concept	●●●	●●●	●●●	●●●	●●●	●●●	●●●	●●●
Typical load profile	●●●	●●●	●●●	●●●	●●●	●●●	●●●	●●●
Control variable	●●●	●●●	●●●	●●●	●●●	●●●	●●●	●●●
Control horizon and latency	●○○	●○○	●○○	●○○	●○○	●○○	●○○	●○○
Operative continuity	●●●	●●●	●●●	●●●	●●●	●●●	●●●	●●●
Operative steps	●●○	●●○	●●○	●●○	●●○	●●○	●●●	●●○
Output flexibility	○○○	○○○	●●●	●●●	○○○	○○○	●●●	●●●
Bivalence	○○○	○○○	○○○	○○○	○○○	○○○	○○○	○○○
Buffer capability	●●●	●●●	●●●	○○○	●●●	●●●	○○○	○○○
Redundancy	○○○	○○○	○○○	●●●	○○○	○○○	●●●	●●●
Operative Shiftability	●●●	●●●	○○○	●●●	●●●	●●●	○○○	○○○
Interruptible	●●●	○○○	○○○	○○○	○○○	●●●	○○○	○○○
Task flexibility	○○○	●●●	●●●	●●●	○○○	●●●	●●●	○○○
Routing flexibility	○○○	○○○	○○○	●●●	○○○	○○○	●●●	○○○
Manufacturing principle	●●○	●●○	●●●	●●○	●●●	●●●	●●○	●●○
Production method	●●●	●●●	●●●	●●●	●●●	●●●	●●●	●●○
Working shift model	●●●	●●●	●●○	●●○	●●●	●●○	●●○	●●○
Production planning horizon	●○○	●○○	●○○	●○○	●○○	●○○	●○○	●○○
Change of manufacturing orders	●○○	●○○	●●●	●○○	●○○	●○○	●○○	●○○
Change horizon	●○○	●○○	●○○	●○○	●○○	●○○	●○○	●○○
Product-based divergences	●●○	●●○	●●●	●●●	●●○	●●○	●●●	●●○
Multiple energy carriers	○○○	○○○	○○○	○○○	○○○	○○○	○○○	○○○
Labor flexibility	●○○	●○○	●●●	●●●	●○○	●○○	●●●	○○○
Relevant Costs	●○○	●○○	●○○	●○○	●○○	●○○	●○○	●○○

[1] Role legend: ○○○Irrelevant, ●○○Relevant, ●●○Influential, ●●● Crucial.

Table 3. Level of influence of the physical, operative and production characteristics in the existence of technical EFMs [1].

Characteristic Name	Technical EFM Categories						
	Adaptation of Operation Parameters	Operation Interruption	Adjustment of the Operational Sequence	Inherent Energy Storage	Bivalent Operation	Dedicated Energy Storage	Energy Carrier Exchange
Technical unit	●●●	●●●	●●●	●●●	●●●	●●●	●●●
Industrial system layout	●●●	●●●	●●●	●●●	●●●	●●●	●●●
Power rating	●●○	●●○	●●○	●●○	●●○	●●○	●●○
Operative time	●●○	●●○	●●○	●●○	●●○	●●○	●●○
Control concept	●●●	●●●	●●●	●●●	●●●	●●●	●●●
Typical load profile	●●●	●●●	●●●	●●●	●●●	●●●	●●●
Control variable	●●●	●●●	●●●	●●●	●●●	●●●	●●●
Control horizon and latency	●○○	●○○	●○○	●○○	●○○	●○○	●○○
Operative continuity	●●○	●●●	●●○	●●○	●●○	●●○	●●○
Operative steps	●●○	●●○	●●●	●●○	●●○	●●○	●●○
Output flexibility	●●●	○○○	●●●	●●●	○○○	●●●	○○○
Bivalence	○○○	○○○	○○○	○○○	●●●	○○○	●●●
Buffer capability	○○○	●●●	○○○	●●●	○○○	●●●	○○○
Redundancy	○○○	○○○	○○○	○○○	●●●	○○○	○○○
Operative Shiftability	○○○	●●●	●●●	●●●	○○○	○○○	○○○
Interruptible	○○○	●●●	○○○	○○○	○○○	○○○	○○○
Task flexibility	○○○	○○○	○○○	○○○	○○○	○○○	○○○
Routing flexibility	○○○	○○○	○○○	○○○	○○○	○○○	○○○
Manufacturing principle	●●○	●●○	●●○	●●○	●●○	●●○	●●○
Production method	●●○	●●○	●●○	●●○	●●○	●●○	●●○
Working shift model	●●○	●●○	●●○	●●○	●●○	●●○	●●○
Production planning horizon	●○○	●○○	●○○	●○○	●○○	●○○	●○○
Change of manufacturing orders	○○○	○○○	○○○	○○○	○○○	○○○	○○○
Change horizon	○○○	○○○	○○○	○○○	○○○	○○○	○○○
Product-based divergences	●●○	●●○	●●○	●●○	●●○	●●○	●●○
Multiple energy carriers	○○○	○○○	○○○	○○○	●●●	○○○	●●●
Labor flexibility	○○○	●●●	●●●	○○○	○○○	○○○	○○○
Relevant Costs	●○○	●○○	●○○	●○○	●○○	●○○	●○○

[1] Role legend: ○○○Irrelevant, ●○○Relevant, ●●○Influential, ●●● Crucial.

The identification of EFMs is hence a cross-analysis of the manifestation of each crucial characteristic and the tested category of organizational or technical EFM. Crucial characteristics can either support the existence of the specific EFM-category and hence be positive. They can hinder the existence of a specific EFM-category, being negative, and excluding the availability of this category in the analyzed industrial system. Finally, they can be unclear on which additional analysis or information is needed to accept or discard the existence of the specific EFM-category in the system. This additional information usually involves further discussion with the relevant stakeholders in the production facility.

The partial consideration of the characteristics, in case the complete list is not known, will reduce the analysis reference framework and then the type of EFMs identified and their EFP, as described in Section 3.1.

All the EFM-categories on which all crucial characteristics are positive are considered the prospective EFMs. After validation with relevant stakeholders in the facility, the next step is their characterization, where the influential and relevant characteristics are considered.

4.7. Characterization of the Validated Energy Flexibility Measures

The characterization of an EFM intends to define its scope. For this purpose, several groups of parameters, referred to as dimensions, have been defined that constitute the EFM characterization framework. The goal of this framework is to standardize the description of EFMs, facilitating their evaluation, modelling, implementation and management.

The proposed characterization framework consists of four different dimensions: functional dimension, performance dimension, temporal dimension and economic dimension. The parameters that constitute each dimension are explained in the following sub-sections.

4.7.1. Functional Dimension

This dimension serves to contextualize the EFM on the industrial system and the respective factory. The parameters that constitute the functional dimension are:

- Industrial system description: definition of the industrial system on which the EFM takes place. The description should at least include a system's layout including all energy-consuming components, their performance data and an outline of the energy and material flows through which the system interacts with the other systems in the factory.
- EFM category: category of the identified prospective EFM based on the general categories defined in Table 1.
- Operative concept: a description of how the identified EFM induces a change of state in the industrial system.
- Adjustment factor and adjustment relationship: the adjustment factor is the independently controlled variable(s) that induces the change of state in the industrial system. The adjustment relationship describes the correlation between the adjustment factor and the rate of energy demand. Usually, a mathematical function describes this relationship in the form of a correlation, i.e., linear, polynomial or step.
- Amount and type of the modes of operation (MO): the MOs describe the operative states the EFM might induce in the industrial system. The modes of operation might be holding if only one operative step is induced on the system per activation of the EFM or modulating if the EFM induces more than one operative state per activation. The amount and type of the MOs are determined, among other characteristics, by the typical load profile of the industrial system, particularly its operative clusters.
- Execution level: highest level across the control pyramid on which the virtual part of an EFM will take place. The execution level usually tends to be the level on which the system is controlled.

4.7.2. Temporal Dimension

This dimension groups the time-related parameters that characterize the EFM. The parameters that constitute the temporal dimension are:

- Active Duration, Δt_{active}: temporal element of the EFP, it comprises the minimum and maximum period on which the EFM is active, meaning the duration on which the industrial system operates under the EFM-induced operative state(s).
- Planning Duration, $\Delta t_{planning}$: minimum and maximum period necessary to plan the activation of an EFM. This parameter responds to the operative continuity of the industrial system. In the case of industrial systems belonging to the MA technical unit is also majorly influenced by the production planning horizon and change horizon, both are determined at the planning level of the EMC technical unit. The planning duration can take place before or after the occurrence of the triggering event of the EFM.
- Perception Duration, $\Delta t_{perception}$: minimum and maximum period between the occurrence of a triggering event and the perception of this event by the control architecture in the EMC technical

unit. The value of this parameter depends on the nature of the triggering event and the control latency in relevant systems in the EMC technical unit. The nature of the triggering event relates to the implementation objective of energy flexible operation as defined in Section 3.

- Decision Duration, $\Delta t_{decision}$: minimum and maximum period ranging from the perception of a triggering event, t_0, to the decision on the activation of the EFM. The performance, particularly the latency, of the systems that constitute the supervisory level of the EMC technical unit determine this parameter.
- Shift Duration, Δt_{shift}: minimum and maximum period covering the change in the operative state. This parameter is usually a function of the latency in the control concept of the industrial system. Nonetheless, it might be influenced by its operative stages and operative continuity.
- Activation Duration, $\Delta t_{activation}$: minimum and maximum period covering from the perception of a triggering event to the achievement of the EFM-induced operative state. It can be understood as the addition of the perception, decision, planning (if it is performed after the triggering event) and shift duration. Their calculation is relevant because it will quantify the overall interval between the triggering event and the fully active EFM. The calculation formula for the Activation Duration is presented in Equation (1).

$$\Delta t_{activation} = \Delta t_{perception} + \Delta t_{decision} + \Delta t_{planning} + \Delta t_{shift} \tag{1}$$

- Deactivation Duration, $\Delta t_{deactivation}$: minimum and maximum period between the end of the active duration of the EFM and the return of the industrial system to its original operative state. As it was the case for the shift duration, this parameter depends on the control concept of the industrial system and its control horizon.
- Regeneration Duration, $\Delta t_{regeneration}$: minimum and maximum period that must elapse before an EFM can be activated again after it has been deactivated. The regeneration duration can be understood as the necessary time to bring stability to the material and energy flows altered by the activation of an EFM.
- Validity, V: parameter outlining the fraction of the operative time of the industrial system on which the EFM will be available for activation. This parameter is defined by the type and amount of operative steps of the industrial system and therefore the validity should include a reference to the specific operative step on which the EFM is available [38].
- Activation Frequency, $N_{activation,T}$: the activation frequency parameter quantifies the maximum number of times an EFM can be executed over a specific period, T, usually a calendar year. Although it might be affected by other externalities, it might be calculated using the ratio between the product of the validity and the period, T, and the complete duration of the execution of an EFM. Equation (2) describes its calculation. The activation frequency should be referenced to the active duration for which it was calculated.

$$N_{activation,T} = \frac{V_T}{\Delta t_{activation} + \Delta t_{active} + \Delta t_{deactivation} + \Delta t_{regeneration}} \tag{2}$$

Figure 6 shows a summary of the different durations in the temporal dimension of an EFM. In the figure, a representative consumption increase EFM is presented.

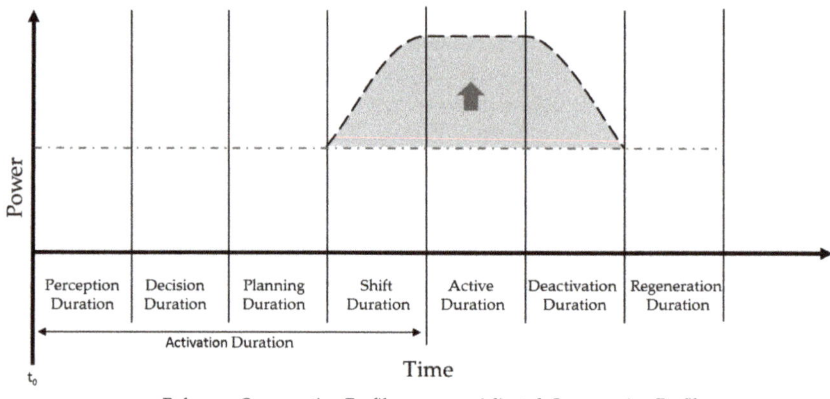

Figure 6. Duration parameters describing the comp lete execution of an EFM (Adapted from Reference [23], own illustration).

4.7.3. Performance Dimension

The performance dimension groups the characterization parameters of the EFM related to the change in the rate of energy demand. These parameters are:

- Flexibility Type: describes the direction on which the operative state will be changed by the activation of each of the MOs of the EFM. The possible flexibility types are:

  - Load increase ($\uparrow$): increase in the energy demand rate compared to the reference consumption profile. The increase can involve just an increase in the rate consumption or the complete switch-on of the industrial system. In a load increase, there is no consumption compensation requirement. Therefore, the activation of an EFM of this type will constitute an overall increase in energy consumption of the system.

  - Load decrease ($\downarrow$): reduction in the energy demand rate compared to the reference consumption profile. Similarly, like the increase, the renunciation can involve both a reduction of energy consumption and a complete switch-off of the influenced industrial system. In a load renunciation, consumption compensation is also not required. Therefore, the activation of this type of an EFM will constitute an overall decrease in energy consumption.

  - Bidirectional ($\uparrow\downarrow$): the ability of the EFM to offer both a load increase and renunciation. Nonetheless, once activated in either direction this type of EFM will not require a compensation of the altered energy consumption.

  - Consumption shift ($\leftrightarrow$): temporary rearrangement of the energy consumption, increase or decrease, with proportional compensation. The consumption shift is backwards when consumption is shifted to an earlier point in time. Inversely, it will be forward if it is postponed to a later point in time. A special case of load shift is "valley-filling" where the tasks that generate the consumption profile are broke down and rearranged at different points in time, thus reducing peak-consumption. In any of the consumption shift cases, the net energy consumption will stay constant despite activating the EFM. The different flexibility types are typified in Figure 7.

- Flexible Power, ΔP_{flex}: the power delta of the EFP, it describes the maximum difference of rate of energy demand between the reference operative state and the EFM-induced operative state. The unit for this parameter is usually kW_{flex}.

- Flexible Energy Carrier: this parameter, defines the energy carrier or carriers influenced by the activation of the EFM. Usually, as previously introduced, the focus, due to its attractiveness, is on the electrical energy consumption. Nonetheless, at least, for the Bivalent Operation and Energy Carrier Exchange EFM-categories, another energy carrier is also influenced.
- Flexible Energy, $E_{flex,T}$: the average amount of energy that could be adapted through the activation of an EFM over a specific period, T, typically a year. The flexible energy consists of the product of the average flexible power, the active duration and the retrieval frequency for this active duration, as presented in Equation (3). The unit for this parameter is usually MWh_{flex} and it must be referenced to the active duration for which it is calculated.

$$\Delta E_{flex,T} = \Delta P_{flex} * \Delta t_{activ} * N_{activation,T} \tag{3}$$

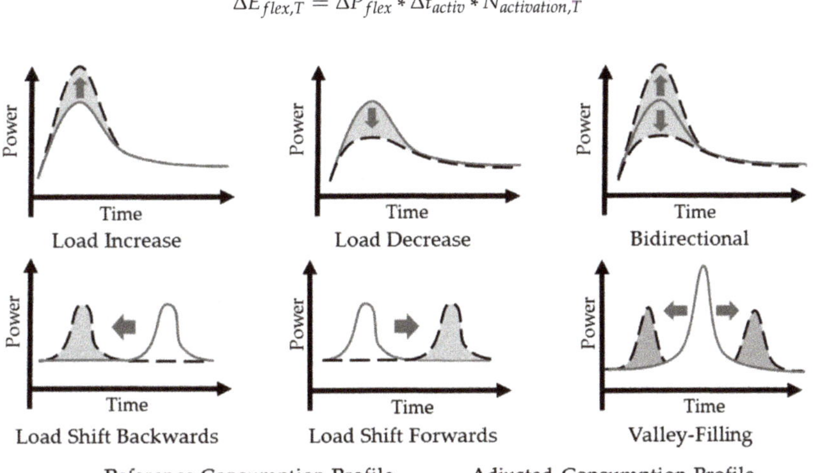

Figure 7. Different flexibility types of EFMs (Adapted from References [39,40], own illustration).

4.7.4. Economic Dimension

This dimension comprehends all the parameters related to the costs of the implementation and execution of an EFM. The parameters that constitute the economic dimension are:

- Investment Costs, $C_{investment}$: fixed, one-time expenses incurred to implement an EFM. Simply put the expenses necessary to bring the EFM to an operative status. The investment costs can be tangible including further development of component technology, further development of the IT-infrastructure and strengthening of the proprietary energy distribution infrastructure. These costs can also be intangible, like those associated with, the acquisition of software tools, hiring of third-party services or personnel training among others.
- Activation Costs, $C_{activation}$: ongoing expenses related to the activation of the EFM. These expenses are only incurred when the EFM is executed and hence are a function of the activation frequency. Examples include increased material, energy or labor costs due to the adaptation of the operative cycle of the industrial system and potential opportunity costs due to the activation of the EFM.
- Maintenance Costs, $C_{maintenance,T}$: ongoing expenses to keep the availability of the EFM over a specific time span, T, typically a calendar year. These costs are activation-independent. Therefore, they will stay unaffected by the activation frequency of the EFM. Examples include the hiring of third-party services to trade in energy markets and additional component wear and tear costs associated with energy flexible operation.
- Expected payback period, $\tau_{payback}$: the expected period, typically given in years, on which the EFM is expected to reach a break-even point or the point on which the revenues associated with

the EFM offset its costs. The company's management usually defines the expected payback period. Normally it obeys to their historical approach to factory-upgrade investments.

- EFM specific cost, $c_{flex,T}$: cost summary indicator of the EFM, it represents the cost of the EFM by a unit of flexible energy over a specific period (T). It is calculated through the formula presented in Equation (4).

$$c_{flex,T} = \frac{\frac{C_{investment}}{\tau_{payback}} * k + C_{maintenance,T} + C_{activation} * N_{activation,T}}{\Delta E_{flex,T}} \tag{4}$$

where k represents the temporal conversion factor between $\tau_{payback}$ and T.

Once all the different parameters across the four dimensions have been determined, the EFM has been fully characterized and the economical EFP can be determined.

4.8. Calculation of the Economical and Viable EFP of the Characterized EFMs

As previously mentioned, during this step, the calculated flexible power and active duration constitute the practical EFP of the industrial system, assuming all the relevant characteristics of the industrial system and the production facility have been considered. To calculate the economical EFP, the expected gross revenues, as a function of the intended implementation objective of the EFM, have to be estimated. An exact formula for the calculation of the gross revenues depends on the targeted implementation objective, as defined in the first step of the presented methodology. Generally, the gross revenues constitute the monetary savings achieved by the activation of the EFM when compared to the reference operation of the industrial system. Once the gross revenues, R_{flex}, have been calculated, the EFM specific gross revenues, $r_{flex,gross,T}$, constitute the ratio between the revenues for the specific period, T, and the flexible energy for the same period, as presented in Equation (5).

$$r_{flex,gross,T} = \frac{R_{flex}}{\Delta E_{flex,T}} \tag{5}$$

The difference between $r_{flex,gross,T}$ and $c_{flex,T}$ will provide the specific net revenues, $r_{flex,net,T}$, of the EFM for the period T, as presented in Equation (6).

$$r_{flex,net,T} = r_{flex,gross,T} - c_{flex,T} \tag{6}$$

The $r_{flex,net,T}$, will define the economic feasibility of the EFM on its current configuration. A negative or equal to zero $r_{flex,net,T}$ will indicate that the costs are too high. Therefore, the scope of the EFM needs to be optimized. This usually refers to reducing the activation costs by altering the flexible power, active duration or activation frequency of the EFM. If a cost-reduction is not possible, the EFM is deemed economically unfeasible and needs to be rejected. When the $r_{flex,net,T}$ is positive, the EFM will be economically feasible. Nonetheless, the scope of the EFM can be revisited to pursue the maximization of $r_{flex,net,T}$.

The resulting EFP, once the $r_{flex,net,T}$ of the EFM is maximized, constitutes the economical EFP. The final step will be to evaluate the scope and financial benefits of the EFM and weight it against other comparable investments, i.e., energy efficiency measures or other EFMs, and then decide on its implementation. This decision might further delimit the scope of the EFM, hence constituting the viable EFP.

At this moment, the EFM is completely identified and characterized and it can be grouped with the other identified EFMs across the facility, thus establishing the EFM catalogue of the facility.

5. Application of the Proposed Methodology

In the last section, the different steps to identify and characterize EFMs in industrial systems were described in detail. In this section, a representative example of an EFM using the described methodology is presented. The example belongs to an EFM identification analysis performed on an existing production facility. The scope of the analysis was limited to the calculation of the practical EFP. Therefore, the final step of the methodology is not presented in this example.

5.1. Delimitation of the Available Industrial Systems

The analyzed facility is a machinery assembly facility. All five technical units, as depicted in Figure 1, are present in the facility. The analysis, nonetheless, focused on the MA, TBS and EM technical units, employing the data from EMC technical unit as energy consumption records were available for the industrial systems in these technical units. The MA technical unit consists of three manufacturing segments, a press shop, a paint shop and an assembly production line. The TBS technical unit includes three industrial systems a compressed air system, a chilled water air-conditioning system, and a gas-fired hot water system. The EM technical unit consists of natural gas and electricity supplied to the facility. The latter supported by a photovoltaic array and two gas-fired combined heat and power (CHP) engines. The main objective for the implementation of energy flexibility in the facility is the intelligent response to the volatility of energy prices.

All the industrial systems were analyzed first for suitability and then to identify and characterize their available EFMs. The analysis extended over a period of two working weeks, once all available inputs were collected. In the following subsections, the application of the proposed methodology for a specific available industrial system, the chilled water air-conditioning system, is described.

5.2. Determination of the Relevant Physical Characteristics of the Chilled Water Air-Conditioning System

The physical characteristics of the chilled water air-conditioning system are:

1. Technical Unit: The system belongs to the TBS technical unit of the production facility.
2. Industrial system layout: the chilled water air-conditioning system consists of a 7/12 °C chilled water (CHW) circuit that provides room cooling for a production hall and an on-site data center. The cooling output is provided by three water-cooled, screw-driven mechanical chillers (CHWDX) and two hot water, single-effect absorption chillers (CHWAB) plus a free cooling module (CHWFC). The heat abatement of these units is performed by a 32/37 °C cooling water (CW) circuit with three cooling towers and three pumps. The hot water for the CHWABs is usually fed from the two CHP engines on-site but through minor modifications might be sourced from the 95/60 °C hot water (HW) system onsite. The cooling is delivered through a series of air handling units to the production hall and a data center that supports the production activities. Due to their air-quality-specific operation pattern, the air-handling units were not considered in the analysis of this system. For analysis purposes, the hot water loop for the CHWABs is assumed as an energy carrier entering the system. Therefore, the HW generation sources were not analyzed. The layout of the chilled water system is depicted in Figure 8.
3. Power rating and cooling output of the cooling units: The power rating and cooling output of the cooling generation units are summarized in Table 4.
4. Power rating and output of the other energy-consuming components: The power rating and design output of the other relevant energy-consuming components, pumps and cooling towers, in the system, are presented in Table 5.
5. Operative Time: The system operates 24/7 in stand-by mode, going into active operation when there is a cooling demand. Therefore, its maximum operative time is limited to the working shifts in the production facility.
6. Control Concept: The cooling demand is a function of the ambient temperature on-site. The system is controlled at the supervisory level through a SCADA architecture that monitors the air return temperature in the air-handler units and the return water temperature in the chilled water circuit. The current control concept prioritizes the operation of the CHWDX units for cooling supply. These mechanical chiller units are activated sequentially, based on the return water

temperature. Activation priority is given to CHWDX-3, due to its better performance at partial loads. The other mechanical chiller units, CHWDX-1 and CHWDX-2, are rotated to guarantee equalized running time among them. The absorption units, CHWAB-1 and CHWAB-2, are mostly activated in-junction with two combined heat and power (CHP) engines on site. The CHP engines are activated for peak-shaving purposes in the factory. The absorption chiller units are also used to provide redundancy to the mechanical chiller units. The free-cooling module, CHWFC-1 gains priority activation, when the ambient temperature drops below 10 °C.

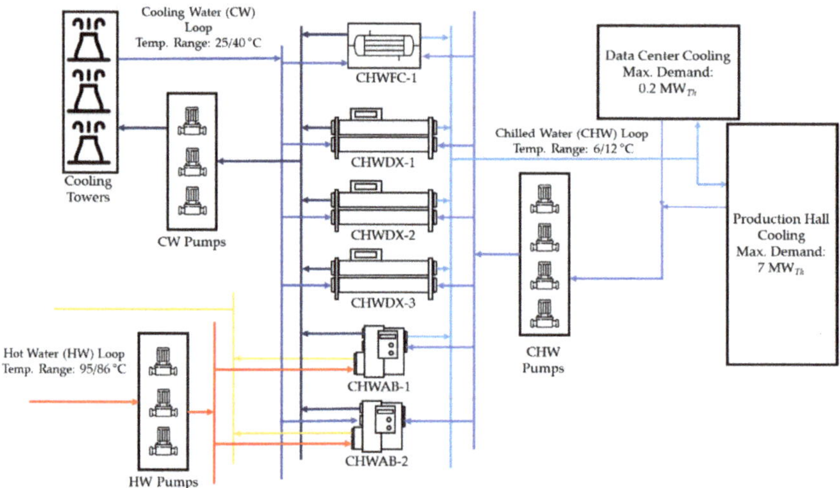

Figure 8. Schematic of the chilled water system.

Table 4. Technical specifications of the generation units in the Chilled Water System.

Designation [1]	Type	Cooling Output	Power Rating
CHWDX-1	Mechanical Chiller	2814 kW$_{Th}$	484 kW$_{Elec}$
CHWDX-2	Mechanical Chiller	2814 kW$_{Th}$	484 kW$_{Elec}$
CHWDX-3	Mechanical Chiller	2575 kW$_{Th}$	479 kW$_{Elec}$
CHWAB-1	Absorption Chiller	1653 kW$_{Th}$	6.9 kW$_{Elec}$
CHWAB-2	Absorption Chiller	1653 kW$_{Th}$	6.9 kW$_{Elec}$
CHWFC-1	Free-cooling Module	1250 kW$_{Th}$	0 kW$_{Elec}$

[1] Designation used in Figure 8.

Table 5. Technical specification of other components in the chilled water system.

Designation [1]	Output	Power Rating (kW$_{Elec}$)
Cooling towers	Flow: 43–260 L/s Range: 40/25 °C (WBT: 21 °C)	30
CW Pumps	Flow: 250 L/s Head: 90 kPa	37.3
HW Pumps	Flow: 65 L/s Head: 50 kPa	5.6
CHW Pumps	Flow: 90 L/s Head: 100 kPa	15

[1] Designation used in Figure 8.

5.3. Suitability of the Chilled Water System for Energy Flexible Operation

The results of the suitability analysis for energy flexible operation of the system are presented in Table 6. Regarding its controllability: as the system is used for air temperature conditioning, it is state variable, outdoor temperature-dependent and hence can be classified as a Co1. As there is high redundancy in the system and it belongs to the TBS technical unit, its criticality is estimated at a Cr3. This because a change of state in the system is neutral for process continuity as long as the demand is met. Finally, the interdependence is given as In1, as the system counts just the inherent thermal inertia across in the chilled-water-piping grid and the conditioned rooms.

Table 6. Suitability analysis of the chilled water system (left: score, right: radar graph).

Criteria	Level	
Controllability: State variable dependent	1	
Criticality: Neutral influence	3	
Interdependence: Inherent decoupling capabilities	1	
Overall Score: Moderate suitability	**3**	

5.4. Determination of the Relevant Operative Characteristics of the Chilled Water Air-Conditioning System

The relevant operative characteristics of the chilled water air-conditioned system are:

- Typical load and output profile: There is a three-year data record of the cooling consumption in the factory on a 15 min basis. The data record also includes the cooling output and the electrical consumption of the components in Tables 4 and 5. Employing the silhouette analysis and the K-means algorithm, as previously described, on the data record, values were clustered and an average cooling output and electrical input profile per cluster were calculated. As the measurements followed a normal distribution, their spread for each 15 min period was calculated using the two standard deviations over (2σ) and below (-2σ) the mean. The results are presented in Figure 9, the average consumption profile is color-highlighted and the range (±2 standard deviations) is shown in grey.
- Control Variable: As mentioned the system operates continuously, ramping up and down the different cooling generation units as a function of the return water temperature.
- Control horizon and latency: The ramping up and down of the system to a new operative state lasts between 5 and 10 min. The control components present a latency under five milliseconds.
- Operative Continuity: the system presents a discontinuous operative continuity.
- Operative steps: Each of the cooling units ramp up and down in single steps depending on the number of cooling circuits they present. The CHWDXs present 2 circuits, hence 2 operative steps, and the CHWABs present a single one, as does the CHWFC.
- Output Flexibility: the aforementioned cooling circuits in each of the cooling generation units provide the output flexibility.
- Bivalence: Due to the different functioning principle of the CHWDX and CHWFC, the system can be considered as presenting bivalence.
- Redundancy: As previously mentioned, the CHWAB act as redundancy for the CHWDX units. The other components in the CW and HW circuits present 2N + 1 redundancy, while the pumps in the CHW circuit present 3N + 1 redundancy.

The system, as already mentioned, does not present any buffering capability, is not shiftable or interruptible and has no routing or task flexibility.

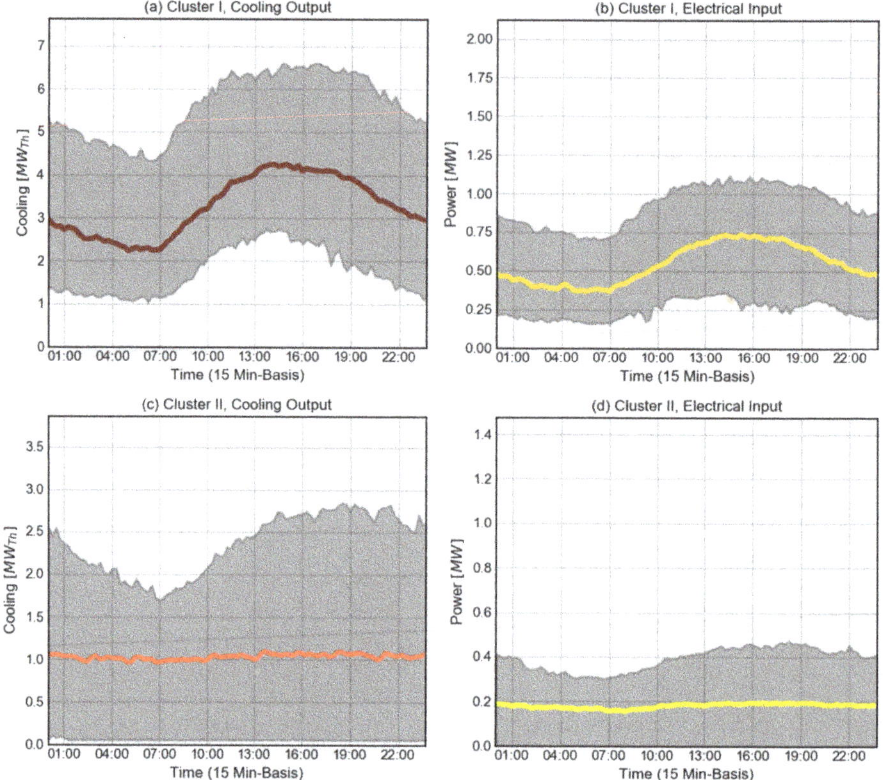

Figure 9. Typical cooling output and electrical input profiles for the chilled water system, (**a**) Operative cluster I, Cooling Output, (**b**) Operative Cluster I, Electrical Input, (**c**) Operative Cluster II, Cooling Output, (**d**) Operative Cluster II, Electrical Input.

5.5. Determination of the Relevant Production Characteristics of the Production Facility

The relevant production characteristics of the production facility on which the system finds itself are:

- Manufacturing Principle: MTS
- Production Method: Flow Processing
- Working Shift Model: 3 shifts (8 h long), 5 days a week, 50 weeks per year
- Production planning horizon: Weekly
- Product-based divergence in energy consumption: None
- Multiple Energy Carriers on Site: Relevant to this system are natural gas and electricity.

The production characteristics not mentioned are irrelevant for this system, except for the relevant costs, which, as sensible information from the company, cannot be disclosed. Additional to provided information it is important to mention that the facility is located in Central Europe and, therefore, the outdoor temperature ranges from −14 to 42 °C.

5.6. Identification of Prospective EFMs in the Chilled Water Air-Conditioning System

The analysis of the different characteristics explains the existence of two operative clusters in the system as shown in Figure 9. Cluster I, the first operation profile, comprises working days and outdoor temperatures over 10 °C, where the cooling demand increases and no free-cooling option is available. Cluster II, the second operation profile covers working days below 10 °C, where the cooling demand is relatively reduced and a considerable part of the cooling demand is supplied by CHWFC-1. Additionally, Cluster II also covers non-working days, where there a base cooling demand, mainly for the data center. This was inferred from an analysis of the days in a calendar year on which each of the determined operative clusters will be active.

The analysis of the chilled water system using a cross-analysis matrix based on Tables 2 and 3 and evaluating each crucial characteristic as either positive or negative showed the availability of two different EFMs in the analyzed system:

1. Adaptation of resource allocation: The adaptation of resource allocation EFM focuses on the possibility of switching between the CHWDX chillers and the CHWAB chillers while supplying the cooling consumption of the facility. Due to the considerable difference in the power rating between chiller types and hence in their electrical EER, the rotation of the absorption and mechanical chiller units induce a change in the electrical input of the chilled water system.
2. Dedicated Energy Storage: The installation of CHW storage to supply the totality or a share of the cooling demand, at specific periods.

The availability of both prospective EFM was validated with the energy managers in the facility. Due to expected high investment costs of the dedicated energy storage, this EFM was deemed unattractive.

5.7. Characterization of the Validated EFM in the Chilled Water Air-Conditioning System

The characterization framework defines the scope of the validated EFM. The following tables determine or quantify the different parameters and hence the different dimensions as described in Section 4.7. Table 7 describes the functional dimension of the EFM.

Table 7. Functional Dimension of the identified EFM [1].

Parameter	Description					
System Description	CHW air-conditioning system to supply space cooling.					
EFM Category	Adaptation of resource allocation					
Operative Concept	Switching between types of cooling generation units to either increase (↑), by prioritizing the usage of mechanical chillers, or decrease (↓), by prioritizing the usage of the absorption chillers, the electrical demand of the system					
Adjustment Factor	Ramp-Up and Down of the specific chiller units.				Adjustment Relationship	Step (On/Off)
MO Amount	MO-1	Cluster I, ↑	MO-3	Cluster II, ↑	MO Types	Holding
	MO-2	Cluster I, ↓	MO-4	Cluster II, ↓		
Execution Level	Supervisory Level					

[1] MO: Mode of Operation.

In Table 7, the system description, EFM category and operative concept come from the analysis performed in the previous steps. The adjustment factor responds to the control variable in the system, the ramping up and down of cooling generation units. Four different modes of operation (MO) have been defined for the EFM. The division is based on the number of clusters identified in the analysis of the typical load profile and the ability of the EFM to induce an increase (↑) or a decrease (↓) in the electrical consumption. All of these modes of operation are defined as holding because when activated, they only induce one operative state in the system. As the control of the system is performed via a SCADA system the execution level is set on the supervisory level. As expected, the functional dimension responds directly to the physical and operative characteristics of the analyzed industrial system. Table 8 presents the temporal dimension of the identified EFM.

Table 8. Temporal Dimension of the identified EFM [1].

Parameter	Value (Min-Max)				Remarks [2]
Active Duration, Δt_{Active}	12 min–8 h				Valid for all four MOs
Planning Duration, $\Delta t_{Planning}$	0				Valid for all four MOs
Perception Duration, $\Delta t_{Perception}$	15 min–24 h				Valid for all four MOs
Decision Duration, $\Delta t_{Decision}$	<1 min				Valid for all four MOs
Shift Duration, Δt_{Shift}	5–10 min				Valid for all four MOs
Activation Duration, $\Delta t_{Activation}$	10.5 min–24.2 h				Valid for all four MOs
Deactivation Duration, $\Delta t_{Deactivation}$	5–10 min				Valid for all four MOs
Regeneration Duration, $\Delta t_{Regeneration}$	12 min				Valid for all four MOs
Validity, V	MO-1	69%	MO-3	31%	
	MO-2	69%	MO-4	31%	
Activation Frequency, $N_{Activation,T}$	MO-1	709^2	MO-3	323^2	
	MO-2	709^2	MO-4	323^2	

[1] MO: Mode of Operation. [2] Calculated for Active Duration, Δt_{Active}, equal to 8 h.

The Active Duration minimum is restricted to avoid compressor short cycling (>5 cycles/h), which might cause the operative failure of the cooling generation units. The maximum Active Duration is limited to one working shift in the facility as an analysis of the typical profiles showed that MOs in the system can change form one shift to the other. The wide range in the Active Duration supports the intended implementation objective as the duration of price volatility can extend over several hours. The Planning Duration is set to zero as no planning is necessary, in both cases, the system can execute its task, provide cooling, without interruption. The Perception Duration depends on the specific market on which electricity is being purchased and hence ranges from 5 min for intra-day handling to 24 h for day-ahead handling. The Decision Duration is considered automatized and hence it is defined by the latency of the components in the EMS. The Shift Duration responds to the ramp-up and down duration of the different cooling generation units. The Activation Duration aggregates the ramping-up of the EFM and it is calculated using Equation (1). The major element deciding the Activation Duration is the Perception duration and hence the identified EFM presents a very high capability to quickly react to price volatility and hence achieved the intended implementation objective. The Deactivation Duration mirrors the Shift Duration and the Regeneration Duration corresponds to the short-cycling avoidance requirement in the cooling generation units. Both are also relatively short, allowing for the EFM to be used to respond to subsequent electricity price variations. The Validity responds to production characteristics of the facility and hence to the operatives clusters determined by the typical profile analysis. The Activation Frequency is calculated using Equation (2).

Table 9 presents the performance dimension of the identified EFM. The given load increase (↑), ΔP_{flex}, values quantify the maximum and average difference between the typical electrical input and the necessary electrical input if the typical cooling output is satisfied by only using the mechanical chillers. The load reduction (↓), ΔP_{flex}, on the other hand, quantify the maximum and average difference between this typical electrical input and the necessary electrical input if the typical cooling output is primarily supplied using the absorption chillers. In reality, the ΔP_{flex} is dynamic and hence a function of the state of operation of the system. The state of operation will depend on the instantaneous cooling demand, in turn, a function of both the outdoor temperature and the level of production in the facility. The given values are hence a static approximation to the dynamic ΔP_{flex}, value.

As previously hinted in the functional dimension, the EFM presents a bidirectional flexibility type without a need for later compensation. The different ΔP_{flex} responds to the different cooling demand of the typical profiles in the facility. A reduction in the cooling demand, MO-3 and MO-4, diminishes the flexible power. Moreover, the flexible power of a load increase is considerably lower than that of a load reduction as in the reference operation, the CHWDXs have operative priority, and hence are already supplying a portion of the cooling demand. The most attractive MO is MO-2, an electrical consumption reduction during production and with an ambient temperature over 10 °C. This MO can achieve, on average, a reduction of 88% of the electrical consumption of the system for a period of up to 8 h.

Table 9. Performance Dimension of the identified EFM.

Parameter	Value			
Flexibility Type	Bidirectional ($\uparrow\downarrow$)			
Maximum and Average Flexible Power, $\Delta P_{flex,max}$ ($\Delta P_{flex,avg}$) (kW$_{Flex}$)	MO-1 MO-2	145.2 (117.2) 498.9 (440.3)	MO-3 MO-4	42.9 (25.7) 145.2 (117.2)
Flexible Energy Carrier(s)	Electricity/Hot Water (<95 °C)			
Average Flexible Energy [1] $E_{flex,avg,year}$ (MWh$_{Flex}$)	MO-1 MO-2	665.1 2498.7	MO-3 MO-4	66.5 303.3

[1] Calculated for Active Duration, Δt_{Active}, equal to 8 h.

Regarding the additional characterization parameters, the flexible energy carriers respond to the operative principle of the cooling generation units in the system and the flexible energy is calculated using Equation (3). As the physical and operative characteristics of the chilled water system and, the production characteristics of the facility are considered to calculate the ΔP_{flex} and the Δt_{Active}, they represent the practical EFP.

Finally, in Table 10, the economical dimension of the EFM is presented and the calculation reasoning behind each parameter is described.

Table 10. Economic Dimension of the identified EFM.

Parameter	Value			Description
Investment Costs, $C_{investment}$	33,599.25 €			Minor modifications in the system piping and acquisition of new components for the EMS.
Activation Costs, $C_{activation}$	0.00 €			No activation costs are considered for the EFM. The costs of each energy carrier are excluded as activation costs as they are used in the net revenue analysis.
Maintenance Costs, $C_{maintenance,T}$	53,019.56 €			Due to the added rotation, additional maintenance costs have to be accounted for the chiller units.
Expected payback period, $\tau_{payback}$	3 years			Defined by the company based on industry standards for these investments.
EFM specific cost [1] $c_{flex,T}$ (€/MWh$_{Flex}$)	MO-1 MO-2	96.6 25.7	MO-3 965.6 MO-4 211.8	Calculated using Equation (4).

[1] Calculated for Active Duration, Δt_{Active}, equal to 8 h.

As can be inferred from the descriptions the implementation of the EFM will only represent investment and activation costs. The investment costs relate to additional infrastructure to have a constant supply of HW in the facility, and additional IT-infrastructure to allow the reaction to dynamic electrical prices. The maintenance costs relate, mainly to additional operative hours of the CHWABs which present relatively high maintenance costs, due to their operative principle. Although a relatively high payback period is given, it is clear that MO-3 and MO-4 are prohibitively expensive, based on the average price of electricity in the EU which is approximately 100 €/MWh [41].

5.8. Calculation of the Economical and Viable EFP of the Identified EFM

As mentioned before, the final step of the proposed methodology, the calculation of the economic and viable EFP, based on the net revenues of the EFM was not a part of the performed analysis. Nonetheless, as can be inferred, the gross revenues are dependent on the variability of the price paid for electricity and will be specific to each MO. For MO-1 and MO-3, load increase revenues are achieved if the electricity price is lower than the average price paid for the combination of electricity and hot water as energy inputs. This consideration will further limit the active duration and the activation frequency of these MOs, hence reducing flexible energy and increasing their specific costs. These considerations hint that these MOs might not be economically attractive for the company to activate. Nonetheless, they will be practically available if the EFM is implemented. On the other hand, for MO-2 and MO-4 load decrease, the revenues are reached if the price of generating HW is lower than that for electricity. This consideration will also reduce the active duration and the activation frequency of these MOs. Nonetheless, due to the considerable low specific cost of MO-2, and its high validity this might constitute a very attractive EFM overall for the company.

The economic EFP will hence constitute the flexible power and active duration in which the EFM generates revenues on each MO. The MOs that do not produce revenues should not be further considered. In the case of the viable EFP, the company has to make decisions on the active duration and activation frequency they intend for the EFM, weighing potential risks or negative consequences in the facility's performance, i.e., in its energy efficiency, which was not a part of this analysis.

6. Discussion

The initial application of the methodology provided several insights that are discussed in this section. Under ideal conditions, the definition available industrial systems, Step 1, will only respond to the grouping energy consuming components in industrial systems and their categorization in technical units. Nonetheless, as detailed monitoring of all energy-consuming loads is not yet a standard in the industrial sector, a very relevant aspect in the decision of which industrial systems will be analyzed is the available data records of their energy consumption and their output. The selection of implementation objectives is relevant to provide an end goal to the analysis, however, it is frequent that once the EFMs are identified and characterized new implementation objectives become relevant.

The physical characteristics of the industrial systems, determined in Step 2, give a general view of the system and its operation and can lead to an initial understanding of the energy flexibility capabilities of the system. Nevertheless, excessive reliance on these characteristics might be misleading. During the application of the methodology, it was the case that initially thought available EFMs were deemed unavailable by the operative characteristics of the system or the production characteristics of the facility.

The suitability analysis, conducted in Step 3, allows sorting among the available industrial systems and reduce the duration of the analysis particularly when the production facility is very complex and hence constituted by a large number of industrial systems. Nonetheless, its qualitative nature demands caution as a wrong assessment of any of the three criteria might discard suitable industrial systems. In the cases where a determination was not clear, the consideration of operative characteristics of the industrial system, which normally occurs in the subsequent step, significantly helped the analysis.

In contrast to the physical characteristics, the operative characteristics of the industrial system and the production characteristics of the facility, determined in Steps 4 and 5, play a sorting role in either supporting or discarding the availability of each EFM-Category, as conducted in Step 6. Particularly in very wide encompassing EFMs categories, like dedicated energy storage, which initially seem to be available for all sorts of industrial systems.

The determination of the different parameters in the characterization framework in, Step 7, is intrinsically dependent of the nature of each industrial system and therefore is considerably difficult to standardize, here the experience of the person conducting the analysis, the thoroughness of the surveyed system and facility characteristics and, the input of relevant stakeholders from the production facility proved vital to obtain realistic values.

Similarly, the calculation of the economical and viable EFP, in Step 8, is very case-specific and only general guidelines can be given regarding how this step should be conducted.

In general, the application of the proposed methodology shows that it is not able to replace the accuracy of modelling the industrial system to simulate its operation under energy flexible operation, as described in References [15,38] among others. As explained, the methodology relies on typical profiles of energy consumption and patterns of operation. As these profiles and patterns are a simplification of the actual dynamic operation of an industrial system, the performance of EFMs once implemented will diverge from the provided characterization. Nonetheless, the methodology presents considerable value, as it pinpoints the industrial systems suitable for energy flexible operation, from the large list of available industrial systems in a typical production facility. Moreover, it systematically identifies and characterizes the specific actions that induce energy flexible operation in these systems in the form EFMs, which is not only novel but provides a key input for the modelling, evaluation, implementation and management of the energy flexibility capabilities of the industrial system. Subsequent modelling of the industrial system acts then as a supplement, focusing on improving the accuracy of the values of the characterization parameters and being used as a prognosis tool to plan the management of the EFMs.

Additionally; the initial results also show that the methodology is promising but can be improved by improving the tools to establish the typical operative patterns of industrial systems. The accuracy of the results is highly dependent on the approach used to establish these patterns. It is hence crucial to examine thoroughly the available machine learning algorithms on data mining and clustering to find the best fitting for the task. These algorithms provide extremely relevant insights towards understanding how energy consumption is affected, particularly by the operative characteristics of the system and the production characteristics of the facility. Therefore, the most fitting algorithms and their optimal usage will facilitate the identification of EFMs and provide more accurate quantification of their characterization parameters.

7. Conclusions and Outlook

The paper presents a methodology to identify and characterize energy flexibility measures in the industrial systems that constitute a production facility. The methodology is meant to be the basis of an industrial energy audit focusing on the topic of energy flexibility and hence providing vital information for enterprises to implement and exploit the energy flexibility capabilities of their production facilities. The proposed methodology follows a similar procedure than the current standards in industrial energy auditing aimed to improve industrial energy management and identify energy efficiency measures [28,29]. As those standards, and as previously stated in the requirements, the proposed methodology needed to be systematic, agile, current operation friendly, applicable to the plethora of industrial systems and its outcomes needed to be relevant for the industrial stakeholders. The methodology starts by establishing the available industrial systems in the facility. Allowing the definition of different system boundaries depending on the morphology of the analyzed production facility, and hence adapting to the heterogenous nature of industrial systems. The fact that the expected implementation objectives from energy flexible operation are incorporated in the methodology provides a clear end goal for the analyzed production facilities, and hence prioritizes outcomes to the specific company needs, providing relevancy to its outcomes. The suitability analysis allows focusing only on those relevant industrial systems, reducing the analysis duration contributing to its agility. This acts as a counterpart to the "big-consumers" approach usually used in energy efficiency auditing which might be misleading in the case of energy flexibility. The analysis of the physical and operative characteristics of the industrial system and the production characteristics of the facility allows considering the current operative nature of the analyzed industrial systems, guaranteeing its affinity with the current operation approach. Moreover, it provides a more agile approach to analyze the dynamic nature of industrial systems than building a dedicated system model. Overall, the methodology is systematic as it follows a linear approach where decisions are made following previously defined criteria and allowing a multi-level analysis of the industrial systems to identify the available EFMs. EFM-categories are analyzed and only discarded under specific techno-economic considerations, not on biased assumptions. The creation of the characterization framework, that consistently delimits the scope of each EFM, facilitates the subsequent evaluation, implementation and management.

The methodology is currently being implemented to identify and characterize EFMs in several production facilities within the framework of the second phase of Kopernikus-project "SynErgie".

The results are expected to be used to evaluate the benefit-based performance of each EFM to then prioritize and facilitate their implementation [42]. The characterization parameters of the EFMs will also be used as input in the simulation of the production facility under energy flexible operation using digital twinning modelling. Moreover, the outcomes of the proposed methodology will also be used to develop energy management and optimization strategies for the analyzed production facilities. Continuous improvement of the methods and tools described in this article is expected as more production facilities are audited.

Author Contributions: A.T. and F.H. worked on the conceptualization of the methodology. A.T. developed the general structure and wrote the first draft of the manuscript. F.H. supported the review and editing of the manuscript. A.T. and F.H. worked together in the validation of the methodology, visualization and the application in practical examples. A.S. provided the idea, supervised the work of A.T., helped to review the article and was instrumental in the funding of this work. All authors have read and agreed to the published version of the manuscript.

Funding: This research was funded by the German Federal Ministry of Education and Research (BMBF), Grant No. 03SFK3G1.

Acknowledgments: The authors gratefully acknowledge the financial support of the Kopernikus-project "SynErgie" by the Federal Ministry of Education and Research (BMBF), and the project supervision of the project management organization Projektträger Jülich (PtJ).

Conflicts of Interest: The funders had no role in the design of the study; in the collection, analyses, or interpretation of data; in the writing of the manuscript, or in the decision to publish the results.

Acronyms

The following acronyms were used in this publication:

AS	Auxiliary Systems technical unit
AGV	Automated Guided Vehicles
ATO	Assembly to Order manufacturing principle
CHP	Combined Heat and Power
CHW	Chilled Water
CHWAB	Chilled Water Absorption Chiller
CHWDX	Chilled Water Mechanical Chiller
CHWFC	Chilled Water Free-Cooling Module
Co	Controllability
Cr	Criticality
In	Input/output Interdependence
CW	Cooling Water
DR	Demand Response
DSEF	Demand Side Energy Flexibility
EEP	Institute for Energy Efficiency in Production
EER	Energy efficiency ratio
EFM	Energy Flexibility Measure
EFP	Energy Flexibility Potential
EMC	Energy and Manufacturing Control technical unit
EM	Energy and Media technical unit
EMS	Energy Management System
ERP	Enterprise Resource Planning System
ETO	Engineering-to-Order manufacturing principle
GHG	Green House Gases
HMI	Human-Machine Interfaces
HVAC	Heating, Ventilation and Air-conditioning systems
HW	Hot Water

IEF	Industrial Energy Flexibility
IPA	Institute for Manufacturing Engineering and Automation
IRENA	International Renewable Energy Agency
MA	Manufacturing technical unit
MES	Manufacturing Execution System
MTO	Make-to-Order manufacturing principle
MTS	Make-to-Stock manufacturing principle
MO	Modes of Operation of the energy flexibility measure
NDC	Nationally Determined Contribution
PLC	Program Logic Controllers
PM	Production Machines
SCADA	Supervisory control and data acquisition system
SSEF	Supply Side Energy Flexibility
TBS	Technical Building Services technical unit
VRE	Variable Renewable Energy Sources
WBT	Wet-Bulb Temperature
WS	Workstations

References

1. International Renewable Energy Agency. *Global Energy Transformation: A Roadmap to 2050 (2019 Edition)*; International Renewable Energy Agency: Abu Dhabi, UAE, 2019; ISBN 978-92-9260-121-8.
2. International Renewable Energy Agency. *Global Renewables Outlook: Energy Transformation 2050*; International Renewable Energy Agency: Abu Dhabi, UAE, 2020; ISBN 978-92-9260-238-3.
3. International Renewable Energy Agency. *Power System Flexibility for the Energy Transition, Part 1: Overview for Policymakers*; International Renewable Energy Agency: Abu Dhabi, UAE, 2018; ISBN 9789292600891.
4. Cochran, J.; Miller, M.; Zinaman, O.; Milligan, M.; Arent, D.; Palmintier, B.; O'Malley, M.; Mueller, S.; Lannoye, E.; Tuchy, A. *Flexibility in 21st Century Power Systems*; National Renewable Energy Laboratory: Golden, CO, USA, 2014.
5. Bradley, P.; Leach, M.; Torriti, J. A review of the costs and benefits of demand response for electricity in the UK. *Energy Policy* **2013**, *52*, 312–327. [CrossRef]
6. Publications Office of the European Union. *Energy Balance Sheets*, 2019 ed.; Publications Office of the European Union: Luxembourg, 2019; ISBN 9276087141.
7. EU-28: Breakdown of Industrial Energy Use 2015|Statista. Available online: https://www.statista.com/statistics/859496/renewable-energy-consumption-by-source-european-union-eu-28/ (accessed on 9 May 2020).
8. Sauer, A.; Abele, E.; Buhl, H.U. (Eds.) *Energieflexibilität in der Deutschen Industrie: Ergebnisse aus dem Kopernikus-Projekt—Synchronisierte und Energieadaptive Produktionstechnik zur Flexiblen Ausrichtung von Industrieprozessen auf eine Fluktuierende Energieversorgung (SynErgie)*; Fraunhofer Verlag: Stuttgart, UAE, 2019; ISBN 9783839614792.
9. Strbac, G. Demand side management: Benefits and challenges. *Energy Policy* **2008**, *36*, 4419–4426. [CrossRef]
10. Munuera, L. Demand Response (Tracking Report). 2020. Available online: https://www.iea.org/reports/demand-response (accessed on 3 August 2020).
11. Virag, A.; Jezdinsky, T. Flexible Industrial Demand Assessment. Case Studies Summary; Public Version. 2018. Available online: http://www.industre.eu/downloads/ (accessed on 4 August 2020).
12. Wiendahl, H.-P.; Reichardt, J.; Nyhuis, P. *Handbook Factory Planning and Design*; Springer: Heidelberg, Germany, 2015; ISBN 9783662463901.
13. Kiran, D.R. *Production Planning and Control. A Comprehensive Approach*; Butterworth-Heinemann: Amsterdam, The Netherlands, 2019; ISBN 9780128183649.
14. Posselt, G. *Towards Energy Transparent Factories*, 1st ed.; Springer: Berlin, Germany, 2016; ISBN 9783319208688.
15. Beier, J. *Simulation Approach towards Energy Flexible Manufacturing Systems*; Springer: Cham, Switzerland, 2017; ISBN 978-3-319-46639-2.
16. Cannistraro, G.; Cannistraro, M.; Cannistraro, A.; Galvagno, A.; Trovato, G. Evaluation on the convenience of a citizen service district heating for residential use: A new scenario introduced by high energy efficiency systems. *IJHT* **2015**, *33*, 167–172. [CrossRef]

17. Cannistraro, G.; Cannistraro, M.; Cannistraro, A.; Galvagno, A.; Trovato, G. Technical and Economic Evaluations about the Integration of Cotrigeneration Systems in the Dairy Industry. *IJHT* **2016**, *34*, S332–S336. [CrossRef]

18. Sauer, A.; Weckmann, S. Industrial Smart Grids–Ein Beitrag für ein Nachhaltiges Energiesystem. In *CSR und Digitalisierung*; Hildebrandt, A., Landhäußer, W., Eds.; Springer: Berlin, Germany, 2017; pp. 209–226, ISBN 978-3-662-53201-0.

19. Schnell, G.; Wiedemann, B. (Eds.) *Bussysteme in der Automatisierungs- und Prozesstechnik Grundlagen Systeme und Trends der Industriellen Kommunikation*, 6th ed.; Vieweg + Teubner: Wiesbaden, Germany, 2006; ISBN 9783834891082.

20. Reinhart, G.; Reinhardt, S.; Graßl, M. Energieflexible Produktionssysteme: Einführungen zur Bewertung der Energieeffizienz von Produktionssystemen. *WT Werkstattstech. Online* **2012**, *102*, 622–628.

21. Federal Energy Regulatory Commission; U.S. Department of Energy. *Implementation Proposal for the National Action Plan on Demand Response*; U.S. Department of Energy: Washington, DC, USA, 2011.

22. Zancanella, P.; Bertoldi, P.; Kiss, B. *Demand Response Status in EU Member States*; Publications Office of the European Union: Luxembourg, 2016.

23. Verein Deutscher Ingenieure (Ed.) *Energieflexible Fabrik Grundlagen*; VDI 5207; Blatt 1; VDI-Gesellschaft Produktion und Logistik (GPL): Düsseldorf, Germany, 2019.

24. Schulze, C.; Blume, S.; Siemon, L.; Herrmann, C.; Thiede, S. Towards energy flexible and energy self-sufficient manufacturing systems. *Procedia CIRP* **2019**, *81*, 683–688. [CrossRef]

25. Dufter, C.; Guminski, A.; Orthofer, C.; von Roon, S.; Gruber, A. (Eds.) Lastflexibilisierung in der Industrie–Metastudienanalyse zur Identifikation relevanter Aspekte bei der Potenzialermittlung. In Proceedings of the IEEWT 2017—10. Internationale Energiewirtschaftstagung, Vienna, Austria, 16 February 2017.

26. Popp, R.S.H.; Zaeh, M.F. Determination of the Technical Energy Flexibility of Production Systems. *AMR* **2014**, *1018*, 365–372. [CrossRef]

27. Grassl, M. *Bewertung der Energieflexibilität von Produktionssystemen//Bewertung der Energieflexibilität in der Produktion, Dissertation*; Technical University Munich: Munich, Germany, 2015; ISBN 978-3-8316-4476-6.

28. DIN Deutsches Institut für Normung e. V. *Energy Audits. Part 3: Processes*; DIN Deutsches Institut für Normung e. V.: Berlin, Germany, 2014; DIN EN 16247-3:2014-08.

29. ISO International Standards Organization. *Energy Audits. Requirements with Guidance for Use*; ISO International Standards Organization: Geneva, Switzerland, 2014; ISO 50002:2014.

30. Abele, E.; Schraml, P.; Moog, D. Electric Load Management on Machine Tools. *Procedia CIRP* **2016**, *55*, 164–169. [CrossRef]

31. Abele, E.; Schraml, P.; Beck, M.; Flum, D.; Eisele, C. Two Practical Approaches to Assess the Energy Demand of Production Machines. In *Eco-Factories of the Future*; Thiede, S., Herrmann, C., Eds.; Springer International Publishing: Cham, Switzerland, 2019; pp. 127–146. ISBN 978-3-319-93729-8.

32. DIN Deutsches Institut für Normung e. V. *Referenzmodell zur Charakterisierung der Energieflexibilität von Industrieunternehmen*; DIN Deutsches Institut für Normung e. V.: Berlin, Germany, 2018; DIN SPEC 91366:2018-04.

33. Wang, Y.; Chen, Q.; Kang, C.; Xia, Q. Clustering of Electricity Consumption Behavior Dynamics, toward Big Data Applications. *IEEE Trans. Smart Grid* **2016**, *7*, 2437–2447. [CrossRef]

34. Richard, M.-A.; Fortin, H.; Poulin, A.; Leduc, M.-A.; Fournier, M. Daily load profiles clustering: A powerful tool for demand-side management in medium-sized industries. In Proceedings of the ACEEE Summer Study on Energy Efficiency in Industry, Denver, CO, USA, 15–18 August 2017; pp. 160–171.

35. Yilmaz, S.; Chambers, J.; Patel, M.K. Comparison of clustering approaches for domestic electricity load profile characterisation—Implications for demand-side management. *Energy* **2019**, *180*, 665–677. [CrossRef]

36. Chapman, S.N. *The Fundamentals of Production Planning and Control*; Pearson/Prentice Hall: Upper Saddle River, NJ, USA, 2006; ISBN 9780130176158.

37. Westkämper, E. *Einführung in die Organisation der Produktion*; Springer: Berlin/Heidelberg, Germany, 2006; ISBN 978-3-540-30764-8.

38. Schott, P.; Sedlmeir, J.; Strobel, N.; Weber, T.; Fridgen, G.; Abele, E. A Generic Data Model for Describing Flexibility in Power Markets. *Energies* **2019**, *12*, 1893. [CrossRef]

39. Grunewald, P.; Diakonova, M. Flexibility, dynamism and diversity in energy supply and demand: A critical review. *Energy Res. Soc. Sci.* **2018**, *38*, 58–66. [CrossRef]

40. Lund, P.D.; Lindgren, J.; Mikkola, J.; Salpakari, J. Review of energy system flexibility measures to enable high levels of variable renewable electricity. *Renew. Sustain. Energy Rev.* **2015**, *45*, 785–807. [CrossRef]
41. Industrial Electricity Prices across Europe 2018|Statista. Available online: https://www.statista.com/statistics/267068/industrial-electricity-prices-in-europe/ (accessed on 9 May 2020).
42. Tristan, A.; Emde, A.; Reisinger, M.; Stauch, M.; Sauer, A. Energieflexibilität im Industrial Smart Grid. *WT Werkstattstechnik Online* **2019**, *109*, 301–306.

Publisher's Note: MDPI stays neutral with regard to jurisdictional claims in published maps and institutional affiliations.

Review

Review of Control and Energy Management Approaches in Micro-Grid Systems

Abdellatif Elmouatamid [1,2,*], Radouane Ouladsine [1], Mohamed Bakhouya [1], Najib El Kamoun [2], Mohammed Khaidar [2] and Khalid Zine-Dine [3]

[1] College of Engineering and Architecture, LERMA Lab, International University of Rabat, 11100 Sala ElJadida, Morocco; radouane.ouladsine@uir.ac.ma (R.O.); mohamed.bakhouya@uir.ac.ma (M.B.)
[2] STIC Laboratory, Faculty of Sciences, CUR"EnR&SIE", Chouaïb Doukkali University, 24000 El Jadida, Morocco; elkamoun.n@ucd.ac.ma (N.E.K.); khaidar.m@ucd.ac.ma (M.K.)
[3] Faculty of Science, Mohammed V University, 10000 Rabat, Morocco; khalid.zinedine@um5.ac.ma
* Correspondence: abdellatif.elmouatamid@uir.ac.ma; Tel.: +212-649-966-652

Abstract: The demand for electricity is increased due to the development of the industry, the electrification of transport, the rise of household demand, and the increase in demand for digitally connected devices and air conditioning systems. For that, solutions and actions should be developed for greater consumers of electricity. For instance, MG (Micro-grid) buildings are one of the main consumers of electricity, and if they are correctly constructed, controlled, and operated, a significant energy saving can be attained. As a solution, hybrid RES (renewable energy source) systems are proposed, offering the possibility for simple consumers to be producers of electricity. This hybrid system contains different renewable generators connected to energy storage systems, making it possible to locally produce a part of energy in order to minimize the consumption from the utility grid. This work gives a concise state-of-the-art overview of the main control approaches for energy management in MG systems. Principally, this study is carried out in order to define the suitable control approach for MGs for energy management in buildings. A classification of approaches is also given in order to shed more light on the need for predictive control for energy management in MGs.

Keywords: control approaches; energy management; optimization method; objective function; control constraints

Citation: Elmouatamid, A.; Ouladsine, R.; Bakhouya, M.; El Kamoun, N.; Khaidar, M.; Zine-Dine, K. Review of Control and Energy Management Approaches in Micro-Grid Systems. *Energies* **2021**, *14*, 168. https://dx.doi.org/10.3390/en14010168

Received: 3 November 2020
Accepted: 11 December 2020
Published: 31 December 2020

Publisher's Note: MDPI stays neutral with regard to jurisdictional claims in published maps and institutional affiliations.

1. Introduction

Proper management of energy flow in MG (Micro-grid) systems must be carried out in order to improve the global performance of the system, to minimize the cost of the electrical bill, and to extend the lifetime of its components (e.g., converters, batteries, fuel cells). In general, energy management (EM) approaches involve an objective function, which could be used to maximize the efficiency of the hybrid RES system and to minimize energy consumption while improving the consumers' quality of services. For instance, an EM control strategy that considers only the availability of the electricity can be developed to switch, at each time, from RESs (renewable energy sources) to storage devices or to the utility grid without considering the electricity price or the profitability of the system. In other cases, control strategies can interact with the generators by limiting the power generation. The aim is to ensure the electrical quality of services and, consequently, minimize the profitability of the installation. However, despite the ability of these strategies to reach the defined objective, they might decrease the performance of other criteria, such as the batteries' lifetime, the system's installation cost, and profitability.

Actual commercial inverters provide high-performance energy balance by interconnecting RESs, energy storage systems, and the utility grid, taking into consideration only a single-objective function. This later is mainly implemented in order to increase the availability of the electricity for building's loads. With a limited configuration, the inverter can use batteries or the TEG at any moment without taking into account other constraints, such as the electricity cost and the C/D (charge/discharge) cycle of the batteries. For instance, high and frequent cycles of the C/D cycle of batteries could decrease their performance while reducing the system's profitability. EM strategies that are deployed in the actual inverters use "if-else" statements to perform real-time decisions. For instance, the defined setpoint values (i.e., control inputs) cannot be adjusted according to predictive variations of RESs production, load demand, and battery SoC (state of charge). Such EM strategies are considered as "passive strategy" in their decisions and actions [1]. Control strategies incorporating multiple-objective functions are therefore required for efficient energy management (i.e., ensuring electricity availability) while taking into consideration operational constraints (e.g., costs, reliability, and flexibility). In fact, "active strategies" for EM should be developed in order to adapt the setpoint values accordingly. These strategies could use intelligent and predictive control techniques together with recent IoT/Big-data technologies (e.g., data monitoring, data analysis, data mining, machine learning) for efficient EM in hybrid RES systems. In this work, control structures and strategies from the literature are presented by highlighting their advantages and drawbacks in the context of MG for smart buildings.

2. Control Architectures

In hybrid energetic systems or MG systems, distributed and hybrid RES generators (e.g., PV (photovoltaic) panels and wind turbines) are used to produce clean energy (e.g., solar, wind), while energy storage systems are installed to compensate the fluctuation between RESs generation and load consumption. These hybrid systems can either operate on grid-connected or standalone modes depending on desired and fixed objectives. However, while the penetration of these distributed generators is continuously growing, new energy management approaches are required for their seamless integration within existing electricity network. Table 1 presents resent literature works concerning the deployment of hybrid systems. As highly stated in Table 1, batteries are the most commonly used devices for energy storage.

Table 1. Survey through collection of EM (energy management) for the hybrid MG (Micro-grid) system.

Ref.	Grid	DG	PV	WT	Biomass	FC	Hydrogen	Battery	Diesel	Super-Capacitor	EV	Performance Evaluation
[2]		✓	✓	✓		✓		✓				The Multi-Objective Particle Swarm Optimization algorithm is used to improve electric energy utilization in remote areas. Simulation results are presented.
[3]			✓	✓		✓		✓				The development of a methodology for modeling and optimally sizing a hybrid system of RESs and two energy storage devices (hydrogen and batteries). Simulation results are presented.
[4]		✓	✓			✓	✓		✓			The Crow search algorithm is used to optimize and size a hybrid system. Two constraints are considered to minimize the total net cost: Loss of power supply probability and renewable energy portion. Simulation results are presented.
[5]	✓		✓	✓				✓				The operation of a grid-connected hybrid PV-wind system is performed using a standalone inverter capable of working in grid-connection mode and standalone mode. Experimental investigations are presented.

Table 1. *Cont.*

Ref.	Grid	DG	PV	WT	Biomass	FC	Hydrogen	Battery	Diesel	Super-Capacitor	EV	Performance Evaluation
[6]								✓		✓	✓	The work proposed a real-time EM control strategy combining wavelet transform, neural network, and fuzzy logic methods. Experimental results exposed that the power variation and the peak power of the battery pack have been successfully suppressed.
[7]						✓		✓		✓	✓	An intelligent control strategy is developed for a hybrid energy storage system, composed of fuel cell, battery, and super capacitor. Multi-input/multi-output state-space model is used to perform the study. Simulation results are presented.
[8]	✓	✓	✓	✓				✓				A multi-objective optimization problem, over a receding control horizon, is used for energy storage dispatch and sharing of renewable energy resources in a network of grid-connected MG. The multi-objective optimization is formulated as a lexicographic program to allow preferential treatment of multiple MG. Simulation results are presented.
[9]	✓	✓	✓	✓	✓			✓				An economic linear programming model is developed with a sliding-time-window to assess design and scheduling of biomass, combined heat and power-based MG systems. Simulation results are presented.
[10]	✓	✓	✓	✓				✓				Distribution network including RESs is studied for optimal dispatch model of mixed-power generation by considering the charging/discharging scheduling of battery. Bee-colony-optimization method is proposed to solve the daily economic dispatch of MG systems. Simulation results are presented.
[11]		✓				✓	✓	✓				A combined sizing and EM methodology is proposed and formulated as a leader-follower problem. The leader problem focuses on sizing and aims at selecting the optimal size for the MG components. It is solved using a genetic algorithm. Simulation results are presented.
[12]			✓	✓	✓				✓			A strategy for the optimal management of a multi-good standalone MG integrated with RES is investigated. The proposed approach is defined through an EM model able to determine the schedule of each programmable unit to fulfil the community needs at the lowest operation cost. Simulation results are presented.
[13]	✓	✓	✓	✓		✓		✓			✓	Electrical vehicles are used for peak shaving and load curve correction in a MG system. The deployed methods deal with the simultaneous scheduling of electrical vehicles and reactive loads in order to minimize operation cost and emission in presence of RES in MG system. Simulation results are presented.
[14]		✓	✓			✓		✓				A power management system is presented to manage the power output from RES, fuel cell, and batteries with delivery of hydrogen from an electrolyzer. The deployed strategy handles the source effectively by considering the limited lifecycle of storage devices. It eliminates the need for a dump load in the MG when the storage devices are charged to the maximum capacity. Simulation results are presented.

Therefore, the deployment of an energy management approach should be able to enhance the dynamic response of distributed energy resources under different operating conditions and maximize the usage of RES power generation while ensuring the stability when one or more sources are connected or disconnected into/from the system. In this way, different approaches from literature have been proposed for EM (Table 2). As shown in Table 2, the most suitable control strategies could be selected according to fixed constraints and objective functions. These control strategies can be classified into three main categories: Centralized, decentralized, and hierarchal control, as mentioned in Figure 1. These control strategies are presented in the rest of this section.

Table 2. Survey through collection of EM for the hybrid MG system.

Ref.	Main Objective	EM Approach	MG Scale	Control Structure
[3]	A methodology for modeling and optimally sizing a hybrid system for renewable energy considering two energy storage devices: Hydrogen and batteries.	Wavelet transform, Neural network and Fuzzy logic (FL)	Large	Not specified
[4]	A method is developed to size an off-grid PV/diesel/FC hybrid energy system in order to optimize the number of system components with respect to the cost minimization of the installation.	Crow search algorithm	Large	Hierarchical
[8]	An EM method is deployed in a MG system containing energy storage devices and renewable energy based distributed generators in grid-connected MG. In the studied approach, the neighboring MG share the capacity of their distributed resources and energy storage devices aiming at reducing the operational costs.	Lexicographic programming, Linear programming, Receding horizon control	Large	Hierarchical
[9]	A deterministic constrained optimization and stochastic optimization approaches to estimate the uncertainties in biomass-integrated MG supplying both heat and electricity. The work developed an economic linear programming model with a sliding time window to assess design, scheduling of biomass-combined power and heat-based MG systems.	Linear programming model with a sliding time window	Small	Decentralized
[10]	A MG energy management strategy by considering RES integration into the distribution network. The time-of-use, other technical constraints, and an enhanced bee colony optimization is proposed to solve the daily economic dispatch of MG systems.	Enhanced bee colony optimization	Small	Centralized
[11]	Authors proposed a combined EM and sizing methodology, formulated as a leader follower problem. The leader problem focuses on sizing and aims at selecting the optimal size for the MG components. The problem is solved using a genetic algorithm and the follower problem is formulated as a unit commitment problem and is solved with a mixed integer linear program.	Mixed integer linear program	Small	Centralized
[14]	Authors proposed an EM approach to divert excess energy of PV to the electrolyzer.	Linear Programming	Small	Centralized
[15]	An analysis of energy management system of a MG using a robust optimization taking the uncertainties of wind power and solar power generations and energy consumption into consideration.	Agent-based modelling	Large	Decentralized

Table 2. *Cont.*

Ref.	Main Objective	EM Approach	MG Scale	Control Structure
[16]	An algorithm for EM system of a MG using multi-layer ant colony approach pointing on determining the optimum point of operation for local distributed energy generation with least electricity production cost. The studied algorithm has the capability of analyzing the constraints related to economic and technical aspects of the problem.	Multi-layer ant colony approach	Medium	Not specified
[17]	A method known as contingency-based energy management for a system of MGs. A stochastic optimization is proposed according to various scenarios of the contingencies.	Contingency-based energy management	Large	Hierarchical
[18]	A fuzzy EM approach is deployed to smooth the power flow of a MG containing heat and power unit. The aims is to use the surplus of electrical power of the MG for storing in electrical energy storage systems and ensuring the water temperature of the thermal storage system in the desired value in order to supply residential buildings.	Fuzzy energy management strategy	Medium	Not specified
[19]	A model predictive control technique to determine the optimal operation of the MG system using an extended horizon of evaluation and recourse. The EM problem is decomposed into Unit Commitment and Optimal Power Flow problems in order to avoid a mixed-integer non-linear formulation.	Model predictive control	Large	Centralized
[20]	Authors present an EM system to minimize the daily operating cost of a MG and maximize the self-consumption of the deployed RES by selecting the best setting for a central battery storage system based on a defined cost function.	Convex Programming, Model Predictive, and Rolling Horizon	Medium	Hierarchical
[21]	The operating cost of MG is minimized, while considering droop controlled active and reactive power dispatch of AC side MG as a constraint.	Mixed integer nonlinear programming	Small	Centralized

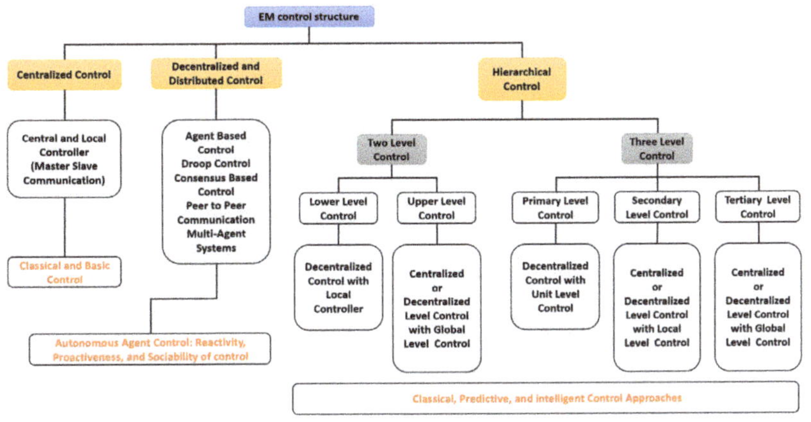

Figure 1. Control structure for energy management in MG systems.

2.1. Centralized Control

Centralized control approaches use a single central controller (CC), which is characterized by a high-performance computing unit and a secure communication infrastructure in order to manage different entities of the system (e.g., RESs, storage systems, TEG). Each entity uses a local controller (LC) in order to communicate and directly interact with the CC. Moreover, using recent communication and computing technologies (e.g., IoT, Big-Data), the CC is able to monitor, collect, and analyze real-time data. This allows all entities to collaborate with the central EM controller while ensuring a flexible MG operation in both grid-connected and standalone mode (Figure 2). The CC collects data, such as RES energy production, energy consumption pattern, the energy price from market operators, and weather conditions, and then executes the optimal and efficient system's control.

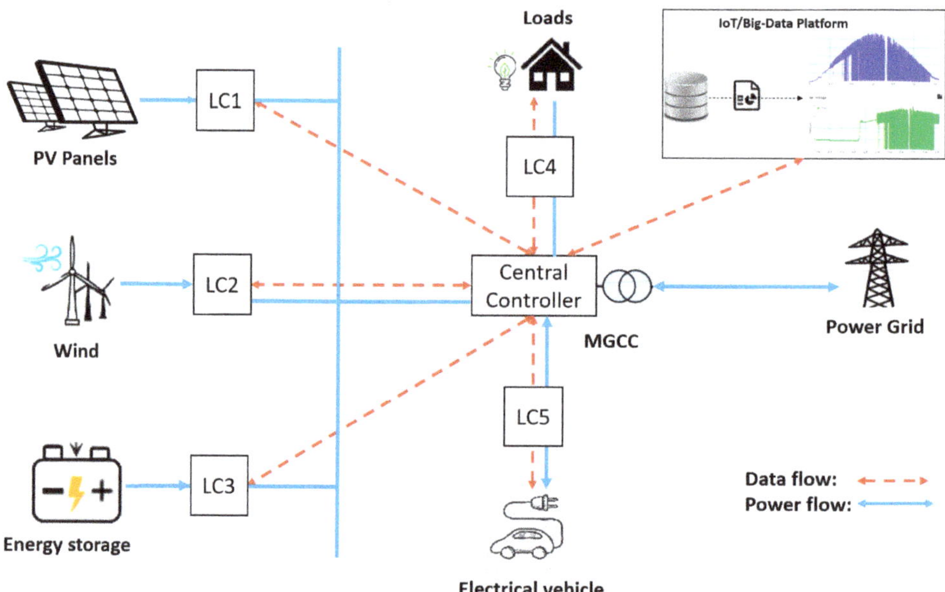

Figure 2. Centralized control structure.

Numerous research works have developed and deployed centralized EM strategies. For instance, the authors of [22] proposed a centralized controller in order to optimize the operation of MG by maximizing the production of distributed RESs generators while establishing back-and-forth energy transfer with the main utility grid. The efficiency of the proposed solution on MG system was investigated by considering a typical case network operating under various market policies and spot market prices. Moreover, the authors of [19] developed a centralized EM system for a standalone MG system based on the model predictive control method in order to reduce the computational loads. In fact, the studied problem was solved iteratively by nonlinear programming (NLP) and mixed integer linear programming (MILP) techniques. Other centralized control strategies are summarized in Table 2. However, despite the ease of implementing the centralized strategies, they have shown their limits, especially when dealing with large-scale hybrid systems [23].

2.2. Decentralized Control

Unlike centralized strategies, in decentralized control, each entity is considered autonomous using a LC. This means that groups of entities are controlled separately by a leader. In literature, the terms 'decentralized' and 'distributed controls' are often used in place of each other [24,25]. The distributed control can be considered as a decentralized control in which LCs use local measurements, such as frequency and voltage values, to elect the leader entity. They are also allowed to share information with neighbors. For a distributed control, LCs do not only use local measurements but also are able to send and receive required information to other LCs [26]. In decentralized control approaches, limited local connections are required and the control decisions are made based only on local measurements (Figure 3). It does not require a high-performance computing unit and a high-level connectivity [27].

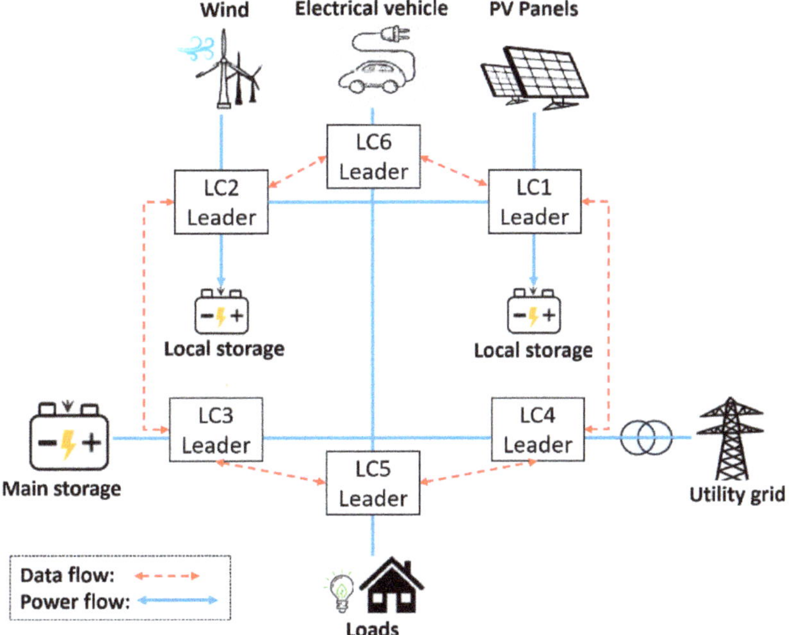

Figure 3. Decentralized control structure.

As depicted in Figure 3, each LC operates individually on managed energy sources, storage systems, and loads without central control. The control decisions are determined locally based on local measurements, which are shared among controllers using peer-to-peer communication.

However, monitoring, processing, and data visualization is considered critical in order to coordinate various distributed controllers and achieve a global operation goal. This process is standardized by the norm IEC-61968 for a single-building energy management system and by IEC-61850 for interoperability between building MG systems [28,29]. Depending on the communication network availability, the decentralized control can be classified into three operation modes: (i) Fully dependent, in which the distributed controllers generate local control decision while communicating information with each other via a CC; (ii) partially independent, in which LCs communicate with each other and share information with the CC in order to generate central decisions; and (iii) fully independent, in which the distributed controllers communicate directly with each other and independently from the CC [30]. However, despite the flexibility of these operational modes, the decentralized control structure presents low performance compared to centralized control [25,31–33]. This is due to the low response time and the incomplete information about the total MG system installation.

2.3. Hierarchical Control

Hierarchical control is mainly proposed for SG (smart grid) systems. In fact, the extended geographic areas of these systems and the extensive communication and computation requirements make the implementation of fully centralized approaches a difficult task. At the same time, higher coupling between the different LCs requires a maximum level of coordination, which cannot be achieved by decentralized control structures. However, a compromise between the fully centralized and decentralized control structures is realized by providing hierarchical control structures [34,35] according to three control levels: Primary, secondary, and tertiary, as depicted in Figure 4.

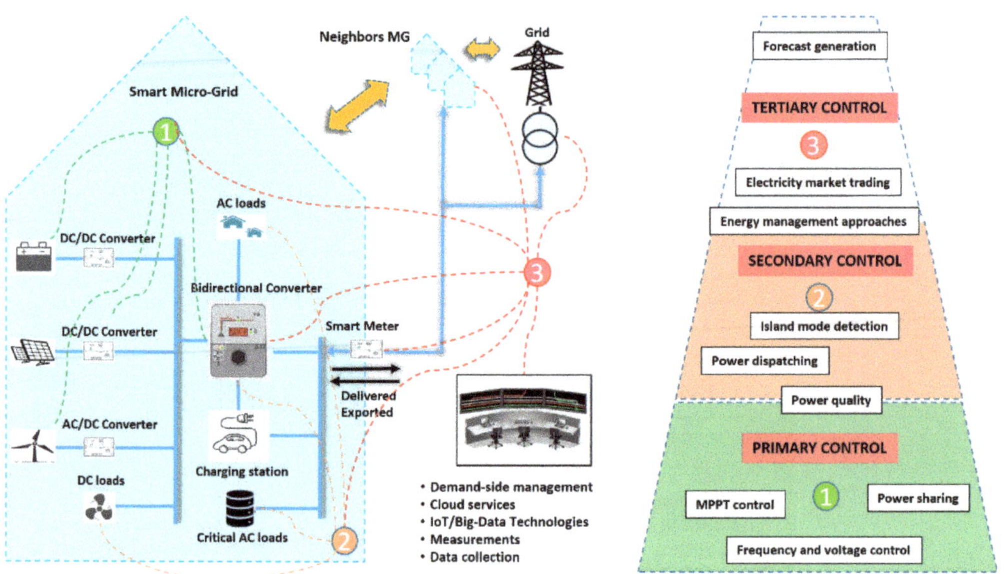

Figure 4. Hierarchical control structure.

The primary control level stabilizes the voltage and frequency generated from each source in order to respect the limits required by the standards [36–38]. In addition, the primary control level detects the operating mode of MG systems, offering the ability to operate in grid-connected and standalone modes [39]. For the secondary control level, the MG voltage and frequency are restored after system's load variation. The aim is to ensure and enhance the power quality within the required standards values, allowing the synchronization between the MG systems and the main electrical network [40].

The main objectives of tertiary control are the power flow control in the grid-connected mode, ensuring then the optimal operation in both modes like capacitance and inductance [41]. Figure 5 includes the structures of each level of the hierarchical control. The control levels differ in the response time frame speed in which they operate as well as the infrastructure requirements, especially for the communication, which is normalized by the standards IEC 61850-7-420 and EN13757-4 [36]. The hierarchical control can be implemented in parallel in both centralized and distributed structure. The advantages and disadvantages of each control structure are presented in Table 3.

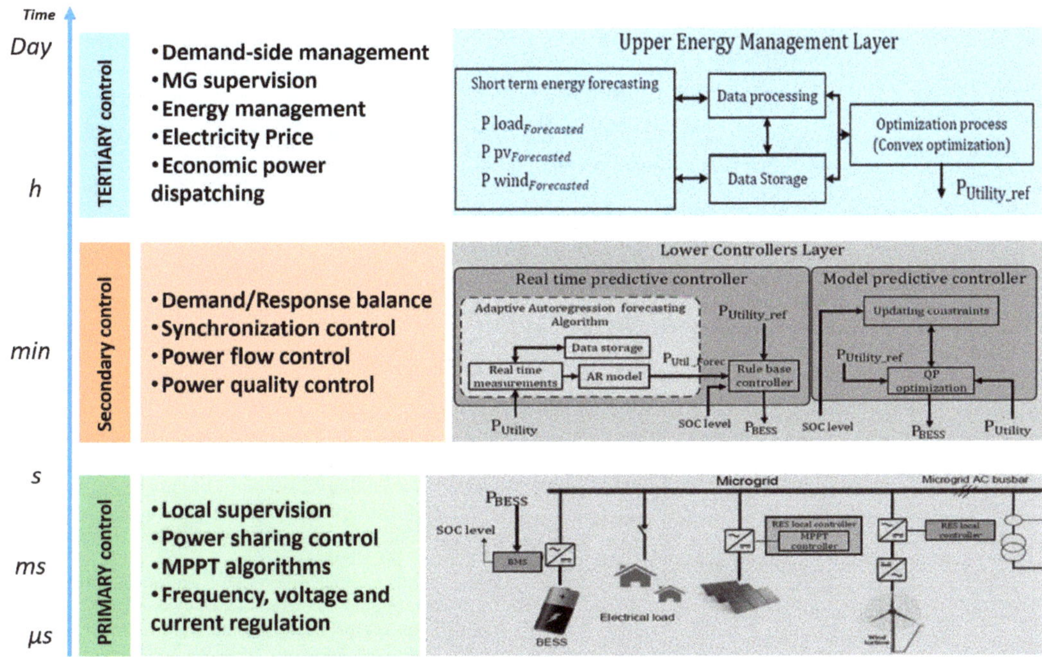

Figure 5. Hierarchical control levels.

Table 3. Control architectures for hybrid system, advantages and inconveniences.

EM	Advantages	Inconveniences
Centralized	<ul><li>Strong controllability and real-time observability of the whole MG system;</li><li>Provides strong supervision and wide control of the whole system;</li><li>Mature and established approaches for control of many systems;</li><li>Suitable for small size MG systems where the collected information is performed by low bandwidths communication [42];</li><li>Suitable for the internal control in MG system;</li><li>Global optimization of all entities of the same MG;</li><li>Offers high-performance computing unit and a secure communication infrastructure;</li><li>Holds the control strategy that considers the MG entirely and depends on the simple architecture of the system to build a global knowledge making the EM control easier to be deployed;</li><li>Straightforward implementation, the CC allows economic implementation and it is easy to maintain;</li><li>Optimal decision is guaranteed.</li></ul>	<ul><li>The failure of the CC affects the whole system operation;</li><li>Heavy computation burden is a technical barrier for the deployment;</li><li>Not well designed to support plug-and-play functionalities of a large number of entities;</li><li>Need a high level of connectivity due to the direct interaction of each entities with the central;</li><li>Requiring high processing unit for the CC;</li><li>More prone to failures since only one unit regulates the voltage and leads to reduce life spam of Battery bank stack [43];</li><li>Poor scalability and responsible for shorter battery life [44];</li><li>Since all information is collected and handled at one CC, the computational burden increases making the control less effective for real-time communication requirements;</li><li>Reliability is degraded for the whole system.</li></ul>

Table 3. *Cont.*

EM	Advantages	Inconveniences
Decentralized	• Distributed processing system with autonomous control capability; • Peer-to-peer nodes communication, allowing greater flexibility of operation, and avoiding single-point failure; • Higher reliability due to the redundancy of controllers and communication; • Distributed generators are controlled by independent controllers through their local variables offering redundancy communication link; • Insufficient information about other entities of the MG systems; • Droop control strategy is usually used to avoid circulating currents between the converters without the use of digital communication link; • Avoiding single-point failure, enhancing the expandability, and allowing greater flexibility of operation; • High privacy for the entities and less amount of information; • Reduction of the computational need and releasing the traffic on the communication network; • Reduces computational burden and increases reliability and robustness; • Easy realization of plug-and-play functionality.	• Incomplete information about the overall MG status; • Voltages and currents average regulation requires more data transmission through the MG; • Local optimization in EMS is not able to provide a global solution for operating cost minimization of the total MG; • The distributed processing does not guarantee global optimal results for the whole MG system; • A high complexity of implementation compared to centralized and hierarchical control; • Load dependency problem, responsible for the circulating currents in distributed generators, accuracy of load sharing can be achieved with the compromise of deviation in the voltages compared to their rated values; • Unsuitability for non-linear loads due to harmonics and inability to achieve coordinated performance of multiple components with different characteristics, and poor transient performance; • Requires effective synchronization and strong communication to achieve synchronicity; • Requires fast periodical reconfiguration.
Hierarchical	• More suitable for DC MG systems; • The voltage and the current are regulated locally by the source converters; • Flexible regulation of the system voltage within acceptable intervals; • Economic power dispatch among the converters, between the MG, the utility grid as well as the neighboring-MG; • Synchronous generators with the same frequency for all over the grid; • The operation constraints are dispatched to different levels reducing the processing time; • Improving the current mismatches among the controllers; • Combining the previous control structures; • Optimal decision is possible.	• The distributed generators should participate in voltage regulation and frequency control; • Some generators operate in limited power mode while supplying only the power planned by the electricity market; • The distributed generators are responsible for adjusting the differences between the planned demand and the actual load. Therefore, the demand should be forecasted to plan correctly the output of the generators; • Adjacent layers coordination is required; • There is no transfer of information and energy if there is a communication fault in the upper layer; • Fewer computation burdens.

3. Control Strategies

The deployment of more than one energy source in MG systems requires the use of efficient control strategies/approaches for managing energy flow. This requires the development and deployment of EM systems. EM systems should be able to effectively coordinate energy sharing and trading among all electrical networks while supplying loads according to the operational conditions and economic constraints with secure, reliable, and efficient power system operation. In fact, optimization techniques for D/R, demand-side management, and power quality management are needed to achieve different EM system objectives while satisfying multiple constraints, such as electricity price minimization and occupants' comfort maximization, as mentioned in Figure 6.

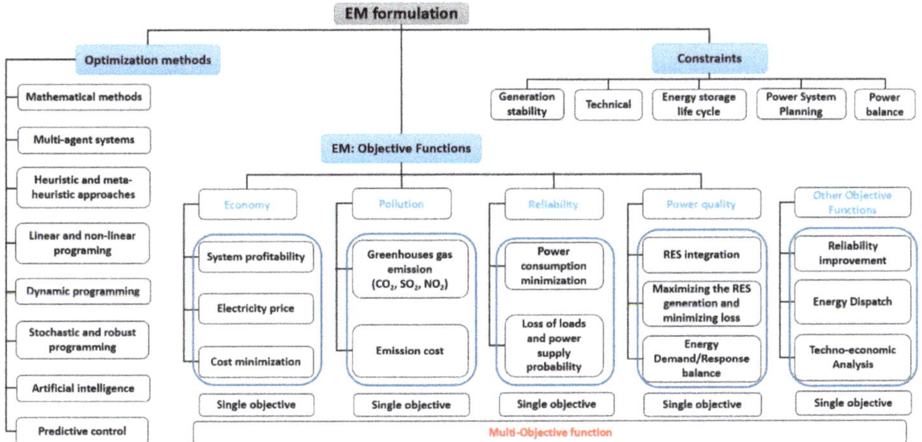

Figure 6. Objective functions constraints, and optimization methods for optimum operation of MG systems.

The concept of EM system is not new and began with the first electrical network, known as "Energy Control Center." In the past decade, the electrical network has been developed and new challenges have been evolved. Consequently, new ICTs (information and communication technologies) have been deployed in order to improve the electrical power sector.

The EM system was also developed to be renamed as a SCADA-EMS (supervisory control and data acquisition-energy management system), which is charged to deploy various control techniques like services control, distributed management systems, and demand-side management [33]. With the deployment of RESs, the EM system should be capable of creating an energy balance between the variable demand and the stochastic RES generation in an efficient manner. It could have a control center, which is capable of supervising, monitoring, managing, and optimizing the operation of distributed generators, diversified consumers, and the transport/distribution facility of the electricity. Actually, the EM system is not limited to the classical control objective, but has been developed to work for real-time applications, predictive control scheduling, and transmission security management.

Several approaches have been proposed and have used diversified objective functions and constraints together with optimization methods for efficient energy management, as depicted in Figure 6.

3.1. Objective Functions and Constraints

The deployment of EM control strategies specifies the main objective functions, which could be related to the operation cost, pollution, reliability, and power quality [11,45–47]. For instance, the main aim of using economic objective functions is to minimize the electricity price. Different formulations have been studied for cost minimization in MGs. For instance, the authors of [48] an EM strategy for electricity cost minimization in residential MG, which was constituted by multiple households with distributed energy resources. This EM strategy considered predefined purchasing/selling decisions, at each time slot, for reducing the electricity cost as well scheduling decisions for the shifted loads. The authors of [49] formulated the cost minimization as a dynamic economic load dispatch problem. A metaheuristic algorithm was introduced and compared with other methods, such as the differential evolution algorithm, genetic algorithm, and particle swarm optimization. The authors of [50] proposed an optimal strategy by evaluating the performance of different hybrid MG systems. A mathematical model was studied for sizing the component of the MG in order to meet the lowest possible cost while maximizing load demand under varying weather conditions. The obtained results presented the optimal configuration for MG system components to achieve the lowest cost of energy and net present cost. In addition, the dynamic analysis showed that, in order to reduce the voltage-drop during disturbances, it is essential to carefully install the sources in the buses connected to high energy demand. The authors of [20] presented an EM system to minimize the daily operating cost of a MG while maximizing the self-consumption of the deployed RES by selecting the best setting for a central battery storage system according to a defined cost function. A simple comparison was made to show the advantages of two different layer controllers: The rolling horizon predictive controller and modem predictive controller. The experimental results showed the performance of the proposed strategy to work in real-time with high accuracy. The yearly RES self-consumption and the yearly operation cost of the MG were calculated with and without the rolling horizon, showing the utility of the method to minimize the cost. Another interesting work was presented by the authors of [51], who introduced an optimization model for managing a residential MG which contained RESs and a charging spot with a "vehicle-to-grid" system. In this EM system, not only were energy costs considered, but battery installation costs were also introduced in the system minimization.

The deployment of EM approaches, which consider the pollution factor as an objective function, take time to validate, since the whole procedure should consider the life cycle of the different deployed equipment. In fact, every new energy source technology which is promoted as being "renewable" or "sustainable" is subject to an energy balance analysis in order to calculate the net energy yield. The energy analysis does not only consider the data for present generation systems, but also the data for the probable improvements in production and energy system technology [52]. The equivalent CO_2, generated during the fabrication of each component, should be calculated and compared to the equivalent energy which is generated during its life cycle. We consider that this energy is generated by traditional sources in order to estimate the equivalent CO_2 emission and that, by comparing these two elements of CO_2 generation, the profitability of the system concerning the pollution objective can be defined. For example, the authors of [53] studied the life cycle of the balance system component of 3.5 MW_p multi-crystalline PV installation. The life cycle and the boundary conditions were calculated for each component of a PV installation (e.g., PV metal support, aluminum frames). The authors of [52] presented estimations of the energy requirements for manufacturing PV systems and evaluated the energy balance for an example of PV system applications. The work investigated the effects of the future developments in PV generation technology in order to assess the long-term predictions of PV system as a candidate for a sustainable energy supply and for CO_2 mitigation. The authors considered the energy payback time to estimate the CO_2 mitigation potential and concluded that 90% of greenhouse gas emissions during the PV system life cycle are caused by the energy used during system manufacturing and not during the system operation.

Like economic and pollution aspects, the term 'reliability' covers different aspects concerning the system operation cost, profitability, fails and maintenance, and productivity. Consequently, as mentioned above, RESs have a significant cost and consume a lot of energy in their fabrication. In order to maximize the profitability and system's reliability, the production of these sources should be maximized. Therefore, the main aim is to maximize the use of renewable energy generation, minimizing the loss of energy, keeping the storage energy system at a good state of health, and ensuring a safety and efficient supply of energy to the loads. In this way, the authors of [54] presented an electricity market strategy for reliability enhancement of islanded multi-MG systems. A techno-economical objective function was deployed to account the profit of MG owners and to enhance the reliability of the system as well. Distribution functions were used for the probabilistic modeling of RESs and loads, and an electricity market strategy was proposed to improve the profit of the MG owners. However, the power quality, particularly the power loss, is still a main issue for the system's reliability. Therefore, several works have proposed suitable EM methods and control techniques to minimize the power loss in MG systems. For instance, the authors of [55] integrated a MG with static synchronous compensator controller in order to ensure the higher power flow with enhanced voltage profile and reduced power loss. They concluded that the static synchronous compensator controller raises the capacity of the distribution line and contributes to voltage profile improvements and power loss reduction. Similar works have considered the concept of power loss minimization, such as those presented by the authors of [56–58]. Several objective function can be considered for the deployment of the EM strategies. The reliability improvement is a noticeable task in modern power systems due to its direct influence on the electricity price and more precisely social safety [59]. The auhors of [59] studied an approach for optimal operation of distribution networks. A hybrid algorithm (Grey-Wolf Optimizer and Particle Swarm Optimization) was proposed to solve the proposed multi-objective function. The results were compared with those presented in literature works to demonstrate the powerful of the proposed algorithm. A beneficial literature work for multi-objective EM was improved by the authors of [60], who studied a multi-objective EM in an MG system. Techno-economic analysis and energy dispatch were presented for standalone and grid-connected MG infrastructure with hybrid RESs and storage devices.

After defining the system's constraints and objective functions, suitable optimization methods are required to accordingly ensure the exchange of power flow between the installed RES/storage and the MGs on the one hand, and between MGs and the utility grid on the other hand. The rest of this section is dedicated to an overview of main methods from literature.

3.2. Optimization and Control Methods

Numerous research works have been carried out for MG control according to system's topologies, structures, and operation modes [33,61,62]. For example, optimization and control methods should manage the stochastic nature of the installed RES generators by ensuring a reliable supply of power to consumers while keeping the storage system, electricity bill, and occupants' comfort at the acceptable operation conditions. Figure 7 presents a proposed classification of the MG control methods commonly used in MG operations. A brief description of each method is presented in the rest of this section. Furthermore, various steps should be specified, as depicted in Figure 8, for EM in MG.

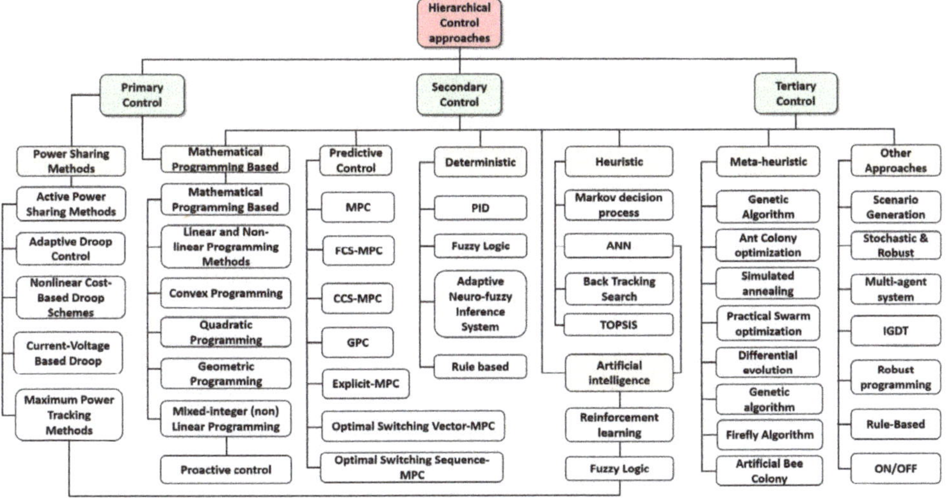

Figure 7. Control approaches for energy management systems.

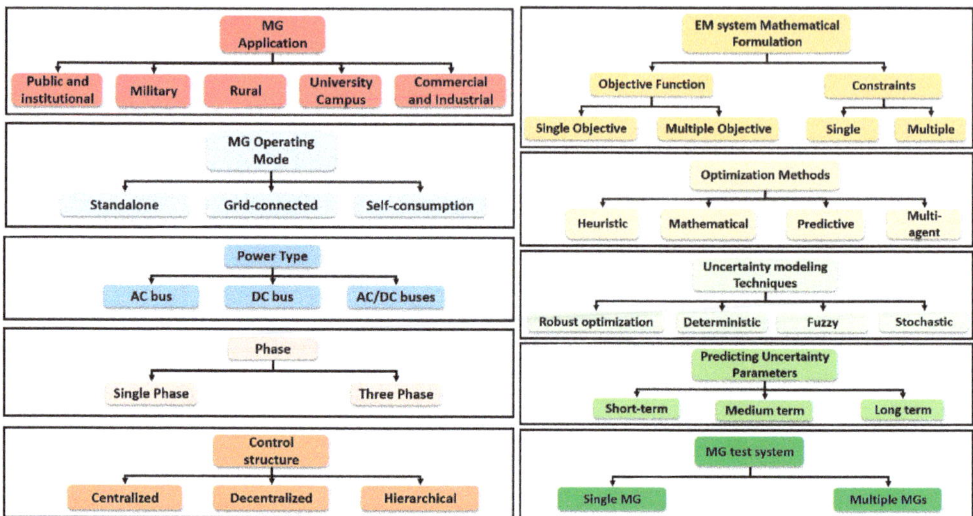

Figure 8. MG and EM system specification and underlying construction steps.

3.2.1. Predictive Control Methods

Recently, predictive control approaches have been proposed for advanced systems control according to defined constraints with the aim of developing predictive controllers for efficient energy flow in MG systems. These controllers could forecast future actions and decisions, but they require forecasted inputs' values (e.g., power consumption/production). With recent progress in IoT and Big-data technologies, together with ML, it now possible to deploy sensors for gathering contextual data [63]. These data could be processed and used for predicting n-step-ahead values. Therefore, the forecasted values are the main inputs for generating the most suitable and future actions by predictive control approaches [64,65].

MPC and GPC are the well-known approaches, having the capabilities of predicting future events and forecasting right control decisions accordingly. In fact, they have the ability to incorporate optimization mechanisms, which makes it possible to integrate system's constraints and disturbances in forecasted control decisions. For instance, the GPC is widely used in advanced control applications, such as in EM and buildings' automation systems [66,67]. For example, the authors of [68] introduced a home EM system for battery storage and PV systems. For the optimal operation strategy, the proposed planning was expressed as a stochastic mixed-integer nonlinear programming. The power generated by the PV system was considered as an uncertain parameter and modeled by a probability distribution function. The battery storage system was used to store energy during off-peak/low-cost hours and discharge energy during on-peak/high-cost hours. However, the main limitation of this EM strategy was the passive reaction of the system with the cost and the peak demand variability. It was programmed by a fixed time interval that presented predefined periods of on-peak and high-cost and was not defined by an active function for the interactive variability of the cost and the electricity demand. Moreover, the authors of [67] proposed an adaptive and dynamic optimization technique based on the stochastic MPC approach. The proposed EM approach was applied for distributed energy resources scheduling problem for a set of smart homes with different sources of energy. Its aim was to address the uncertainty and variability issues of the PV power generation. This study was designed for large-scale smart houses by taking into consideration their cooperation with their neighbors. Another interesting work was presented by the authors of [69], who proposed an EM system using an MPC, where a simple state-space model was used for the performance modeling of a MG system. This work considered the RES power production and the consumption as measured disturbances parameters for the EM system. Therefore, the storage systems and the cost were modeled as constraints for the MG system, which were solved by the state-space equations. In addition, other works have been presented in the literature which have referred to the optimal control of RES in MG systems considering hybrid storage systems, as detailed by the authors of [70]. The authors of [71] used the MPC for optimal control of distributed energy resources with a battery storage system. A mixed-logical framework was applied to model the deployed household system. In other works, the MPC was used for EM of MG systems that were connected to the charging station for electrical vehicles [72–74]. The authors of [72] used an algorithm based on the MPC model for the economic optimization of an MG laboratory. The laboratory contained a hybrid storage system composed of hydrogen storage and battery bank with a connection to the utility grid and a charging station for electric vehicles. A hierarchical control structure was proposed together with the MPC method, which operated at different timescales. The proposed methods operated on the first level to maintain the MG stability and on the second level in order to perform the management of electricity purchase and sale to the utility grid, manage the use of energy storages, and maximize the use of RESs. The presented results showed the reliable operation of the proposed control algorithm to manage the MG system. The authors of [73] proposed an optimal EM approach based on the MPC controller for the MG with external agents, including battery storage system and fuel cell electric vehicles. The MPC problems were solved by a mixed-integer quadratic programming. The Mixed Logic Dynamic framework was used to model the plant, and the operation and degradation costs were included in the objective function. The proposed approach considered the best time period in to recharge/refuel the vehicle, finding lower prices for the recharge of the vehicle battery or the refueling of the vehicle fuel cell if they were planned before the day-ahead market session. Therefore, generic MPC models were introduced by the authors of [75,76] for economic optimization in MG systems. The authors of [75] presented mathematical optimization models of residential energy hubs. The model can be readily integrated into household automation systems and EM systems to improve their effectiveness and reduce the total energy costs and emissions while considering their preferences and comfort. Mathematical models of major household demands have been developed. The authors of [76] developed an MPC approach to optimize an MG system's operation. A mixed-integer-linear framework was illustrated, which included economic dispatch, energy storage, unit commitment, and grid interaction. The cost was addressed and parameterized in detail in the problem formulation. The experimental results were presented, showing the performance of the proposed approach to save money compared to the current practice.

It is worth noting that the MPC family was proposed for electronic power, especially power converter control. The GPC is one of the CCS-MPC (Continuous Control Set MPC) methods that calculate a continuous control command in order to generate the desired output of the power converter. The CCS-MPC models have a lower computational cost than the other existing methods, such as the FCS-MPC (Finite Control Set), OSV-MPC (Optimal Switching Vector), and OSS-MPC (Optimal Switching Sequence) [77]. It can be used for long predictive horizon problems by calculating the control actions beforehand and then limiting the online computation burden. Mainly, the calculation time is the main factor for the deployment of MPC control families. In past decades, the development of computing units and the integration of ICTs and ML algorithms for power electronic applications has encouraged the use of predictive control for the power converter. For instance, the authors of [78,79] used an FCS-MPC for the current control of three-phase inverter. The authors of studied this in [80] for a multiphase inverter, the authors of [81,82] for a multilevel inverter, and the authors of [83,84] for a matrix converter. For more details, we refer readers to an interesting review, which is related to predictive control applications in power electronics [85]. These approaches offer the possibility to integrate multiple-objective functions and constraints with the possibility of integration in the different control levels. Mainly, with the integration of the new ICT, the predictive control can be developed to present high performance for control command and action predictions. In addition, the use of ML algorithms to forecast the control input parameters offers more reliability and flexibility to the predictive control approaches.

3.2.2. Classical Approaches

Many EM optimization approaches are based on classical approaches, such as mixed-integer linear and nonlinear programming. These approaches can be considered as efficient methods for MG systems control according to the specified objective and constraints. For instance, the authors of [86] proposed a MG EM system for power sharing, power trading with the main grid, continuous run, and on/off mixed mode based on the linear programming optimization method. In this study, the on/off mode was solved by a MILP solution approach, which optimized the operation of MG with respect to the operation mode of the main grid, fuel cell, and energy storage system. The authors of [87] developed a real-coded genetic algorithm and a MILP-based method to schedule the unit commitment and economic dispatch of MG units. The work considered the voltages limits, equipment loadings, and unit constraints in its formulation, and the proposed algorithm deployed a flexible set of sub-functions and intelligent convergence behavior, as well as diversified searching approaches and penalty methods for constraint violations. At the same, a method was investigated to deal with the constraints of MILP algorithm in handling the nonlinear network topology constraints. Another interesting work was presented by the authors of [88], who proposed an MILP-based approach for managing electrical and heat demands in a multiple MG environment. The proposed strategy considered different energy converters and storages, distributed energy generators, and electricity/heat storage units for an optimal scheduling of MG, including technical and economic ties between electricity and natural gas systems. The deployed algorithm was developed based on AC power flow, while the deployed model respected reactive power and voltage security constraints, allowing the MG system to minimize the operation cost. Moreover, several other works have been presented using these approaches. For example, the authors of [21] minimized the operating cost of MG using MINLP, while considering, as a constraint, droop controlled active and reactive power dispatch of AC side MG. The authors of [89] proposed an EM approach for MG under an operation system of transformer nominal operation and voltage security. Three objective functions, customer benefits, load leveling, and network losses, were studied.

Generally, the objective function and constraints deployed in linear programming methods are linear functions with whole-valued and real-valued decision variables. This family of approaches is often used for system analysis and optimization, as it presents a flexible and powerful method for solving large and complex problems, such as distributed generation and MG systems.

Dynamic programming methods are used to solve more complex problems that can be sequenced and discretized. The studied problems are usually fragmented into sub-problems that are optimally solved, while the obtained solutions are superimposed to develop an optimal solution for the original problem [90]. Therefore, rule-based methods are generally used to implement the EM system because they do not require any future data profile to make a decision, thus making them more suitable for real-time applications. For example, the authors of [91] presented a rule-based EM system in which a rule-based algorithm was used to implement the priority of RES usage and manage the power flow of the proposed MG components. A nature-inspired optimization algorithm was used to optimize the MG system's operations for long-term capacity planning. The main goal of the proposed objective function was to minimize the cost of energy in MG systems as well as the deficiency of power supply probability. Other works have proposed rule-based methods to control and optimize the energy flow in MG systems. For example, the authors of [92] developed a control algorithm to provide power compatibility and EM for different resources in the MG. A real-time control system was used to experimentally validate the hybrid system in the MG. The results showed that the proposed approach provided stable operation of the MG subsystems under various power generation and consumption conditions. The authors of [93] studied a method to build the optimal EM for MG- connected system, which included the energy trading cost with the main grid and the battery aging cost. The authors used a dynamic programming algorithm to minimize the cash flow of the system while maximizing the power supply from the main grid.

Like other classical methods, dynamic programming algorithms can be considered as mathematical optimization methods, which can be used to simplify a complicated problem to simpler sub-problems for being solved in a recursive manner. They are able to provide optimal decisions. However, they require high computational costs, which make them difficult to implement in embedded devises.

3.2.3. Heuristic and Metaheuristic Approaches

Heuristic and metaheuristic approaches are used in many disciplines, such as in telecommunications and transportation systems. Recent studies have developed EM approaches for MG systems. For instance, the authors of [94] introduced a heuristic method for the optimal operation and EM of DC MG systems. The studied problem was formulated in the form of a single-objective optimization problem by focusing only on cost minimization. The authors of [95] proposed a metaheuristic based system by integrating the Harmony search algorithm and the enhanced differential evolution. To ensure that the power consumption did not exceed a fixed threshold value during peak periods, multiple knapsacks were used, and the proposed system outperformed the existing metaheuristic techniques in terms of cost and peak-to-average ratio. The authors of [96] proposed an economical model for energy storage system together with a real coded-genetic algorithm model for MG systems operating in a grid-connected mode. The developed algorithm maximized the present cost of energy storage system over its lifespan based on its capital, energy arbitrage revenue, operation cost, and maintenance cost. The authors of [97] proposed an optimal EM system for a grid-connected MG system based on the genetic algorithm, which considered the electricity price, power consumption, and uncertainty of RES generation. The work showed that particle swarm optimization method is more efficient in term of finding the best solution of the studied optimization function in comparison with genetic algorithm and combinatorial particle swarm optimization. A deterministic EM problem was solved by the authors of [98] via the multi-period gravitational search algorithm. The authors of [99] used a multi-objective particle swarm optimization algorithm to solve the EM system problem, which was considered as a multi-objective problem. However, the authors of [13,100] solved the EM system problem as a single-objective problem using particle swarm optimization-based algorithms. A metaheuristic approach for MG configuration in green data centers was presented by the authors of [101]. An optimization model was presented that considered the electricity costs and greenhouse gas emissions associated with all components of the MG systems, as well as their interactions. The model was applied to a real scenario of a data center with a given load demand in a specified environment. The authors calculated the degradation costs and the operational cost based on a system lifetime of 20 years. The developed model ensured good-quality MG configurations for different tradeoffs of cost and sustainability. Another work, presented by the authors of [102], combined an intelligent expert system fuzzy logic and a metaheuristic algorithm Grey-Wolf Optimizer. The proposed approaches solved the economic and environmental optimization problems of the MG systems by considering the uncertainties of RES and fluctuation in the power demand. In addition, a monitoring technique was developed with the fuzzy system to evaluate the input parameters to control the battery charge/discharge cycle, taking into account the economic aspect of the Grey-Wolf Optimizer optimization problem. The battery storage system operated by tracking the local generation costs of the installed MG and the total costs of the battery storage, which increased the possibility of charging the storage system at low costs during off-peak times. A metaheuristic home energy management system was studied by the authors of [103]. The authors evaluated the performance of the home energy management system using three metaheuristic optimization techniques: Bacterial foraging optimization, the Harmony search algorithm, and Enhanced deferential evolution. The objectives were to minimize the energy consumption, electricity cost, and reduction in peak-to-average ratio while maximizing user comfort. The obtained results showed that a tradeoff between user comfort and cost exists for the control constraints. In terms of cost, the results showed that the Harmony search algorithm performs better among other techniques. Another new interesting work, presented by the authors of [104], used a metaheuristic-based vector-decoupled algorithm to balance the control and operation of a hybrid MG system in the presence of stochastic renewable energy sources and the electric vehicle charging structure. The proposed control method ensured the stability of both frequency and voltage levels during the high-pulsed demand conditions and severe conditions of islanding operation mode together with the variability of RESs production. The presented results exposed the effectiveness and robustness of the proposed method to manage the real and reactive power exchange between the installed DC and AC buses of the MG within acceptable voltage and frequency variability.

Generally, heuristic optimization approaches use exploratory methods, in a reasonable time, to solve the optimization problems. However, they are unable to assure optimality of the obtained results [105]. The metaheuristic approaches are efficient and popular methods that are used for control and EM in the MG system. Several works in the literature that have analyzed the performance of these approaches. In some works, the metaheuristic control has been coupled with other control approaches in order to benefit from the performance of both approaches [106,107].

3.2.4. Artificial Intelligent Methods

Artificial neural networks are examples of artificial methods. They are considered as stochastic methods, which could be used to solve optimization problems for system having random variables. For MG systems, RESs have a variable nature caused by the weather conditions, which affect the power generation. As example, the authors of [108] presented an expert system for EM in MG systems using neural networks in order to predict the power generation of the installed RESs. The authors of [109] proposed a mathematical model for a smart load management in a standalone MG system. The studied loads were modeled by neural networks, and a predictive control was used to manage the energy according to predicted load variation. The authors of [110] presented an EM system for an MG system connected to the utility grid with the main objective of maximizing the use of renewable energies while minimizing the carbon emission. Two neural networks were used to model the proposed EM system using evolutionary adaptive dynamic programming and learning concepts. For the deployed neural networks, one was used for the management strategy and the other was used to check the optimal system's performance. The authors of [111] used a neural network to control a bidirectional rectifier/inverter. A dynamic programming algorithm was implemented and was trained using back propagation through time. The deployed neural networks showed a high ability to trace rapidly changing reference commands for frequency and voltage and satisfied control requirements for a faulted power system. The neural network controller used in this work was performed and studied under typical vector control conditions. The authors of [112] proposed a Lagrange-programming neural networks method for an efficient control and management of MG system with the main objective to minimize the overall cost of MG. In this work, the load was classified into different categories of controllable load, thermal load, price sensitive load, and critical load, while variable neurons and Lagrange neurons were combined to obtain optimal scheduling of MG operation. Mainly, neural networks can control, optimize, and identify system's parameters in online or offline applications. Unlike the previous approaches, neural networks can solve problems with nonlinear data in large-scale MG systems because of their ability to solve the system's stability via self-learning and prediction capabilities [113,114].

MAB control approaches are generally used in MGs because they are decentralized while allowing multiple interacting agents to follow their specified rules and goals and to perform autonomously dedicated functions [115]. The principal element of MAB methods is the agent, which can be a virtual or physical entity situated in a specified system (e.g., buildings, MG). It is capable of autonomously reacting depending on the changes of the system's environment [42,116]. The authors of [117] applied a comprehensive description about different optimization techniques to EM and a comparison with other techniques was realized including MAB. The authors of [118] presented an EM based on the differential evolution algorithm, developed in JADE (Java Agent Development Environment) for grid outage. The proposed MAB approach showed its efficiency in minimizing the load's uncertainty as well as the generation costs from the intermittent nature of RES generation. The approach also considered the price variation in the utility grid, and the critical loads were considered while selecting the best solution. The authors of [119] proposed a fault-tolerant multi-agent control approach for coordinated energy and comfort management in integrated buildings and MG systems. Several cooperative agents were presented and trained in order to reach a global coordination, to satisfy related constraints, and to meet the system's objectives. The integrated buildings and MG systems were mathematically formulated as a multi-objective optimization problem, which was solved under different operating conditions. Other interesting research works, which have considered the MAB control approaches for EM in MG systems, are presented by the authors of [120–122]. Multi-agent systems offer the opportunity to implement more than basic control. They have three key features, namely reactive, proactive, and social abilities. From their characteristics, the agent technology is promising for the implementation of flexible, scalable, and distributed systems [123,124]. The usage of MAB method is rapidly growing in power systems, especially for EM in MG systems. MABs, combined with system modeling, make the arrangements of MG units autonomously directed making the scheme more intelligent and protective. The deployment of MAB control in the MG system considers each agent as an intelligent unit, which can communicate with their neighboring agents in a collaborative way to determine future control actions to achieve the common objective. The communication with neighboring agents requires the deployment of advanced ICTs in order to benefit from the advantage of such approaches.

Ant Colony Optimization (ACO) is one of the more commonly used methods for EM in MG systems due to its flexibility for specified constraints, low computational time and complexity, and ease of implementation. This classical method is inspired by the behavior of real ants to search for good solutions to a given optimization problem. It is a simple computational agent that converts the optimization problems into the problems of finding the shortest path on a weighted graph. The authors of [125] used an AOM method for EM in demand side management. The authors first designed an EM controller model using multiple knapsack problem and applied an ACO approach to obtain a viable solution for the designed objective function. By simulation. the authors attempted to justify that the ACO works efficiently in terms of electricity bill reduction and the minimization of peak-to-average ratio while considering user satisfaction. Another ACO method was developed by the authors of [126], who investigated a combined cost optimization scheme in order to minimize both operational cost and emission levels while satisfying the MG's load demand. The proposed technique was compared with two other techniques, Lagrange and Gradient, to evaluate the proposed method performance. Mainly, other optimization methods based on AI have been used in the literature for EM and optimization problems. Particle Swarm Optimization was presented by the auhors of [127] for EM fuzzy controller design in dual source propelled electric vehicles. A systemic analysis of the power in energy storage was established by a mathematical model of EM problem.

Despite the efficiency of the abovementioned methods, still real-time and predictive control approaches are required for intelligent energy management in smart MG systems.

3.2.5. Other Interesting Approaches

One of the more interesting approaches for EM is proactive control. The principal of this approach is a mixed-integer optimal control problem that can be presented as a mixed-integer nonlinear programming problem [128]. The problem consists of finding optimal rules for a set of binary and continuous control variables that minimize the future predictable cost of the system over the time horizon. The proactive control is an "operation-oriented measures" scheme that makes the system capable of dealing with the unfavorable condition for the system operation. The authors of [129] presented an MG proactive control approach to manage the adverse impacts of extreme windstorms. When alerts were received for the forecasted windstorm, the approach found a conservative schedule of MG with the minimum number of vulnerable branches in service while the total load was served. The conservative schedule ensured the MG normal operation prior to the windstorm while reducing the MG vulnerability at the event arrival. This method increased the benefits for generation reschedule, conservation voltage regulation, network reconfiguration, and optimal parameter settings of droop-controlled units. The authors of [130] discussed unified resilience evaluation and the operational enhancement approach, including a procedure for assessing the impact of severe weather on power systems. The proposed approach aimed to mitigate the cascading effects that may occur during weather emergencies. Another work, presented by the authors of [131], studied the installation of a battery energy storage system with a PV system in a hierarchical trans-active EM approach in order to reduce consumer's electricity bills. A cost-benefit analysis approach was developed for proactive houses which combined PV units and battery storage systems. The developed control algorithm controlled the charge/discharge cycle of the battery based on an economic benefit analysis in real-time electricity rate and battery cost to give an exact idea of returns and yearly savings to consumers on their investment. The performance of this method can be enhanced when a proactive system is managed using predictive approaches. The authors of [101] compared reactive feedback control and Model Predictive Control in terms of energy consumed, energy error, and management effort for a given data center. The work proposed a feedback control strategy based on the data center model in order to optimize the quality of service, the energy consumed, and the management effort. It is perceived from the literature that the concept of proactive control for energy management in MG systems is rarely used. The concept is very interesting for control-based predictive decisions. Due to the development of information and communication technologies, especially microcontrollers, proactive control can be improved in future researches for EM in MG systems. The method is capable of making the system more preferment with the existing disturbances system operation.

Another interesting control approach is the FL. Like neural networks, the FL method is considered as one of the nonlinear techniques that are used for power regulation with power electronics-based converters. This intelligent control consists of a fuzzifier, rule evaluator, and a defuzzifier, while a set of rules known as rule-based and database is considered for the control strategy deployment. Mainly, the FL method is used to control space vector PWM based three-phase rectifier and is used with intelligent techniques-based Droop-Control to manage multiple distributed energy DC-MG systems [132]. For instance, the authors of [133] proposed a voltage control technic using an FL-based centralized controller with gain scheduling control for DC-MG with an electric-double-layer-capacitor as energy storage. A fuzzy-based control strategy, proposed by the authors of [134,135], is capable of determining small voltage and frequency steps regulations to improve the performance of Droop-Control by diminishing the mismatch in the common bus without heavy communication links. This work considered the frequency and voltage as uncoupled variables and then corrected each one separately by considering that the voltage is a local variable and the frequency is a global variable of the system. The proposed fuzzy method changed the frequency and the voltage reference value in the droop equation of the Voltage Sources Inverters to correct its variation. The authors of [102] used FL and a metaheuristic algorithm known as Grey-Wolf optimization to optimize the interconnection between multiple MG systems. The main aims of this method were to minimize both the costs for the generator units and the emission levels of the fossil fuel sources. Several works have studied the use of FL for energy management in MG systems. The authors of [136] deployed a mode transition strategy to smooth the mode variation and a fuzzy controller was used to determine the operation mode of coupled MG system with 20 different grid-connected and standalone MG systems. The FL was also considered as a deterministic algorithm for frequency and voltage regulation in both primary and secondary control levels and was characterized by low computational cost and easiness of implementation. In the literature, FL is the most deterministic approaches used together with PI controller. Some FL methods can be classified as AI methods.

4. Comparison of Control Approaches for MG Systems

The choice of an EM approach is an essential requirement for the reliable and stable operation for MG system. Depending on the characteristics of the deployed system (e.g., topologies, operation modes, structure), an EM can be selected. However, the deployment of an approach does not signify that the others are not reliable, and the studied constraints and the fixed objective of the control strategy are the main issue in order to identify the utility of the deployed method. In the rest of this section, the advantages and the disadvantages of different control techniques are presented (see Table 4).

Table 4. Brief comparison of control approaches.

Control Approach	Application	Advantage	Disadvantage
Model predictive control [85,137,138]	• Reliable for power sharing between MG and the utility grid • Hybrid AC/DC coupled MG	• Robust against uncertainty • Power smoothing • Multiple control objective and constraint functions are implemented for the same control strategy • Optimal control	• Requiring the use of advanced ICTs • Control parameters information should be defined in advance
Adaptive droop [139,140]	• Hybrid system of RESs • Parallel DC/DC converter • Heavy loading conditions	• The different operation modes eliminate the overload conditions between generator unites, storage devices, and utility grid; • Minimizing circulating current.	• Difficult to select the proper voltage levels • Generating interconnection resistances between the installed converter and requiring information about the DC bus • Control parameters should be known in advance.
Artificial neural networks [141,142]	• Distributed power generation units • Multiple MG system interconnection	• The approach can control, optimize, and identify the system's parameters in online or offline applications • Solve problems with nonlinear data approaches in large-scale systems in MG • Solve the system's stability and fault tolerance via self-learning and prediction	• Complexity of the model structure • Experimental interpretation of the model is difficult (black boxes) • Difficult to determine the best network structure in case of adding or raising units from the MG topology • Possibility only on stable system structure
Distributed cooperation control [143–145]	• The control is optimal for DC-MG system • Improving voltage levels for DC-MG	• Flexible, robust, and, extensible • Optimal coordination control and improved voltage profile	• Less security for the communication system • Frequency response nature cannot be visualized

Table 4. *Cont.*

Control Approach	Application	Advantage	Disadvantage
Conventional droop [146,147]	• Reliable for DC-MG • Linear loads • Inductive transmission lines	• Easy implementation for the primary control	• Voltage regulation is not ensured • The voltage drops across the bus resistance, causing a current sharing degradation • Active and reactive power bandwidth variation of the controllers affects the voltage and frequency controls
FL based control [148,149]	• Reliable for primary control • Voltage and frequency regulation	• Improved voltage and frequency regulation and power sharing for multiple MG	• Requiring a high processing unit • Errors methods adopted for the participation function and time-consuming process
Multi-agent-based control [123,150,151]	• Distributed power generation units • Multiple MG system interconnection	• The group of agents can address larger problems than any individual is capable to do in MG system • Redundancy and economies of large scale • The ability to meet global constraints • Flexibility to work in uncertain environments under unforeseen conditions	• Potential for conflicts; need for increased agent sophistication • Short term benefits may not outweigh organization construction costs for the installed MG systems • Requiring a high connectivity between agents and the LC • The agent should operate at the same parameters of the other agents, especially for voltage and frequency regulation

A good approach must consider the stochastic nature of different control parameters, the installation cost, the components lifetime, the distributed resources, and the reliable and safety operation of the MG system. In fact, the deployment of an EM control strategy requires the classification of the whole system into different levels, while each level should operate by coordinating with the other levels from the sources (e.g., maximum power point tracking) to the end consumers, which can be a local consumer or a neighboring MG consumer. Nowadays, smart components are installed for each source and for each MG system, which can cooperate between them due to the new ICTs. Especially, the actual inverters can execute different control strategies from the source power regulation to the interconnectivity to the utility grid or to the neighboring MG. In addition, the inverters can be installed for a large scale of MG systems, creating a cluster of data and electricity exchange, while these inverters could be connected to the internet in order to store the historic data in the cloud. Mainly, the main objective function for each inverter is ensuring continuous power supply to the consumers without considering the lifetime of the battery storage system or the cost of electricity. In this context, the development of an EM control strategy that considers the electricity price variation and minimizes the battery C/D cycle is required. These two issues allow the maximization of the system profitability by minimizing the electricity bill and avoiding a frequent replacement of battery storage in a MG system. The main idea is to develop an intelligent and predictive control strategy that can optimally control the distributed resources in the MG by considering multiple constraints and objective functions at the same time.

5. State of the Art Synthesis and Our Contribution

Control strategies generally use single-objective function procedures (e.g., maximizing the quality of the services). Without considering different operating constraints, these procedures are easier to implement and to deploy in real-sitting scenarios. Moreover, control strategies, which take into consideration only the energy availability within MG components (e.g., energy sources, storage devices, traditional electric grid), could be implemented by simple algorithms. These algorithms implement procedures that switch, at each time, from RES either to storage devices or to the TEG. For instance, actual commercial inverters are able to efficiently manage the interconnection between RESs, energy storage systems, and the utility grid by incorporating a single-objective function. In particular, the MG system's EM takes into consideration only the availability of the electricity for being supplied to buildings loads. The inverter can use either batteries or the utility grid once without taking into account other parameters, such as the actual electricity cost as well as battery C/D cycles. However, in a limited time, high battery C/D cycles could decrease their performance, which impact on the profitability of the system. In other cases, controllers can interact with energy sources generators (e.g., solar, wind) in real-time in order to limit the power generation (LPPT). The aim is to ensure the quality of the electrical services (e.g., frequency, voltage), and consequently, to minimize the profitability of MG system's components. Despite their advantages, they could have negative impacts on the batteries' lifecycle and system's profitability. Therefore, context-awareness principles and predictive analytics could be exploited for developing context-driven control approaches.

The current state of knowledge aims to develop context-driven control approaches for the energy management of MG systems in the context of smart buildings. Mainly, a predictive control approach, named MAPCASTE (Measure, Analyze, Predict, foreCAST, and Execute) [37], is developed and deployed in real-sitting scenarios for energy management in MG systems (see Figure 9). Unlike the control approaches from literature, MAPCASTE considers multiple-objective functions, which take into consideration battery C/D cycles as well as electricity price forecasting [37]. The main aim is to ensure, in an optimal way, the continuous electricity supply from different installed sources (e.g., RESs, batteries, TEG) to building's services. The proposed approach is based on predictive control models, which are able to generate a sequence of future control actions over a prediction horizon.

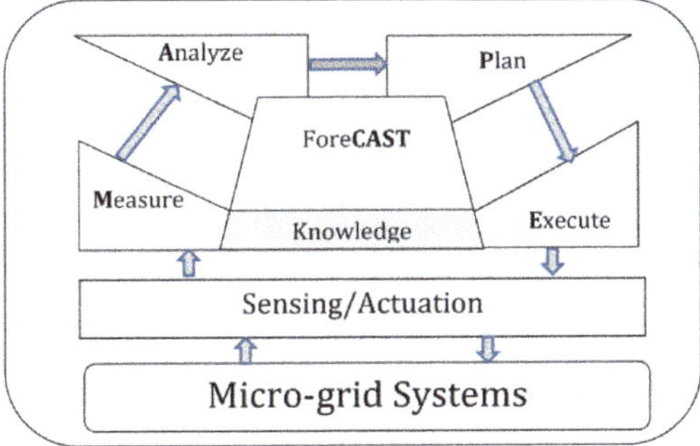

Figure 9. The proposed control approach schemes with operation process.

However, in order to carry out the MAPCASTE, several forecasted inputs values are required, mainly the power production/consumption and batteries SoC. This requires an advanced metering infrastructure, which makes it possible to measure and predict all inputs values. Therefore, an MG was deployed together with an IoT/Big data platform in order to conduct experiments and validate developed models. The deployed MG system contains RESs and battery storage systems, which are connected together with the TEG in order to supply the electrical energy to the building's loads (e.g., lighting, ventilation). The IoT/Big data platform was developed and deployed in order to allow measuring and forecasting RESs power generation, loads consumption, and batteries SoC. Sensing/actuating components with a control card are installed in order to monitor and manage the whole MG system, offering the possibility to test the developed control techniques in real context [37,152]. Moreover, based on this review, ongoing works focus on the development of smart converters. In fact, the actual commercial inverters offer the possibility to manage the power flows between different power sources, loads, energy storage systems, and utility grids with high performance. However, these inverters are limited generally to a single-objective function, the satisfaction of the load demand, without considering other operating constraints, such as the electricity price and the battery state of health. Moreover, the integration of new IoT/Big-data technologies to the actual inverter has improved the performance of the system to control and predict the suitable actions for EM and control. Mainly, the integration of machine-learning algorithms is required to analyze the data and to predict the actions for EM in MG systems. In this way, the development of smart inverter has enhanced the possibility to integrate multiple-objective functions and operating constraints that can be integrated in the EM approaches. Therefore, the deployment of predictive control strategies in real scenarios requires the use of open-access power converter. For that, we are deploying our proper power inverter in order to have the ability to conduct real testing of predictive control strategies with specific constraints and multiple-objective functions. The deployment of smart inverter offers the possibility to create MG networks using IoT/Big-data technologies. In this context, a platform for MG2MG energy and data exchange will be developed based on the predictive control deployed in the smart inverters.

6. Conclusions

The energy management and optimization control in MG systems are becoming a multiple-objective "management/optimization" function to be satisfied by solving simultaneously technical, economic, and environmental problems. Therefore, several approaches (e.g., exact, stochastic, and predictive) have been proposed for energy management. These approaches were chosen based on their practicality, reliability, and resource availability in MG environment. This work reviewed recent research work related to EM in MG systems. In particular, we focused on different control approaches that have been proposed to efficiently operate MG systems, including centralized, decentralized, and hierarchical management structures. A comprehensive description of control and optimization methods was highlighted, particularly to identify the most common and effective method for EM in MG systems. Predictive control was a good candidate, since it integrates optimal control and multivariable processes and is a flexible control scheme that allows the easy inclusion of system constraints and optimization functions. It is robust against uncertainty and power-smoothing problems. Thus, multiple control objective and constraint functions can be implemented for the same control strategy. However, despite the power of these predictive control techniques, their deployment in real-sitting scenarios requires a holistic platform that integrates MG components together with all equipment for measuring and predicting important input data. With recent technological advances in microprocessors, data analysis, and machine learning, predictive control can be seen as a promising alternative for energy management in MG systems.

Author Contributions: Conceptualization, A.E., M.B. and R.O.; Data curation, A.E.; Methodology, A.E., R.O. and M.B.; Software, A.E. and R.O.; Supervision, R.O., M.B., N.E.K., M.K. and K.Z.-D.; Validation, R.O., M.K. and M.B.; Writing—original draft, A.E.; Writing—review & editing, A.E., M.B., N.E.K., M.K. and K.Z.-D. All authors have read and agreed to the published version of the manuscript.

Funding: This research was funded by USAID under the PEER program, grant number 5-398. It is also partially supported by HOLSYS project, which is funded by IRESEN (iresen-inno-projet-2020–2022).

Conflicts of Interest: The authors declare no conflict of interest.

References

1. Higuita Cano, M. Système de Gestion de L'énergie Basé sur L'incertitude Pour un Système Hybride à Sources D'énergie Renouvelable Autonome Avec Production D'hydrogène. Ph.D. Thesis, Université du Québec à Trois-Rivières, Quebec, QC, Canada, 2017.
2. Indragandhi, V.; Logesh, R.; Subramaniyaswamy, V.; Vijayakumar, V.; Siarry, P.; Uden, L. Multi-objective optimization and energy management in renewable based AC/DC microgrid. *Comput. Electr. Eng.* **2018**, *70*, 179–198. [CrossRef]
3. Zhang, W.; Maleki, A.; Rosen, M.A.; Liu, J. Optimization with a simulated annealing algorithm of a hybrid system for renewable energy including battery and hydrogen storage. *Energy* **2018**, *163*, 191–207. [CrossRef]
4. Ghaffari, A.; Askarzadeh, A. Design optimization of a hybrid system subject to reliability level and renewable energy penetration. *Energy* **2020**, *193*, 116754. [CrossRef]
5. Dali, M.; Belhadj, J.; Roboam, X. Hybrid solar-wind system with battery storage operating in grid-connected and standalone mode: Control and energy management—Experimental investigation. *Energy* **2010**, *35*, 2587–2595. [CrossRef]
6. Zhang, Q.; Wang, L.; Li, G.; Liu, Y. A real-time energy management control strategy for battery and supercapacitor hybrid energy storage systems of pure electric vehicles. *J. Energy Storage* **2020**, *31*, 101721. [CrossRef]
7. Sai Chandan, R.; Sai Kiran, T.L.; Swapna, G.; Vijay Muni, T. Intelligent Control Strategy for Energy Management System with Fc/Battery/Sc. *J. Crit. Rev.* **2020**, *7*, 344–348. [CrossRef]
8. Sandgani, M.R.; Sirouspour, S. Priority-Based Microgrid Energy Management in a Network Environment. *IEEE Trans. Sustain. Energy* **2018**, *9*, 980–990. [CrossRef]
9. Zheng, Y.; Jenkins, B.M.; Kornbluth, K.; Træholt, C. Optimization under uncertainty of a biomass-integrated renewable energy microgrid with energy storage. *Renew. Energy* **2018**, *123*, 204–217. [CrossRef]
10. Lin, W.-M.; Tu, C.-S.; Tsai, M.-T. Energy Management Strategy for Microgrids by Using Enhanced Bee Colony Optimization. *Energies* **2015**, *9*, 5. [CrossRef]
11. Li, B.; Roche, R.; Miraoui, A. Microgrid sizing with combined evolutionary algorithm and MILP unit commitment. *Appl. Energy* **2017**, *188*, 547–562. [CrossRef]
12. Mazzola, S.; Astolfi, M.; Macchi, E. A detailed model for the optimal management of a multigood microgrid. *Appl. Energy* **2015**, *154*, 862–873. [CrossRef]
13. Rabiee, A.; Sadeghi, M.; Aghaeic, J.; Heidari, A. Optimal operation of microgrids through simultaneous scheduling of electrical vehicles and responsive loads considering wind and PV units uncertainties. *Renew. Sustain. Energy Rev.* **2016**, *57*, 721–739. [CrossRef]
14. Dash, V.; Bajpai, P. Power management control strategy for a stand-alone solar photovoltaic-fuel cell-battery hybrid system. *Sustain. Energy Technol. Assess.* **2015**, *9*, 68–80. [CrossRef]
15. Kuznetsova, E.; Ruiz, C.; Li, Y.-F.; Zio, E. Analysis of robust optimization for decentralized microgrid energy management under uncertainty. *Int. J. Electr. Power Energy Syst.* **2015**, *64*, 815–832. [CrossRef]
16. Marzband, M.; Yousefnejad, E.; Sumper, A.; Domínguez-García, J.L. Real time experimental implementation of optimum energy management system in standalone Microgrid by using multi-layer ant colony optimization. *Int. J. Electr. Power Energy Syst.* **2016**, *75*, 265–274. [CrossRef]
17. Aghdam, F.H.; Salehi, J.; Ghaemi, S. Contingency based energy management of multi-microgrid based distribution network. *Sustain. Cities Soc.* **2018**, *41*, 265–274. [CrossRef]
18. Arcos-Aviles, D.; Sotomayor, D.; Proaño, J.L.; Guinjoan, F.; Marietta, M.P.; Pascual, J.; Marroyo, L.; Sanchis, P. Fuzzy energy management strategy based on microgrid energy rate-of-change applied to an electro-thermal residential microgrid. In Proceedings of the 2017 IEEE 26th International Symposium on Industrial Electronics (ISIE), Edinburgh, UK, 19–21 June 2017; pp. 99–105.
19. Olivares, D.E.; Canizares, C.A.; Kazerani, M. A Centralized Energy Management System for Isolated Microgrids. *IEEE Trans. Smart Grid* **2014**, *5*, 1864–1875. [CrossRef]
20. Elkazaz, M.; Sumner, M.; Thomas, D. Energy management system for hybrid PV-wind-battery microgrid using convex programming, model predictive and rolling horizon predictive control with experimental validation. *Int. J. Electr. Power Energy Syst.* **2020**, *115*, 105483. [CrossRef]
21. Helal, S.A.; Najee, R.J.; Hanna, M.O.; Shaaban, M.F.; Osman, A.H.; Hassan, M.S. An energy management system for hybrid microgrids in remote communities. In Proceedings of the 2017 IEEE 30th Canadian Conference on Electrical and Computer Engineering (CCECE), Windsor, ON, Canada, 30 April–3 May 2017; pp. 1–4.
22. Tsikalakis, A.G.; Hatziargyriou, N.D. Centralized control for optimizing microgrids operation. In Proceedings of the 2011 IEEE Power and Energy Society General Meeting, Detroit, MI, USA, 24–28 July 2011; pp. 1–8.
23. Warnier, M.; Dulman, S.; Koç, Y.; Pauwels, E. Distributed monitoring for the prevention of cascading failures in operational power grids. *Int. J. Crit. Infrastruct. Prot.* **2017**, *17*, 15–27. [CrossRef]
24. Kermani, M.; Carnì, D.L.; Rotondo, S.; Paolillo, A.; Manzo, F.; Martirano, L. A Nearly Zero-Energy Microgrid Testbed Laboratory: Centralized Control Strategy Based on SCADA System. *Energies* **2020**, *13*, 2106. [CrossRef]
25. Yamashita, D.Y.; Vechiu, I.; Gaubert, J.-P. A review of hierarchical control for building microgrids. *Renew. Sustain. Energy Rev.* **2020**, *118*, 109523. [CrossRef]
26. Pourbabak, H.; Chen, T.; Su, W. Centralized, decentralized, and distributed control for Energy Internet. In *The Energy Internet*; Elsevier BV: Amsterdam, The Netherlands, 2019; pp. 3–19.

27. Senjyu, T.; Kuninaka, R.; Urasaki, N.; Fujita, H.; Funabashi, T. Power system stabilization based on robust centralized and decentralized controllers. In Proceedings of the 2005 International Power Engineering Conference, Singapore, 29 November–2 December 2005; pp. 905–910.
28. Liu, C.C.; McArthur, S.; Lee, S.J. *Smart Grid Handbook, 3 Volume Set*; John Wiley & Sons: Piscataway, NJ, USA, 2016.
29. Aftab, M.A.; Hussain, S.M.S.; Ali, I. ICT Technologies, Standards and Protocols for Active Distribution Network Automation and Management. In *Advanced Communication and Control Methods for Future Smartgrids*; IntechOpen: London, UK, 2019.
30. Celik, B.; Roche, R.; Suryanarayanan, S.; Bouquain, D.; Miraoui, A. Electric energy management in residential areas through coordination of multiple smart homes. *Renew. Sustain. Energy Rev.* **2017**, *80*, 260–275. [CrossRef]
31. Feng, X.; Shekhar, A.; Yang, F.; Hebner, R.E.; Bauer, P. Comparison of Hierarchical Control and Distributed Control for Microgrid. *Electr. Power Compon. Syst.* **2017**, *45*, 1043–1056. [CrossRef]
32. Espín-Sarzosa, D.; Palma-Behnke, R.; Núñez-Mata, O. Energy Management Systems for Microgrids: Main Existing Trends in Centralized Control Architectures. *Energies* **2020**, *13*, 547. [CrossRef]
33. Rathor, S.K.; Saxena, D. Energy management system for smart grid: An overview and key issues. *Int. J. Energy Res.* **2020**, *44*, 4067–4109. [CrossRef]
34. Molzahn, D.K.; Dorfler, F.; Sandberg, H.; Low, S.H.; Chakrabarti, S.; Baldick, R.; Lavaei, J. A Survey of Distributed Optimization and Control Algorithms for Electric Power Systems. *IEEE Trans. Smart Grid* **2017**, *8*, 2941–2962. [CrossRef]
35. Van Nam, D.; Sualeh, M.; Kim, D.; Kim, G.-W. A Hierarchical Control System for Autonomous Driving towards Urban Challenges. *Appl. Sci.* **2020**, *10*, 3543. [CrossRef]
36. Elmouatamid, A.; NaitMalek, Y.; Ouladsine, R.; Bakhouya, M.; Elkamoun, N.; Khaidar, M.; Zine-Dine, K. *A Micro-Grid System Infrastructure Implementing IoT/Big-Data Technologies for Efficient Energy Management in Buildings. Submitted to ATSPES'1 (Advanced Technologies for Solar Photovoltaics Energy Systems)*; Springer: Berlin/Heidelberg, Germany, 2020.
37. Elmouatamid, A. MAPCAST: An Adaptive Control Approach using Predictive Analytics for Energy Balance in Micro-Grid Systems. *Int. J. Renew. Energy Res. (IJRER)* **2020**, *10*, 945–954.
38. Gaiceanu, M.; Arama, I.N.; Ghenea, I. DC Microgrid Control. In *Power Systems*; Springer Science and Business Media LLC: Berlin/Heidelberg, Germany, 2019; pp. 357–380.
39. Prabaharan, N.; Jerin, A.R.A.; Najafi, E.; Palanisamy, K. An overview of control techniques and technical challenge for inverters in micro grid. In *Hybrid-Renewable Energy Systems in Microgrids*; Elsevier BV: Amsterdam, The Netherlands, 2018; pp. 97–107.
40. González-Romera, E.; Romero-Cadaval, E.; Roncero-Clemente, C.; Ruiz-Cortés, M.; Barrero-González, F.; Milanés-Montero, M.-I.; Muñoz, A.M. Secondary Control for Storage Power Converters in Isolated Nanogrids to Allow Peer-to-Peer Power Sharing. *Electronics* **2020**, *9*, 140. [CrossRef]
41. Guerrero, J.M.; Chandorkar, M.; Lee, T.-L.; Loh, P.C. Advanced Control Architectures for Intelligent Microgrids—Part I: Decentralized and Hierarchical Control. *IEEE Trans. Ind. Electron.* **2013**, *60*, 1254–1262. [CrossRef]
42. Sahoo, S.K.; Sinha, A.K.; Kishore, N.K. Control Techniques in AC, DC, and Hybrid AC-DC Microgrid: A Review. *IEEE J. Emerg. Sel. Top. Power Electron.* **2018**, *6*, 738–759. [CrossRef]
43. Kumar, J.; Agarwal, A.; Agarwal, V. A review on overall control of DC microgrids. *J. Energy Storage* **2019**, *21*, 113–138. [CrossRef]
44. Cupelli, M.; Monti, A.; De Din, E.; Sulligoi, G. Case study of voltage control for MVDC microgrids with constant power loads—Comparison between centralized and decentralized control strategies. In Proceedings of the 2016 18th Mediterranean Electrotechnical Conference (MELECON), Lemesos, Cyprus, 18–20 April 2016; pp. 1–6.
45. Cannata, N.; Cellura, M.; Longo, S.; Montana, F.; Sanseverino, E.R.; Luu, Q.L.; Nguyen, N.Q. Multi-Objective Optimization of Urban Microgrid Energy Supply According to Economic and Environmental Criteria. In Proceedings of the 2019 IEEE Milan PowerTech, Milan, Italy, 23–27 June 2019; pp. 1–6.
46. Jiménez-Fernández, S.; Camacho-Gómez, C.; Mallol-Poyato, R.; Fernández-Caballero, J.C.; Del Ser, J.; Portilla-Figueras, A.; Salcedo-Sanz, S. Optimal Microgrid Topology Design and Siting of Distributed Generation Sources Using a Multi-Objective Substrate Layer Coral Reefs Optimization Algorithm. *Sustainability* **2018**, *11*, 169. [CrossRef]
47. Hannan, M.; Tan, S.Y.; Al-Shetwi, A.Q.; Jern, K.P.; Begum, R. Optimized controller for renewable energy sources integration into microgrid: Functions, constraints and suggestions. *J. Clean. Prod.* **2020**, *256*, 120419. [CrossRef]
48. Liu, Y.; Yuen, C.; Hassan, N.U.; Huang, S.; Yu, R.; Xie, S. Electricity Cost Minimization for a Microgrid With Distributed Energy Resource Under Different Information Availability. *IEEE Trans. Ind. Electron.* **2014**, *62*, 2571–2583. [CrossRef]
49. Kamboj, V.K.; Bath, S.K.; Dhillon, J.S. Solution of non-convex economic load dispatch problem using Grey Wolf Optimizer. *Neural Comput. Appl.* **2016**, *27*, 1301–1316. [CrossRef]
50. Haidar, A.M.A.; Fakhar, A.; Helwig, A. Sustainable energy planning for cost minimization of autonomous hybrid microgrid using combined multi-objective optimization algorithm. *Sustain. Cities Soc.* **2020**, *62*, 102391. [CrossRef]
51. Igualada, L.; Corchero, C.; Cruz-Zambrano, M.; Heredia, F.-J. Optimal Energy Management for a Residential Microgrid Including a Vehicle-to-Grid System. *IEEE Trans. Smart Grid* **2014**, *5*, 2163–2172. [CrossRef]
52. Alsema, E.A. Energy Payback Time and CO_2 Emissions of PV Systems. In *Practical Handbook of Photovoltaics*; Elsevier BV: Amsterdam, The Netherlands, 2012; pp. 1097–1117.
53. Mason, J.E.; Fthenakis, V.M.; Hansen, T.; Kim, H.C. Energy payback and life-cycle CO_2 emissions of the BOS in an optimized 3·5 MW PV installation. *Prog. Photovolt. Res. Appl.* **2006**, *14*, 179–190. [CrossRef]

54. Jafari, A.; Ganjehlou, H.G.; Khalili, T.; Bidram, A. A fair electricity market strategy for energy management and reliability enhancement of islanded multi-microgrids. *Appl. Energy* **2020**, *270*, 115170. [CrossRef]
55. Yenealem, M.G.; Ngoo, L.M.H.; Shiferaw, D.; Hinga, P. Management of Voltage Profile and Power Loss Minimization in a Grid-Connected Microgrid System Using Fuzzy-Based STATCOM Controller. *J. Electr. Comput. Eng.* **2020**, *2020*, 1–13. [CrossRef]
56. Maknouninejad, A.; Qu, Z. Realizing Unified Microgrid Voltage Profile and Loss Minimization: A Cooperative Distributed Optimization and Control Approach. *IEEE Trans. Smart Grid* **2014**, *5*, 1621–1630. [CrossRef]
57. Shieh, S.-Y.; Ersal, T.; Peng, H. Power Loss Minimization in Islanded Microgrids: A Communication-Free Decentralized Power Control Approach Using Extremum Seeking. *IEEE Access* **2019**, *7*, 20879–20893. [CrossRef]
58. Barakat, M.; Tala-Ighil, B.; Chaoui, H.; Gualous, H.; Hissel, D. Energy Management of a Hybrid Tidal Turbine-Hydrogen Micro-Grid: Losses Minimization Strategy. *Fuel Cells* **2020**, *20*, 342–350. [CrossRef]
59. Azizivahed, A.; Naderi, E.; Narimani, H.; Fathi, M.; Narimani, M.R. A New Bi-Objective Approach to Energy Management in Distribution Networks with Energy Storage Systems. *IEEE Trans. Sustain. Energy* **2017**, *9*, 56–64. [CrossRef]
60. Murty, V.V.S.N.; Kumar, A. Multi-objective energy management in microgrids with hybrid energy sources and battery energy storage systems. *Prot. Control. Mod. Power Syst.* **2020**, *5*, 1–20. [CrossRef]
61. Ontiveros, L.J.; Suvire, G.O.; Mercado, P.E. A New Control Strategy to Integrate Flow Batteries into AC Micro-Grids with High Wind Power Penetration. *Redox Princ. Adv. Appl.* **2017**, *83*. [CrossRef]
62. Shayeghi, H.; Shahryari, E.; Moradzadeh, M.; Siano, P. A Survey on Microgrid Energy Management Considering Flexible Energy Sources. *Energies* **2019**, *12*, 2156. [CrossRef]
63. Elmouatamid, A.; NaitMalek, Y.; Bakhouya, M.; Ouladsine, R.; Elkamoun, N.; Zine-Dine, K.; Khaidar, M. An energy management platform for micro-grid systems using Internet of Things and Big-data technologies. *Proc. Inst. Mech. Eng. Part I: J. Syst. Control. Eng.* **2019**, *233*, 904–917. [CrossRef]
64. Elmouatamid, A.; Ouladsine, R.; Bakhouya, M.; El Kamoun, N.; Zine-Dine, K.; Khaidar, M. A Model Predictive Control Approach for Energy Management in Micro-Grid Systems. In Proceedings of the 2019 International Conference on Smart Energy Systems and Technologies (SEST), Porto, Portugal, 9–11 September 2019; pp. 1–6.
65. Elmouatamid, A.; Ouladsine, R.; Bakhouya, M.; El Kamoun, N.; Zine-Dine, K.; Khaidar, M. A Control Strategy Based on Power Forecasting for Micro-Grid Systems. In Proceedings of the 2019 IEEE International Smart Cities Conference (ISC2), Casablanca, Morocco, 14–17 October 2019; pp. 735–740.
66. Buyak, N.; Deshko, V.; Sukhodub, I. Buildings energy use and human thermal comfort according to energy and exergy approach. *Energy Build.* **2017**, *146*, 172–181. [CrossRef]
67. Rahmani-Andebili, M.; Shen, H. Cooperative distributed energy scheduling for smart homes applying stochastic model predictive control. In Proceedings of the 2017 IEEE International Conference on Communications (ICC), Paris, France, 21–25 May 2017; pp. 1–6.
68. Hemmati, R.; Saboori, H. Stochastic optimal battery storage sizing and scheduling in home energy management systems equipped with solar photovoltaic panels. *Energy Build.* **2017**, *152*, 290–300. [CrossRef]
69. Bordons, C.; Teno, G.; Marquez, J.J.; Ridao, M.A. Effect of the Integration of Disturbances Prediction in Energy Management Systems for Microgrids. In Proceedings of the 2019 International Conference on Smart Energy Systems and Technologies (SEST), Porto, Portugal, 9–11 September 2019; pp. 1–6.
70. Petrollese, M. Optimal Generation Scheduling for Renewable Microgrids Using Hydrogen Storage Systems. Ph.D. Thesis, Università degli Studi di Cagliari, Cagliari, Italy, 2015.
71. Negenborn, R.; Houwing, M.; De Schutter, B.; Hellendoorn, J. Model predictive control for residential energy resources using a mixed-logical dynamic model. In Proceedings of the 2009 International Conference on Networking, Sensing and Control, Okayama, Japan, 26–29 March 2009; pp. 702–707.
72. Mendes, P.R.; Isorna, L.V.; Bordons, C.; Normey-Rico, J.E. Energy management of an experimental microgrid coupled to a V2G system. *J. Power Sources* **2016**, *327*, 702–713. [CrossRef]
73. Garcia-Torres, F.; Vilaplana, D.G.; Bordons, C.; Roncero-Sanchez, P.; Ridao, M.A. Optimal Management of Microgrids With External Agents Including Battery/Fuel Cell Electric Vehicles. *IEEE Trans. Smart Grid* **2019**, *10*, 4299–4308. [CrossRef]
74. Chen, H.; Chen, J.; Lu, H.; Yan, C.; Liu, Z. A Modified MPC-Based Optimal Strategy of Power Management for Fuel Cell Hybrid Vehicles. *IEEE/ASME Trans. Mechatron.* **2020**, *25*, 2009–2018. [CrossRef]
75. Bozchalui, M.C.; Hashmi, S.A.; Hassen, H.; Canizares, C.; Bhattacharya, K. Optimal Operation of Residential Energy Hubs in Smart Grids. *IEEE Trans. Smart Grid* **2012**, *3*, 1755–1766. [CrossRef]
76. Parisio, A.; Rikos, E.; Tzamalis, G.; Glielmo, L. Use of model predictive control for experimental microgrid optimization. *Appl. Energy* **2014**, *115*, 37–46. [CrossRef]
77. Bordons, C.; Montero, C.G. Basic Principles of MPC for Power Converters: Bridging the Gap Between Theory and Practice. *IEEE Ind. Electron. Mag.* **2015**, *9*, 31–43. [CrossRef]
78. Linder, A.; Kennel, R. Model Predictive Control for Electrical Drives. In Proceedings of the 2005 IEEE 36th Power Electronics Specialists Conference, Recife, Brazil, 16 June 2005; pp. 1793–1799.
79. Rodriguez, J.; Pontt, J.; Silva, C.A.; Correa, P.; Lezana, P.; Cortes, P.; Ammann, U. Predictive Current Control of a Voltage Source Inverter. *IEEE Trans. Ind. Electron.* **2007**, *54*, 495–503. [CrossRef]

80. Gregor, R.; Barrero, F.; Toral, S.; Arahal, M.; Prieto, J.; Duran, M.J. Enhanced predictive current control method for the asymmetrical dualthree phase induction machine. In Proceedings of the 2009 IEEE International Electric Machines and Drives Conference, Miami, FL, USA, 3–6 May 2009; pp. 265–272.

81. Lezana, P.; Aguilera, R.; Quevedo, D.E. Model Predictive Control of an Asymmetric Flying Capacitor Converter. *IEEE Trans. Ind. Electron.* **2008**, *56*, 1839–1846. [CrossRef]

82. Cortés, P.; Wilson, A.; Kouro, S.; Rodriguez, J.; Abu-Rub, H. Model Predictive Control of Multilevel Cascaded H-Bridge Inverters. *IEEE Trans. Ind. Electron.* **2010**, *57*, 2691–2699. [CrossRef]

83. Correa, P.; Rodríguez, J.; Rivera, M.; Espinoza, J.R.; Kolar, J.W. Predictive control of an indirect matrix converter. *IEEE Trans. Ind. Electron.* **2009**, *56*, 1847–1853. [CrossRef]

84. Vargas, R.; Ammann, U.; Hudoffsky, B.; Rodriguez, J.; Wheeler, P. Predictive Torque Control of an Induction Machine Fed by a Matrix Converter with Reactive Input Power Control. *IEEE Trans. Power Electron.* **2010**, *25*, 1426–1438. [CrossRef]

85. Vazquez, S.; Leon, J.I.; Franquelo, L.G.; Rodriguez, J.; Young, H.A.; Marquez, A.; Zanchetta, P. Model Predictive Control: A Review of Its Applications in Power Electronics. *IEEE Ind. Electron. Mag.* **2014**, *8*, 16–31. [CrossRef]

86. Sukumar, S.; Mokhlis, H.; Mekhilef, S.; Naidu, K.; Karimi, M. Mix-mode energy management strategy and battery sizing for economic operation of grid-tied microgrid. *Energy* **2017**, *118*, 1322–1333. [CrossRef]

87. Nemati, M.; Braun, M.; Tenbohlen, S. Optimization of unit commitment and economic dispatch in microgrids based on genetic algorithm and mixed integer linear programming. *Appl. Energy* **2018**, *210*, 944–963. [CrossRef]

88. Shekari, T.; Gholami, A.; Aminifar, F. Optimal energy management in multi-carrier microgrids: An MILP approach. *J. Mod. Power Syst. Clean Energy* **2019**, *7*, 876–886. [CrossRef]

89. Panwar, L.K.; Konda, S.R.; Verma, A.; Panigrahi, B.K.; Kumar, R. Operation window constrained strategic energy management of microgrid with electric vehicle and distributed resources. *IET Gener. Transm. Distrib.* **2017**, *11*, 615–626. [CrossRef]

90. García-Vera, Y.E.; Dufo-López, R.; Bernal-Agustín, J.L. Energy Management in Microgrids with Renewable Energy Sources: A Literature Review. *Appl. Sci.* **2019**, *9*, 3854. [CrossRef]

91. Bukar, A.L.; Tan, C.W.; Yiew, L.K.; Ayop, R.; Tan, W.-S. A rule-based energy management scheme for long-term optimal capacity planning of grid-independent microgrid optimized by multi-objective grasshopper optimization algorithm. *Energy Convers. Manag.* **2020**, *221*, 113161. [CrossRef]

92. Merabet, A.; Ahmed, K.T.; Ibrahim, H.; Beguenane, R.; Ghias, A.M.Y.M. Energy Management and Control System for Laboratory Scale Microgrid Based Wind-PV-Battery. *IEEE Trans. Sustain. Energy* **2017**, *8*, 145–154. [CrossRef]

93. An, L.N.; Quoc-Tuan, T. Optimal energy management for grid connected microgrid by using dynamic programming method. In Proceedings of the 2015 IEEE Power & Energy Society General Meeting, Denver, CO, USA, 26–30 July 2015; pp. 1–5.

94. Papari, B.; Edrington, C.; Vu, T.V.; Diaz-Franco, F. A heuristic method for optimal energy management of DC microgrid. In Proceedings of the 2017 IEEE Second International Conference on DC Microgrids (ICDCM), Nuremburg, Germany, 27–29 June 2017; pp. 337–343.

95. Khan, Z.A.; Zafar, A.; Javaid, S.; Aslam, S.; Rahim, M.H.; Javaid, N. Hybrid meta-heuristic optimization based home energy management system in smart grid. *J. Ambient. Intell. Humaniz. Comput.* **2019**, *10*, 4837–4853. [CrossRef]

96. Chen, C.; Duan, S.; Cai, T.; Liu, B.; Hu, G. Smart energy management system for optimal microgrid economic operation. *IET Renew. Power Gener.* **2011**, *5*, 258–267. [CrossRef]

97. Radosavljević, J.; Jevtić, M.; Klimenta, D. Energy and operation management of a microgrid using particle swarm optimization. *Eng. Optim.* **2015**, *48*, 811–830. [CrossRef]

98. Marzband, M.; Ghadimi, M.; Sumper, A.; Domínguez-García, J.L. Experimental validation of a real-time energy management system using multi-period gravitational search algorithm for microgrids in islanded mode. *Appl. Energy* **2014**, *128*, 164–174. [CrossRef]

99. Aghajani, G.; Shayanfar, H.; Shayeghi, H. Presenting a multi-objective generation scheduling model for pricing demand response rate in micro-grid energy management. *Energy Convers. Manag.* **2015**, *106*, 308–321. [CrossRef]

100. Alavi, S.A.; Ahmadian, A.; Aliakbar-Golkar, M. Optimal probabilistic energy management in a typical micro-grid based-on robust optimization and point estimate method. *Energy Convers. Manag.* **2015**, *95*, 314–325. [CrossRef]

101. Rahmani, R.; Moser, I.; Cricenti, A.L. Modelling and optimization of microgrid configuration for green data centres: A metaheuristic approach. *Future Gener. Comput. Syst.* **2020**, *108*, 742–750. [CrossRef]

102. Ei-Bidairi, K.S.; Nguyen, H.D.; Jayasinghe, S.D.G.; Mahmoud, T.S. Multiobjective Intelligent Energy Management Optimization for Grid-Connected Microgrids. In Proceedings of the 2018 IEEE International Conference on Environment and Electrical Engineering and 2018 IEEE Industrial and Commercial Power Systems Europe (EEEIC/I&CPS Europe), Palermo, Italy, 12–15 June 2018; pp. 1–6.

103. Zafar, A.; Shah, S.; Khalid, R.; Hussain, S.M.; Rahim, H.; Javaid, N. A Meta-Heuristic Home Energy Management System. In Proceedings of the 2017 31st International Conference on Advanced Information Networking and Applications Workshops (WAINA), Taipei, Taiwan, 27–29 March 2017; pp. 244–250.

104. Aljohani, T.M.; Ebrahim, A.F.; Mohammed, O. Hybrid Microgrid Energy Management and Control Based on Metaheuristic-Driven Vector-Decoupled Algorithm Considering Intermittent Renewable Sources and Electric Vehicles Charging Lot. *Energies* **2020**, *13*, 3423. [CrossRef]

105. Khan, A.A.; Naeem, M.; Iqbal, M.; Qaisar, S.; Anpalagan, A. A compendium of optimization objectives, constraints, tools and algorithms for energy management in microgrids. *Renew. Sustain. Energy Rev.* **2016**, *58*, 1664–1683. [CrossRef]
106. Khan, A.H.; Li, S.; Chen, D.; Liao, L. Tracking control of redundant mobile manipulator: An RNN based metaheuristic approach. *Neurocomputing* **2020**, *400*, 272–284. [CrossRef]
107. Grisales-Noreña, L.F.; Montoya, O.D.; Ramos-Paja, C.A. An energy management system for optimal operation of BSS in DC distributed generation environments based on a parallel PSO algorithm. *J. Energy Storage* **2020**, *29*, 101488. [CrossRef]
108. Motevasel, M.; Seifi, A.R. Expert energy management of a micro-grid considering wind energy uncertainty. *Energy Convers. Manag.* **2014**, *83*, 58–72. [CrossRef]
109. Solanki, B.V.; Raghurajan, A.; Bhattacharya, K.; Canizares, C.A. Including Smart Loads for Optimal Demand Response in Integrated Energy Management Systems for Isolated Microgrids. *IEEE Trans. Smart Grid* **2017**, *8*, 1739–1748. [CrossRef]
110. Venayagamoorthy, G.K.; Sharma, R.K.; Gautam, P.K.; Ahmadi, A. Dynamic Energy Management System for a Smart Microgrid. *IEEE Trans. Neural Netw. Learn. Syst.* **2016**, *27*, 1643–1656. [CrossRef]
111. Li, S.; Fairbank, M.; Johnson, C.; Wunsch, D.C.; Alonso, E.; Proao, J.L. Artificial Neural Networks for Control of a Grid-Connected Rectifier/Inverter Under Disturbance, Dynamic and Power Converter Switching Conditions. *IEEE Trans. Neural Netw. Learn. Syst.* **2013**, *25*, 738–750. [CrossRef]
112. Wang, T.; He, X.; Deng, T. Neural networks for power management optimal strategy in hybrid microgrid. *Neural Comput. Appl.* **2017**, *31*, 2635–2647. [CrossRef]
113. Mahmoud, M.S.; Alyazidi, N.M.; Abouheaf, M.I. Adaptive intelligent techniques for microgrid control systems: A survey. *Int. J. Electr. Power Energy Syst.* **2017**, *90*, 292–305. [CrossRef]
114. Roslan, M.; Hannan, M.; Ker, P.J.; Uddin, M. Microgrid control methods toward achieving sustainable energy management. *Appl. Energy* **2019**, *240*, 583–607. [CrossRef]
115. Dou, C.; Lv, M.; Zhao, T.; Ji, Y.; Li, H. Decentralised coordinated control of microgrid based on multi-agent system. *IET Gener. Transm. Distrib.* **2015**, *9*, 2474–2484. [CrossRef]
116. Mao, M.; Jin, P.; Hatziargyriou, N.D.; Chang, L. Multiagent-Based Hybrid Energy Management System for Microgrids. *IEEE Trans. Sustain. Energy* **2014**, *5*, 1. [CrossRef]
117. Khan, M.W.; Wang, J.; Ma, M.; Xiong, L.; Li, P.; Wu, F. Optimal energy management and control aspects of distributed microgrid using multi-agent systems. *Sustain. Cities Soc.* **2019**, *44*, 855–870. [CrossRef]
118. Raju, L.; Morais, A.A.; Rathnakumar, R.; Ponnivalavan, S.; Thavam, L.D. Micro-grid Grid Outage Management Using Multi-agent Systems. In Proceedings of the 2017 Second International Conference on Recent Trends and Challenges in Computational Models (ICRTCCM), Tindivanam, India, 3–4 February 2017; pp. 363–368.
119. Anvari-Moghaddam, A.; Rahimi-Kian, A.; Mirian, M.S.; Guerrero, J.M. A multi-agent based energy management solution for integrated buildings and microgrid system. *Appl. Energy* **2017**, *203*, 41–56. [CrossRef]
120. Karavas, C.-S.; Kyriakarakos, G.; Arvanitis, K.G.; Papadakis, G. A multi-agent decentralized energy management system based on distributed intelligence for the design and control of autonomous polygeneration microgrids. *Energy Convers. Manag.* **2015**, *103*, 166–179. [CrossRef]
121. Kofinas, P.; Dounis, A.; Vouros, G. Fuzzy Q-Learning for multi-agent decentralized energy management in microgrids. *Appl. Energy* **2018**, *219*, 53–67. [CrossRef]
122. Samadi, E.; Badri, A.; Ebrahimpour, R. Decentralized multi-agent based energy management of microgrid using reinforcement learning. *Int. J. Electr. Power Energy Syst.* **2020**, *122*, 106211. [CrossRef]
123. Khan, M.R.B.; Jidin, R.; Pasupuleti, J. Multi-agent based distributed control architecture for microgrid energy management and optimization. *Energy Convers. Manag.* **2016**, *112*, 288–307. [CrossRef]
124. Li, C.; Jia, X.; Ying, Z.; Li, X. A microgrids energy management model based on multi-agent system using adaptive weight and chaotic search particle swarm optimization considering demand response. *J. Clean. Prod.* **2020**, *262*, 121247. [CrossRef]
125. Rahim, S.; Iqbal, Z.; Shaheen, N.; Khan, Z.A.; Qasim, U.; Khan, S.A.; Javaid, N. Ant Colony Optimization Based Energy Management Controller for Smart Grid. In Proceedings of the 2016 IEEE 30th International Conference on Advanced Information Networking and Applications (AINA), Crans-Montana, Switzerland, 23–25 March 2016; pp. 1154–1159.
126. Esmat, A.; Magdy, A.; ElKhattam, W.; Elbakly, A.M.; Magdy, A. A novel Energy Management System using Ant Colony Optimization for micro-grids. In Proceedings of the 2013 3rd International Conference on Electric Power and Energy Conversion Systems, Istanbul, Turkey, 2–4 October 2013; pp. 1–6.
127. Chenghui, Z.; Qingsheng, S.; Naxin, C.; Wuhua, L. Particle Swarm Optimization for energy management fuzzy controller design in dual-source electric vehicle. In Proceedings of the 2007 IEEE Power Electronics Specialists Conference, Orlando, FL, USA, 17–21 June 2007; pp. 1405–1410.
128. Zavala, V.M.; Wang, J.; Leyffer, S.; Constantinescu, E.M.; Anitescu, M.; Conzelmann, G. Proactive energy management for next-generation building systems. *Proc. SimBuild* **2010**, *4*, 377–385.
129. Amirioun, M.H.; Aminifar, F.; Lesani, H. Resilience-Oriented Proactive Management of Microgrids Against Windstorms. *IEEE Trans. Power Syst.* **2018**, *33*, 4275–4284. [CrossRef]
130. Panteli, M.; Trakas, D.N.; Mancarella, P.; Hatziargyriou, N.D. Boosting the Power Grid Resilience to Extreme Weather Events Using Defensive Islanding. *IEEE Trans. Smart Grid* **2016**, *7*, 2913–2922. [CrossRef]

131. Amin, U.; Hossain, J.; Lu, J.; Mahmud, M.A. Cost-benefit analysis for proactive consumers in a microgrid for transactive energy management systems. In Proceedings of the 2016 Australasian Universities Power Engineering Conference (AUPEC), Brisbane, Australia, 25–28 September 2016; pp. 1–6.
132. Diaz, N.L.; Dragicevic, T.; Vasquez, J.C.; Guerrero, J.M. Intelligent Distributed Generation and Storage Units for DC Microgrids— A New Concept on Cooperative Control without Communications Beyond Droop Control. *IEEE Trans. Smart Grid* **2014**, *5*, 2476–2485. [CrossRef]
133. Kakigano, H.; Miura, Y.; Ise, T. Distribution Voltage Control for DC Microgrids Using Fuzzy Control and Gain-Scheduling Technique. *IEEE Trans. Power Electron.* **2013**, *28*, 2246–2258. [CrossRef]
134. Agnoletto, E.J.; Neves, R.; Bastos, R.F.; Machado, R.Q.; Oliveira, V.A. Fuzzy secondary controller applied to autonomous operated AC microgrid. In Proceedings of the 2016 European Control Conference (ECC), Aalborg, Denmark, 29 June–1 July 2016; pp. 1788–1793.
135. De Nadai, N.B.; De Souza, A.C.Z.; Costa, J.G.C.; Pinheiro, C.A.M.; Portelinha, F.M. A secondary control based on fuzzy logic to frequency and voltage adjustments in islanded microgrids scenarios. In Proceedings of the 2017 IEEE Manchester PowerTech, Manchester, UK, 18–22 June 2017; pp. 1–6.
136. Jafari, M.; Malekjamshidi, Z.; Zhu, J.; Khooban, M.H. Novel predictive fuzzy logic-based energy management system for grid-connected and off-grid operation of residential smart micro-grids. *IEEE J. Emerg. Sel. Top. Power Electron.* **2018**, 1391–1404. [CrossRef]
137. Vazquez, S.; Rodriguez, J.R.; Rivera, M.; Franquelo, L.G.; Norambuena, M. Model Predictive Control for Power Converters and Drives: Advances and Trends. *IEEE Trans. Ind. Electron.* **2017**, *64*, 935–947. [CrossRef]
138. Castilla, M.D.M.; Bordons, C.; Visioli, A. Event-based state-space model predictive control of a renewable hydrogen-based microgrid for office power demand profiles. *J. Power Sources* **2020**, *450*, 227670. [CrossRef]
139. Augustine, S.; Mishra, M.K.; Lakshminarasamma, N. Adaptive Droop Control Strategy for Load Sharing and Circulating Current Minimization in Low-Voltage Standalone DC Microgrid. *IEEE Trans. Sustain. Energy* **2015**, *6*, 132–141. [CrossRef]
140. Djebbri, S.; Ladaci, S.; Metatla, A.; Balaska, H. Fractional-order model reference adaptive control of a multi-source renewable energy system with coupled DC/DC converters power compensation. *Energy Syst.* **2018**, *11*, 315–355. [CrossRef]
141. Moon, J.; Park, S.; Rho, S.; Hwang, E. A comparative analysis of artificial neural network architectures for building energy consumption forecasting. *Int. J. Distrib. Sens. Netw.* **2019**, *15*. [CrossRef]
142. Reynolds, J.; Ahmad, M.W.; Rezgui, Y.; Hippolyte, J.-L. Operational supply and demand optimisation of a multi-vector district energy system using artificial neural networks and a genetic algorithm. *Appl. Energy* **2019**, *235*, 699–713. [CrossRef]
143. Huang, P.-H.; Liu, P.-C.; Xiao, W.; El Moursi, M.S. A Novel Droop-Based Average Voltage Sharing Control Strategy for DC Microgrids. *IEEE Trans. Smart Grid* **2015**, *6*, 1096–1106. [CrossRef]
144. Liu, T.; Tan, X.; Sun, B.; Wu, Y.; Tsang, D.H. Energy management of cooperative microgrids: A distributed optimization approach. *Int. J. Electr. Power Energy Syst.* **2018**, *96*, 335–346. [CrossRef]
145. Xing, X.; Xie, L.; Meng, H. Cooperative energy management optimization based on distributed MPC in grid-connected microgrids community. *Int. J. Electr. Power Energy Syst.* **2019**, *107*, 186–199. [CrossRef]
146. Huang, H.-H.; Hsieh, C.-Y.; Liao, J.-Y.; Chen, K.-H. Adaptive Droop Resistance Technique for Adaptive Voltage Positioning in Boost DC–DC Converters. *IEEE Trans. Power Electron.* **2011**, *26*, 1920–1932. [CrossRef]
147. Peyghami, S.; Davari, P.; Mokhtari, H.; Blaabjerg, F. Decentralized Droop Control in DC Microgrids Based on a Frequency Injection Approach. *IEEE Trans. Smart Grid* **2019**, *10*, 6782–6791. [CrossRef]
148. Erdinc, O.; Vural, B.; Uzunoglu, M. A wavelet-fuzzy logic based energy management strategy for a fuel cell/battery/ultra-capacitor hybrid vehicular power system. *J. Power Sources* **2009**, *194*, 369–380. [CrossRef]
149. Baset, D.A.-E.; Rezk, H.; Hamada, M. Fuzzy Logic Control Based Energy Management Strategy for Renewable Energy System. In Proceedings of the 2020 International Youth Conference on Radio Electronics, Electrical and Power Engineering (REEPE), Moscow, Russia, 12–14 March 2020; pp. 1–5.
150. Prinsloo, G.; Dobson, R.; Mammoli, A. Synthesis of an intelligent rural village microgrid control strategy based on smartgrid multi-agent modelling and transactive energy management principles. *Energy* **2018**, *147*, 263–278. [CrossRef]
151. Xiong, L.; Li, P.; Wang, Z.; Wang, J. Multi-agent based multi objective renewable energy management for diversified community power consumers. *Appl. Energy* **2020**, *259*, 114140. [CrossRef]
152. Elmouatamid, A.; NaitMalek, Y.; Ouladsine, R.; Bakhouya, M.; Elkamoun, N.; Zine-Dine, K.; Khaidar, M.; Abid, R. Towards a Demand/Response Control Approach for Micro-grid Systems. In Proceedings of the 2018 5th International Conference on Control, Decision and Information Technologies (CoDIT), Thessaloniki, Greece, 10–13 April 2018; pp. 984–988.

Article

Demand Response Coupled with Dynamic Thermal Rating for Increased Transformer Reserve and Lifetime

Ildar Daminov [1,2,*], Rémy Rigo-Mariani [1], Raphael Caire [1], Anton Prokhorov [2] and Marie-Cécile Alvarez-Hérault [1]

[1] University Grenoble Alpes, CNRS, Grenoble INP, G2Elab, 21 Avenue des Martyrs, 38000 Grenoble, France; Remy.Rigo-Mariani@g2elab.grenoble-inp.fr (R.R.-M.); raphael.caire@g2elab.grenoble-inp.fr (R.C.); marie-cecile.alvarez@g2elab.grenoble-inp.fr (M.-C.A.-H.)

[2] Power Engineering School, Tomsk Polytechnic University, 7, Usov Street, 634034 Tomsk, Russia; antonprokhorov@tpu.ru

* Correspondence: ildar.Daminov@g2elab.grenoble-inp.fr

Abstract: (1) Background: This paper proposes a strategy coupling Demand Response Program with Dynamic Thermal Rating to ensure a transformer reserve for the load connection. This solution is an alternative to expensive grid reinforcements. (2) Methods: The proposed methodology firstly considers the N-1 mode under strict assumptions on load and ambient temperature and then identifies critical periods of the year when transformer constraints are violated. For each critical period, the integrated management/sizing problem is solved in YALMIP to find the minimal Demand Response needed to ensure a load connection. However, due to the nonlinear thermal model of transformers, the optimization problem becomes intractable at long periods. To overcome this problem, a validated piece-wise linearization is applied here. (3) Results: It is possible to increase reserve margins significantly compared to conventional approaches. These high reserve margins could be achieved for relatively small Demand Response volumes. For instance, a reserve margin of 75% (of transformer nominal rating) can be ensured if only 1% of the annual energy is curtailed. Moreover, the maximal amplitude of Demand Response (in kW) should be activated only 2–3 h during a year. (4) Conclusions: Improvements for combining Demand Response with Dynamic Thermal Rating are suggested. Results could be used to develop consumer connection agreements with variable network access.

Keywords: Demand Response; dynamic thermal rating; flexibility; hosting capacity; transformer

Citation: Daminov, I.; Rigo-Mariani, R.; Caire, R.; Prokhorov, A.; Alvarez-Hérault, M.-C. Demand Response Coupled with Dynamic Thermal Rating for Increased Transformer Reserve and Lifetime. *Energies* **2021**, *14*, 1378. https://doi.org/10.3390/en14051378

Academic Editor: Pedro Faria

Received: 6 February 2021
Accepted: 24 February 2021
Published: 3 March 2021

Publisher's Note: MDPI stays neutral with regard to jurisdictional claims in published maps and institutional affiliations.

1. Introduction

Nowadays, the energy industry is facing a transition toward decarbonization, decentralization, digitalization, and democratization. While the share of renewables is growing in many countries, the transport and heating sectors are intensively electrified to further reduce greenhouse gas emissions. For instance, International Renewable Energy Agency (IRENA) [1] estimates that the global share of electrical energy in the final use of total energy will rise from 20% in 2015 to 45% in 2050. The same report [1] assesses that an intensive electrification in buildings, transportation, and industry will reduce the emissions by 25%, 54%, and 16%, correspondingly. Combining these results with positive effects from renewables and energy efficiency measures should keep the global warming below the 2 °C limit, as it is set in recent Paris Agreement [2].

Apart from electrification measures, the electric power demand will rise due to social, economic, and climate change reasons [3]. Then, the accommodation of increased electrical demand will become a challenge for both Transmission System Operator (TSO) and Distribution System Operator (DSO) [4] and should require significant investments [5]. For instance, Det Norske Veritas & Germanischer Lloyd (DNV GL) estimates that worldwide annual expenditures for electrical networks will rise three times: from USD 0.49 trillion in 2016 to USD 1.5 trillion in 2030 [6]. Apart from financial constraints, network reinforcements may face a strong opposition due to social, environmental, political, and regulatory issues, as reported in [7].

The above-mentioned constraints force system operators to use any options for ensuring the load connection on short and mid-term horizons. Such alternatives for network reinforcement consist of using flexibilities from controllable distributed generation, storage, demand-side management [8], or Dynamic Thermal Rating (DTR) [9,10]. For instance, the expensive reinforcement of a primary substation transformer can be postponed by the coordinated control of Distributed Energy Resources (DER) and/or by considering the real thermal rating of power equipment. The main idea of DTR relies on the consideration of actual thermal ratings of equipment rather than those calculated for worse ambient conditions, which are not likely to ever happen. This paper focuses on Demand Response (DR) associated with DTR/thermal modeling of oil-immersed distribution transformers as the low-cost technologies among many possible flexibility options [11].

The researchers investigating DR usually consider a conservative thermal rating of network equipment. Thus, the network capacity may be underused. For instance, Martínez Ceseña et al. [12] demonstrated that small end-users can support the network capacity without sacrificing comfort levels. In [13], the same authors suggested a methodology, estimating a business case of DR for a small multi-energy district in order to support the capacity of a distribution network. Celli et al. [14] proposed a model of flexibility aggregation with a particular focus on DR to address network contingencies. Esmat and Usaola [15] developed an algorithm allowing to minimize the total cost of congestion management and taking into account payback effects. Jiang et al. [16] incorporated interruptible loads into substation capacity planning. Mullen [17] investigated the important interactions between demand-side response, load recovery, peak pricing, and network capacity margins. Once again, the thermal rating in these studies is considered conservatively.

At the same time, the researchers considering DTR/thermal modeling do not take into account the possibility of using flexibilities. For example, Elmakis et al. [18,19] developed a probabilistic approach for defining a transformer capacity based on its loss of life. Sen et al. [20] suggested a methodology for the sizing of a new oil-immersed transformer as a replacement for existing equipment. Bunn et al. [21] estimated the capacity of a distribution transformer to accommodate additional demand without impacting reliability indexes. Kostin et al. [22] estimated the reserve capacity (allowable loading) of urban transformers considering a minimum of relative annual electric power losses. Daminov et al. [23] estimated the reserve capacity of a primary substation by considering the DTR of oil-immersed transformers. Once again, these studies consider DTR without taking advantage of flexibilities.

Finally, the researchers who apply DTR together with DR do not explicitly explain how much load can be interconnected to a substation [24]. As an example, Sousa et al. [25] investigate the use of interruptible contracts for mitigating the emergency operation of power transformers. Teja and Yemula [26] prolonged the transformer life by controlling heating/cooling systems in buildings. Davison et al. [27] estimated the number of consumer connections considering the DR, temperature-sensitive load behavior, and DTR of overhead lines (but not for transformers). Zhou et al. [28] proposed bi-level multi-house energy management to coordinate the residential DR considering a transformer aging. Van Der Klauw et al. [29] proposed smart charging strategies of electrical vehicles and a neighborhood's load profile to mitigate transformer aging. Liu et al. [30] suggested a DR strategy to balance benefits for households and the transformer lifespan. Soleimani and Kezunovic [31] suggested a method that defines a charging schedule of electric vehicles that eventually mitigates the transformer aging and reduces risks of failure. Mohsenzadeh et al. [32] developed smart home management strategies to mitigate transformer loss of insulation life. Brinkel et al. [33] found that transformer reinforcement could lead to higher emissions than operating the existing transformer with lower ratings. Humayun et al. presented a series of papers [34–37] dedicated to the joint application of DTR and DR to increase the transformer utilization. Specifically, in [34,35], the authors proposed an optimization model for the maximal utilization of transformer capacity during contingencies. In [36,37], the authors expanded the scope on network automation (load transfer on near substations) and included all the costs occurring along the transformer lifetime.

Some early studies estimated the transformer reserve without considering neither DR nor DTR. For instance, Salehi and Haghifam [38] applied a genetic algorithm to define the reserve capacity of a substation. In [39], Kannan and Au suggested a probabilistic approach for sizing the distribution transformers. Helmi et al. [40] used the power factor correction capacitors to increase the reserve capacity of power transformers. Thus, the scope of this paper lays in the intersection between three domains: Demand Response, Dynamic Thermal Rating, and the problem of reserve estimations (see Figure 1). Although substantial efforts were made in each domain, there is still a gap in their intersections.

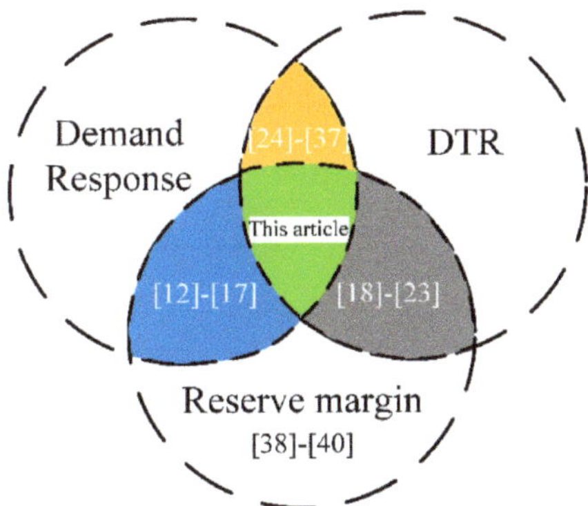

Figure 1. Scope of the paper with regard to the literature survey.

This paper investigates how much and when DR would be required for different reserve margins considering the DTR/thermal constraints of transformers. Only the thermal constraints of transformers and their aging are considered, whereas other limiting factors (e.g., voltage) are ignored. Nevertheless, the paper [41] shows that 78–83% of real network constraints are related to thermal constraints. Conventional approaches assume that DTR has a winding temperature limit of 98 °C, which is a design temperature of winding, rather than a temperature limit—the limit is actually much higher (e.g., 120 °C); see International Electrotechnical Commission (IEC) standard [42]. The question of using design temperature or a temperature limit for transformer loadings was actively discussed in [43–45]. Then, the conventional design temperature approach will be considered as a reference case in the course of this paper.

The considered case study is a Medium Voltage/Low Voltage (MV/LV) substation whose load is represented over a whole year (Figure 2). The objective is the computation of the DR needs (i.e., rated kW and kWh as well as hours of operation) that would allow the connection of additional load over the year. For reserve estimation, we consider strict hypotheses i.e., N-1 conditions with only one operating transformer and a maximum ambient temperature (monthly). The reserve is estimated by adding a constant load along the year, which leads to stronger thermal impacts rather than scaling up the existing load profiles. Thus, the DR design and operation are optimized for different amounts of reserve while keeping transformer temperatures, loading, and normalized aging below the specified limits. In the proposed methodology, we define different intervals along a year with thermal violations and then solve the proposed optimization problem for each interval. Special attention is given to the problem formulation for integrated DR design and management. Especially, a piece-wise linearization (PWL) of the thermal equations is introduced to ensure the convergence for long time intervals. Since transformer equations require minute time resolution and the load data are given in hourly resolution, we suggest an approach to consider different time grids. Finally, two different operating modes for the DR are investigated with "energy shifting" and "energy shedding". Results allow reconstructing the hourly temperature profile over a year and the corresponding transformer aging. The major scientific contributions and outcomes of the paper are as follows:

- A valid PWL model for the thermal model of an oil-immersed transformer is derived, and its adequacy for the problem stated in the paper is confirmed.
- A co-optimization problem for DR design (rated power and energy) and management is formulated and solved with the consideration of thermal and aging limitations over different time horizons and for different reserve margins.
- DTR based on the temperature limit is proved to be more efficient than DTR based on design temperature for use with DR.
- It is proved that if DR is used to shave about 1% of the total energy consumption (a full DR capacity is activated only 2–3 h per year), a transformer reserve can be increased to accommodate 75% more load connection. In these simulations, it is assumed that the thermal state of a transformer is in the quasi-worst situation by using many safety margins: the worst load growth is aligned with the maximal ambient temperature as well as N-1 condition over the whole year. Hence, the calculated DR, which is needed to mitigate the quasi-worst situation, is on the conservative side.

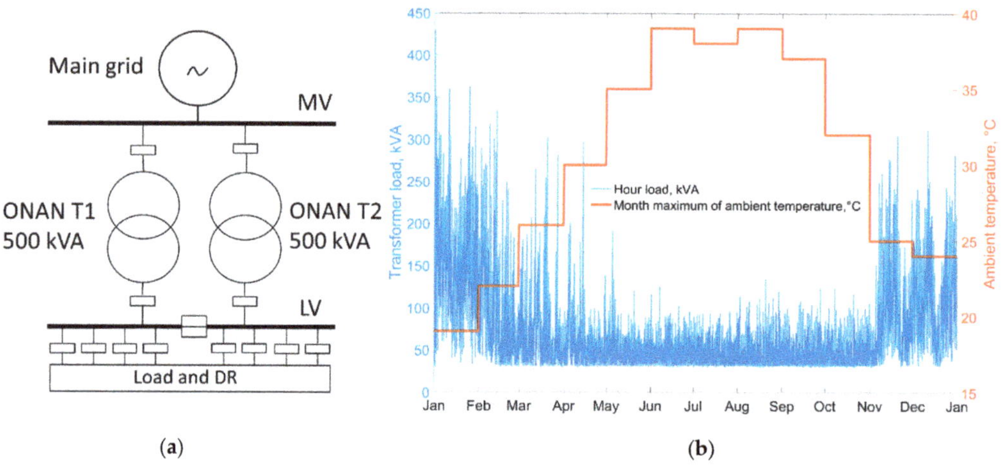

Figure 2. Case study—(**a**) outdoor secondary substation; (**b**) hourly load in kilovolt ampere (kVA) and monthly maximum ambient temperature (θ_t^a) in Grenoble, France.

The paper is organized as follows: Section 2 presents the case study and explains the problem of reserve determination as well as the developed methodology. Section 3 provides details on DR computation (i.e., integrated design and management). Finally, Section 4 provides the main results with validation runs for the DR computation and the optimal design under different reserve constraints before conclusions are drawn in Section 5.

2. End-Users Side Flexibility for Grid Upgrades Deferral

We suggest the reader to start reading from Section 2.1 presenting the case study of this paper. Moreover, this section explains the choice of assumptions allowing us to consider that the calculated DR has safety margins to mitigate the thermal constraints of transformers. Further, Section 2.2 is intended to help the reader understand at what reserve margins it would be sufficient to use DTR only and from what moment it is necessary to apply the DR and DTR. Finally, in Section 2.3, the reader can see an overview of the proposed methodology to find the necessary DR.

2.1. Case Study

The case study (Figure 2a) is an outdoor MV/LV substation with 2×500 kVA distribution transformers both equipped with an ONAN cooling system (Oil Natural Air Natural). Figure 2b shows the considered annual load profile at 1-h resolution representing an aggregated consumption of one hundred houses simulated by physical and behavior approaches [46].

To extract the load profile, authors used a MATLAB application "House load" [47] with all default parameters except the ambient temperature profile (θ_t^a). The updated hourly θ_t^a in Grenoble, France was provided by MeteoBlue for 2019 [48]. To obtain the conservative reserve values, only the historical monthly maximum for the last 30 years was used in thermal simulations (Figure 2b) [23]. Historical maximums of θ_t^a obtained for each month are assumed as a safety margin when simulating the worst thermal state of the transformer for different reserve values. This is especially relevant for global warming with expected constantly rising ambient temperatures. Moreover, this margin can mitigate other errors in thermal characteristics/modeling.

An additional margin in the study comes from the computation of the reserve that assumes an N-1 condition with the loss of one transformer in the considered substation. Thus, if one transformer is out of service, then the remaining distribution transformer 500 kVA can definitely supply all load alone, since the registered peak power is only 430 kVA (86% of nominal rating, 500 kVA). Therefore, the reserve for load connection (in a traditional DSO approach) would be estimated as the difference between the nominal rating and the peak load i.e., $500 - 430 = 70$ kVA (i.e., 14% of nominal rating). Instead of the nominal rating, the DSO can also apply other approaches as static thermal ratings [49,50], seasonal ratings [50], or emergency ratings [51]. Nevertheless, the approach for reserve determination does not change in its nature, as it still represents a simple difference between admissible constant rating and the peak load without consideration of a true thermal state of transformers.

Once thermal modeling is applied, the transformer temperatures (i.e., oil temperature θ_t^o and hot spot temperature θ_t^h) and the Aging EQuivalent (AEQ) over the year can be calculated explicitly using the loading profile P_t^{tr} (in p.u.) and ambient temperature. Current, temperature, and aging, as any variable representing the physical state, have corresponding limits. The choice of those limits can be again used to set important safety margins for thermal modeling. For instance, Table 1 provides current and temperature limits for various loading types: normal cyclic loadings as well as two emergency modes—long-term overloading and short-term overloading. Normal cyclic loading refers to a situation when a transformer is subject to high ambient temperature or higher-than-rated load, but the thermal aging remains the same as for nominal conditions. Long-term emergency loading is a situation when a transformer is subject to elevated temperatures for days or even months, and the short-term emergency loading is a heavy overloading of a transformer for less than 30 min. Due to the temporal nature of emergencies, the IEC standard [42] allows increasing their temperature and current limits.

Table 1. Limits for distribution transformers applicable for different types of loadings [42].

Limits	Loading Type		
	Normal Cyclic	Long-Term Emergency	Short-Term Emergency
P^{tr} (p.u.)	1.5	1.8	2.0
θ^h (°C)	120	140	Not specified
θ^o (°C)	105	115	Not specified
AEQ (p.u.	1	>1 possible	Not specified

In this paper, we use the normal cyclic loading as the strictest limits in the N-1 condition. However, DSO could choose the long-term emergency limits as an alternative for N-1 conditions. This would allow releasing more transformer capacity at the cost of higher risks of overheating and accelerated loss of life. This alternative is not considered in the paper but could be easily integrated without changing the proposed methodology and algorithms.

As stated in [23], the problem of reserve determination is that the load profile of a new consumer cannot be definitely known in advance. Therefore, the transformer's load profile after the connection of new consumers is unknown as well. In addition to the renewable production, leading to the famous duck curve from California, it is believed that the electrification of the heating and transport sector will likely change typical shapes of existing load profiles. In order to add a further "safety margin", the existing shape of a load profile is not increased proportionally in this paper. Instead, the reserve is considered while adding a constant load to the existing load profile all along the representative year [23]. That constant added load profile ensures the worse thermal mode of operation in comparison to any other pattern with the same peak power. Nevertheless, this assumption considers the peak increase only from new load connections but existing consumers, especially industry customers, can also increase the peaks. DSO should still guarantee that consumers can withdraw all power from the distribution network, which is indicated in connection contracts (also known as firm capacity contracts). In fact, consumers usually do not use their full allocation of power, but there is no guarantee that one day, some industrial or commercial consumers will not decide to expand production capacities and then boost the actual power demand up to the subscribed power indicated in connection contracts (of course without informing DSO). In such a case, the above-mentioned assumption can lead to errors in substation peak estimations and even to emergencies. To mitigate such risks, DSO could add a full-subscribed power (i.e., not measured power) of large industrial and commercial consumers to the existing load profile.

2.2. Problem Statement

The problem statement is described in Figure 3 by conducting preliminary thermal studies. These preliminary studies investigate what would be θ^h, θ^o, and AEQ for different reserve margins without using DR (i.e., without taking any measures to decrease the temperatures of transformers). Figure 3 displays the state variables of the transformer as a function of the added constant load from 1% to 100% of a nominal rating of 500 kVA to the existing load profile shown in Figure 2b. The maximal θ^h, θ^o during the year, and corresponding AEQ are estimated for the N-1 condition and computed using the IEC 60076-7 standard [42]. The IEC 60076-7 standard is an internationally recognized loading guide that provides mathematical equations and thermal characteristics of oil-immersed transformers needed to calculate θ^h, θ^o, and AEQ. Specifically, we use the difference equations described in the annex E of the IEC 60076-7 standard. These equations are presented later in Section 3.1 of this paper.

The preliminary thermal studies show that it is possible to connect a constant load of 240 kVA (=0.49 p.u * 500 kVA) in addition to the existing consumption without violating the thermal constraints at 120 °C (see point 1). That 240 kVA reserve is 3.4 times higher than the reserve (70 kVA) calculated with a conventional approach—i.e., the transformer rated power (500 kVA) minus the peak consumption (430 kVA). The 240-kVA reserve is even more significant than the similar reserve obtained for DTR based on design temperature 98 °C (145 kVA, 0.29 p.u. at point 0). Although the use of design temperature in DTR is often claimed to avoid the accelerated aging, we observe that the accelerated ageing occurs only if reserve (a load growth) reaches significant values around 0.70 p.u. (point 3). Thus, the consideration of θ^h_t limit should be preferred over design temperature if the aging limit is explicitly taken into account [23].

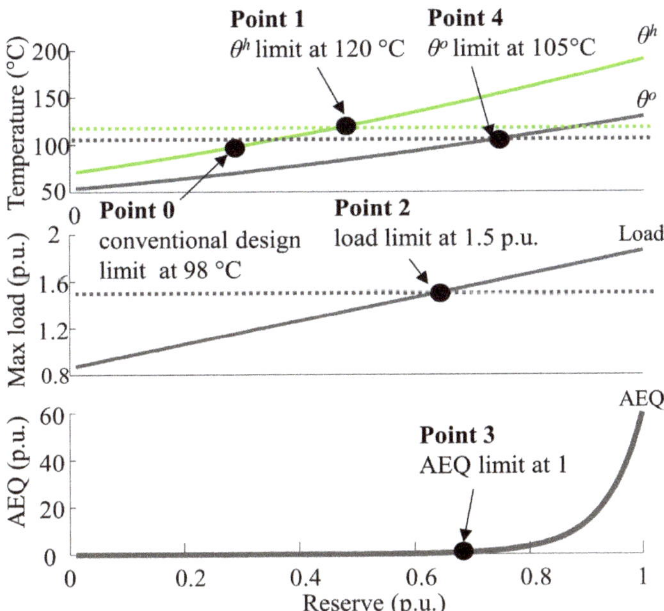

Figure 3. Preliminary thermal results for yearly simulation: load growing from 1% to 100% of nominal rating (reserve in p.u.). The initial loading of the studied distribution transformer is 86% (see the *y*-axis of the middle figure).

Assuming that appropriate DR programs prevent the violation of temperature constraints, more load can be further connected to the transformer until the next limit is reached—current limit at 1.5 p.u. (point 2 at 350 kVA in Figure 3). Starting from this point, DSO should use DR to avoid breaking both θ_t^h and current limits. Thus, from point 2 onward, the reserve can be further increased from 320 to 350 kVA until the next critical point is reached (see point 3 in Figure 3). Starting from this point 3, the AEQ may be higher than the normal annual loss of transformer life. Hence, it is necessary to reduce the θ_t^h even well below its limit to keep the AEQ less than 1 (as it will be shown in Section 4). The last critical point 4 in Figure 3 (reserve of 365 kVA) is a situation when the transformer oil temperature violates its own limit. Finally, an appropriately designed and managed Demand Response should tackle three constraints at the same time—transformer thermal limits (θ_t^o, θ_t^h), current limit, and insulation ageing—which is the objective of the optimization problem introduced in Section 3. Note that heavy computational burden may arise due to the needs for the thermal modeling in accordance with IEC 60076-7 that sets specific requirements for the time step of the load profiles and the temperatures. Indeed, IEC standard [42] requires that the time resolution must be at least two times smaller than the winding time constant, which is usually about 4–10 min. Otherwise, the thermal calculations may lose numerical stability. Practically, this requires that the time step of the thermal model for the transformer does not exceed 2–5 min or may be even less. In this paper, a 1-min resolution is adopted.

It is important to note that DR will be required only a few days per year (e.g., high load, high ambient temperature, or maintenance works). Thus, the DR design and management can be formulated deterministically only for the days when the transformer violates the thermal limits and not for the entire year. Hence, we avoid formulating a large intractable optimization problem. However, we argue that this idea remains valid if the longest interval does not exceed a few days. Figure 4 shows the longest interval with thermal violations considered in the DR optimization problem as a function of the reserve margin (added constant load on the *x*-axis). From Figure 4, we see that the longest interval increases exponentially. For a reserve value of 1.04 p.u. onward, the overheating occurs every day of the simulated year. Hence, nonlinear equations from IEC 60076-7 would lead to the intractability of the optimization problem. Therefore, in Sections 3.3 and 3.4, the paper suggests a linearization of the nonlinear equations of the transformer thermal model.

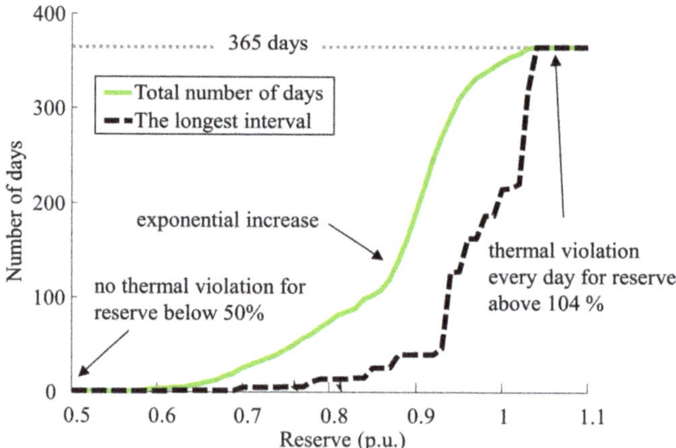

Figure 4. Number of days where thermal limits are reached.

2.3. Methodology

Figure 5 shows the algorithm that allows us to compute the DR required to ensure a given level of reserve.

The workflow consists of three main stages. At the first stage, the annual transformer loading (without DR) is computed with the existing load profile increased by the given reserve margin. Furthermore, the thermal state of the transformer is estimated with the ambient temperature over the whole year. If there are no thermal violations, then no DR is needed, and the algorithm stops here. Hence, another reserve margin can be investigated.

If any overheating is detected, the second stage of the algorithm identifies the interval(s) where transformer temperatures (θ_t^o, θ_t^h) or current limits are violated. The algorithm extracts the loading profile (P_t^l) at every identified interval. Then, the load and ambient temperature profiles over a given interval are considered as inputs for the integrated DR management and design that computes the minimum DR needs to fulfill the operating constraints (see Section 3). Note that the thermal model of the transformer requires initial values for the top oil and hot-spot temperatures at the beginning of the extracted interval. To do that, the optimized transformer loading profile (i.e., with DR management) from previous calculation is used to update the annual load profile from the start of the year until the beginning of the next interval. This cycle repeats until all intervals are investigated and allows us to track the correct initial temperature every time an interval is simulated. The last stage of the algorithm is needed once all intervals are studied and the optimized annual load profile is entirely reconstructed. Finally, the algorithm defines the DR values in terms of power and energy ratings as the maximum DR power and energy computed over the different intervals.

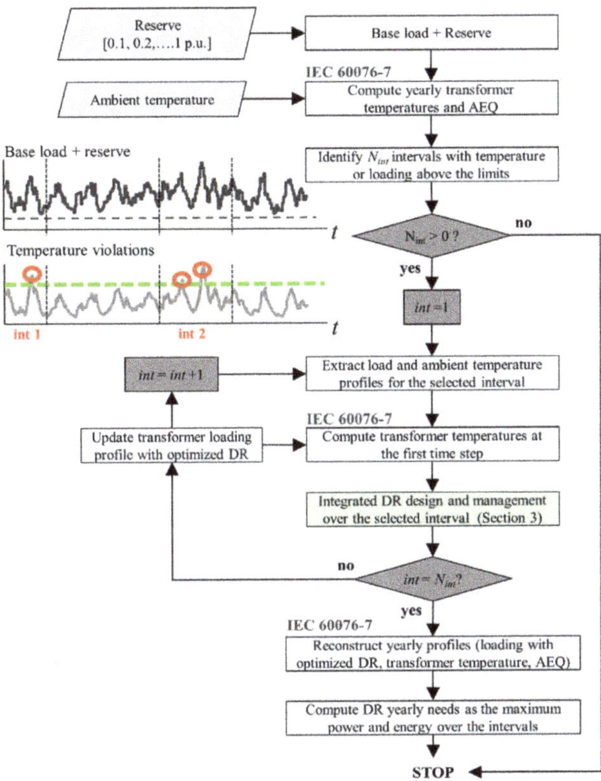

Figure 5. The procedure for finding the needed Demand Response (DR) volume to interconnect the studied reserve.

3. Integrated Design and Management of Demand Response

To understand the thermal modeling of the chosen ONAN transformer, the reader can refer to Section 3.1. Section 3.2 suggests the problem formulation of integrated design and management for DR. As we mentioned earlier, the nonlinear thermal equations of IEC 60076-7 would lead to the intractability of the optimization problem at long intervals. That is why Sections 3.3 and 3.4 explain how nonlinear equations have been linearized in this paper.

3.1. Transformer Thermal Model and Aging

The thermal model of oil-immersed transformers as well as the values for thermal characteristics are derived from the IEC 60076-7 standard [42]. The IEC model allows a discrete representation of the differential equations that govern the thermal behavior of the transformer. The specific thermal characteristics of studied ONAN distribution transformers are given in Table 2. The meaning of each symbol used either in Table 2 or in the next equations is provided in Table 3. Note that the values of Table 2 are very conservative, since IEC 60076-7 was intended to represent the whole transformer fleet by using the same set of thermal characteristics [52]. Generally, the thermal characteristics of transformers are obtained by performing so-called non-truncated heat run tests [53]. Non-truncated heat run tests mean a situation when a constant load is applied to the transformer until reaching the steady-state (constant) temperatures of oil and winding [54]. The reader can see [52,53] for more details on thermal characteristics and the equations of IEC 60076-7 [42].

Table 2. Thermal characteristics of ONAN (Oil Natural Air Natural) distribution transformers at nominal conditions.

x	y	R	τ_o	τ_w	θ_a
0.8	1.6	5	180	4	20
θ^h	$\Delta\theta^{hr}$	$\Delta\theta^{or}$	k_{11}	k_{21}	k_{22}
98	23	55	1	1	2

Table 3. List of the used symbols.

Sets		**Variables**	
		P_t^{DR}	DR flexibility power (kW)
$t \in T$	Generic set of time step (1 min)	P_t^{tr}	Transformer loading (p.u.)
$t^d \in T^D$	Set of time step for dispatch (1 h)	$\delta p_{t,c}^{f1}, \delta p_{t,c}^{f2}$	PWL power in block c at t (p.u.)
$c \in C$	Set of PWL blocks	p^{DRr}	DR flexibility rated power (kW)
		E^{DRr}	DR flexibility rated energy (kWh)
Transformer Parameters			
θ_t^a	Ambient temperature at time t (°C)	x, y	Oil/winding exponent
θ_t^o	Top oil temperature at time t (°C)	R	Loss ratio, hot-spot factor
θ_t^h	Hot-spot temperature at time t (°C)	H	Hot-spot factor
$\Delta\theta_t^{h1}, \Delta\theta_t^{h2}$	Hot-spot to top oil gradients at t (K)	τ^o, τ^w	Oil/winding time constant (min)
$\Delta\theta^{hr}$	Rated hot-spot to oil gradient (K)	k^{11}, k^{21}, k^{22}	Model coefficients
$\Delta\theta^{or}$	Rated top oil temperature rise (K)	K^{trR}	Transformer rated power (kW)
Simulation Parameters			
P_t^l	Load to supply at time t (kW)	$\underline{SOC^{DR}}, \overline{SOC^{DR}}$	Lower/upper state of charge (%)
p_c^{f1}, p_c^{f2}	PWL breakpoints in block c	SOC_0^{DR}	Initial flexibility state of charge (%)
a_c^{f1}, a_c^{f2}	PWL slope in block c	$SOC_{t=T}^{DR}$	Final flexibility state of charge (%)

The IEC model estimates the top oil and hot-spot temperatures over time, θ_t^o and θ_t^h, respectively. As already mentioned, those values depend on the time-series profiles for the ambient temperature (θ_t^a) and the transformer loading (P_t^{tr} in p.u.) as well as the transformer thermal characteristics (Table 2). Specifically, the model includes two nonlinear functions denoted f_t^1 and f_t^2 (1), which are used within equations for θ_t^o and θ_t^h. The equations of IEC standard are ultimately summarized in (2) (for $t > 1$ min) and in (3) (for initialization step i.e., $t = 1$ min).

$$f_t^1\left(P_t^{tr}\right) = \Delta\theta^{or} \times \left(\frac{1 + \left(P_t^{tr}\right)^2 \times R}{1 + R}\right)^x \text{ and } f_t^2\left(P_t^{tr}\right) = \Delta\theta^{hr} \times \left(P_t^{tr}\right)^y \tag{1}$$

$$\begin{cases} \theta_t^o = \theta_{t-1}^o + \frac{1}{k^{11} \times \tau^o}\left(f^1\left(P_t^{tr}\right) + \theta_{t-1}^o - \theta_t^a\right) \\ \Delta\theta_t^{h1} = \Delta\theta_{t-1}^{h1} + \frac{1}{k^{22} \times \tau^w} \times \left(k^{21} \times f_t^2\left(P_t^{tr}\right) - \Delta\theta_{t-1}^{h1}\right) \\ \Delta\theta_t^{h2} = \Delta\theta_{t-1}^{h2} + \frac{k^{22}}{\tau^o} \times \left((k^{21} - 1) \times f_t^2\left(P_t^{tr}\right) - \Delta\theta_{t-1}^{h2}\right) \\ \theta_t^h = \theta_t^o + \Delta\theta_t^{h1} + \Delta\theta_t^{h2} \end{cases} \quad \forall t \in T \tag{2}$$

$$\begin{cases} \theta_{t=1}^o = f^2\left(P_{t=1}^{tr}\right) + \theta_{t=1}^a \\ \Delta\theta_{t=1}^{h1} = k^{21} \times f_{t=1}^2\left(P_{t=1}^{tr}\right) \\ \Delta\theta_{t=1}^{h2} = (k^{21} - 1) \times f_{t=1}^2\left(P_{t=1}^{tr}\right) \end{cases} \tag{3}$$

The thermal model also returns the annual equivalent aging (denoted AEQ) of the transformer insulation for the given ambient temperature and power profiles. The insulation degradation is computed according to (4) with the hot-spot temperature value over the simulated horizon (i.e., $t \in t$) at 1-min resolution here. Note that the aging is normalized with the period duration (the cardinal function $\#T$) and has to remain below 1, which corresponds to a normal degradation with the transformer operating at the design temperature along its estimated lifetime (see the threshold on AEQ violation in the algorithm in Section 2).

$$\text{AEQ} = \frac{1}{\#T} \times \sum_{t \in T} 2^{\frac{\theta_t^h - 98}{6}} \leq 1 \tag{4}$$

3.2. Problem Formulation

The characterization of the DR volume, which is necessary to avoid the transformer overheating, is expressed in the form of a systemic optimization problem. In this optimization problem, the management strategy of the DR (i.e., load power profile modification) is considered along with DR design (sizing). Then, both sizing and management are variables of a single problem. Moreover, dynamic constraints should be considered due to the time dependency of the temperature profiles. The overall problem simply consists of minimizing the DR needs in terms of rated power (P^{DRr} in kW) and the rated capacity (E^{DRr} in kWh). The DR should ultimately fulfill the transformer thermal, aging, and loading constraints given in (5). Additional constraints are introduced to represent the DR operation P_t^{DR} within its bounds (6). This management allows us to modify the transformer loading ($P_t^{tr} \times K^{tr}$) for a given load profile (P_t^l) following the power balance constraint in (7).

$$obj : \min\left(P^{DRr} + \frac{E^{DRr}}{1h} \right) \ s.t. \ \begin{cases} \theta_t^o \leq \overline{\theta^o} \ \forall t \in T \\ \theta_t^h \leq \overline{\theta^h} \ \forall t \in T \\ P_t^{tr} \leq 1.5 \, p.u \ \forall t \in T \\ AEQ \leq 1 \end{cases} \tag{5}$$

$$- P^{DRr} \leq P_t^{DR} \leq P^{DRr} \ \forall t \in T \tag{6}$$

$$P_t^{tr} \times K^{trR} = P_t^l - P_t^{DR} \ \forall t \in T \tag{7}$$

So far, no economic criteria are taken into consideration and no cost is attached to the capacity of the DR flexibility (e.g., cost for storage capacity) and its power (e.g., cost for a battery inverter or backup generator). Note that in practice, this DR flexibility could be of any form and provided by a set of controllable generators, loads, or storage equipment potentially coupled with renewable energy sources. In this work, the DR flexibility is exclusively described in a power and energy domain, and it is modeled similarly to generic storage with a unitary efficiency. Then, additional constraints should be introduced to compute the "virtual state of charge" SOC_t^{DR} and keep it between the bounds (typically 0 and 100%) during the studied interval (8).

Two operating modes are envisioned for the designed DR. At first, when setting similar initial and final state-of-charge values (i.e., SOC_0^{DR} and $SOC_{t=t}^{DR}$) (typically 50%), this DR is managed in an "energy-shifting" mode, with an energy conservation constraint. The "energy-shifting" mode ensures the conservation of the consumed energy, especially through the constraints on the final values for the state of charge. Another operating mode called "energy shedding" is investigated in this paper. In "energy shedding" mode, the DR can be considered as a curtailable load (or an aggregation of curtailable loads). Within the proposed problem formulation, "energy shedding" is simply modeled by setting the "state of charge" at the beginning of the interval (SOC_0^{DR}) to 100% and the $SOC_{t=t}^{DR}$ at the end to 0% and with $P_t^{DR} > 0$ (i.e., equivalent to a DR discharge).

$$\begin{cases} SOC_t^{DR} = SOC_0^{DR} - \sum_{k=1}^{t} \frac{100 \times P_k^{DR} \times \Delta t}{E^{DRR}} \ \forall t \in T \\ \underline{SOC^{DR}} \leq SOC_t^{DR} \leq \overline{SOC^{DR}} \\ SOC_{t=T}^{DR} = SOC_0^{DR} - \sum_{k=1}^{T} \frac{100 \times P_k^{DR} \times \Delta t}{E^{DRr}} \end{cases} \tag{8}$$

Multiplying the state-of-charge constraints on both the left and right hand side by the rated capacity E^{DRr} allows removing the nonlinearity (introduced by the division of operating variable by the design variable). Thus, it allows us to solve the integrated management and sizing problem [55].

3.3. Simplified Formulation for Thermal Constraints

A first set of preliminary tests is intended to solve the proposed problem over a single simulated day using heuristics and meta-heuristics embedded in the MATLAB Optimization Toolbox (e.g., sequential quadratic programming, genetic algorithm). However, those runs suffered from expensive computational times and non-systematic convergence due to the problem size, especially with 1-min resolution for the reference temporal set. Additional complexity is introduced with the computation of the oil and hot-spot temperatures and aging as nonlinear constraints. In order to ensure the convergence within reasonable computational times so that different use cases can be investigated, the problem is linearized by taking advantage of the convexity of the thermal model.

Thus, a conventional piece-wise linearization (PWL) is implemented for the functions introduced in (1). The optimal breakpoints that minimize the linearization error are defined according to the method introduced in [56] and applied for different numbers of PWL segments $c \in C$. Figure 6a displays the obtained results when considering a three-block PWL for the nonlinear functions f^1 and f^2. This PWL allows us to compute the slope coefficients a_c for each function that is further used in the mathematical formulation illustrated for the function f^1 in Figure 6b.

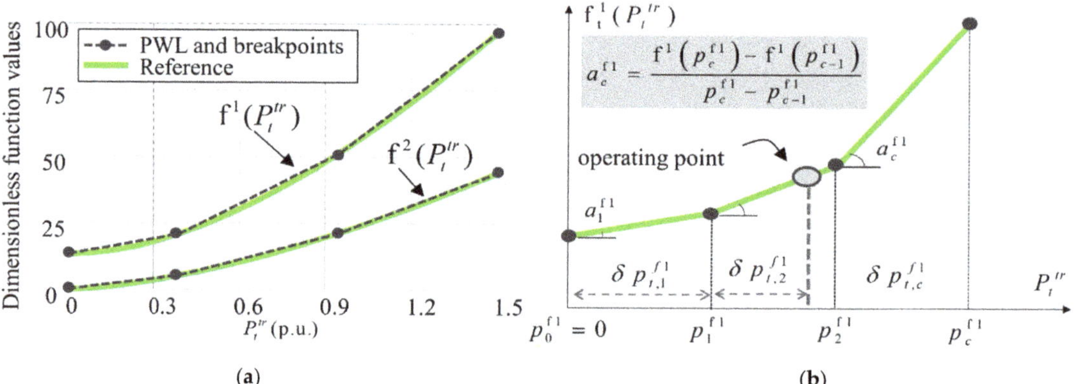

Figure 6. Piece-wise linearization (PWL) process: (**a**) functions fitting and breakpoints; (**b**) mathematical formulation for function f^1.

This typical formulation relies on the introduction of additional variables $\delta P_{t,c}^{f1}$ (similarly for f^2) that represent the contribution of each block c at every time step t. Those semi-definite positive variables are bounded by the breakpoints previously defined, and their summed values should correspond to the transformer power P_t^{tr} at every time step (9). Then, the functions are approximated with the contribution in each block weighted by the corresponding slope coefficient according to (10). Those estimations, for both functions f^1 and f^2, are integrated in the equations for the transformer thermal model (2) and (3), which actually become constraints for the proposed optimization problem—i.e., mathematical constraints rather than physical constraints in Mathematical Language Programming. Since the approximated functions are convex, their PWL formulation should ultimately be introduced in the objective function so that the proper sequence of the activated block (i.e., $\delta P_{t,c}^{f1}$, $\delta P_{t,c}^{f2} > 0$) is ensured for different operating points at every time step. Thus, Equation (11) displays the updated version of the objective. Specific attention is given to the weight α^{pwl} that implicitly performs an arbitrage between the "physical" objective, which is the minimization of the DR flexibility, and the "mathematical" objective that allows the PWL for the convex functions considered. Especially, a high value of α^{pwl} would tend to minimize those functions over the time horizon with temperatures that would remain much lower than the limits and could lead to oversized flexibilities. Preliminary runs and sensitivity analysis for various cases allowed setting a value of $\alpha^{pwl} = 10^{-6}$ so that the resulting DR flexibility is the minimum one. This allows us to reach the worst cases at the temperature limit but never at lower temperatures.

$$
\begin{cases}
\sum_{c \in C} \delta p_{t,c}^{f1} = P_t^{tr} \\
0 \le \delta p_{t,c}^{f1} \le p_c^{f1} - p_{c-1}^{f1}
\end{cases}
\quad \forall t \in T
\tag{9}
$$

$$
f^1(P_t^{lr}) \leftarrow f^1\left(p_0^{f1}\right) + \sum_{c \in C} u_c^{f1} \times \delta p_{t,c}^{f1} \quad \forall t \in T
\tag{10}
$$

$$
obj : \min\left(P^{DRr} + \frac{E^{DRr}}{1h}\right) + \alpha^{pwl} \times \sum_{t \in T}\left(
\begin{array}{l}
f^1\left(p_0^{f1}\right) + \sum_{c \in C} a_c^{f1} \times \delta p_{t,c}^{f1} \\
+f^1\left(p_0^{f2}\right) + \sum_{c \in C} a_c^{f2} \times \delta p_{t,c}^{f2}
\end{array}
\right)
\tag{11}
$$

The optimization problem is written using the YALMIP library in MATLAB [57] and solved with CPLEX 12.0 (IBM, Armonk, New York, NY, USA). CPLEX is preferred here, since the linear programming in MATLAB is not as scalable over 100,000 variables as in CPLEX (in our problem 1 day ≈50,000 variables). Moreover, we observed that for long intervals, MATLAB *linprog* stops prematurely because it exceeds its allocated memory. As the consequence, we saw that on small horizons (one single day of our problem), CPLEX converges at least 25 times faster than MATLAB and continues converging when a number of variables grows at long intervals. As a reminder, the number of variables in the optimization problem becomes an issue for long horizons/reserve margins (see their relation in Figure 4) due to the 1-min time resolution required by the IEC standard [42] (up to 3 million variables for the longest horizons). The reader can see Section 3.5 on how the hour time resolution of P_t^{DR} was used for reducing the problem complexity while keeping temperatures (state variables) in 1-min time resolution in accordance with IEC standard [42].

Validation runs were performed without flexibility, and no thermal limits were identified. Figure 7 displays the results when comparing the output temperatures (top oil and hot spot) with the PWL approximation compared to the reference model.

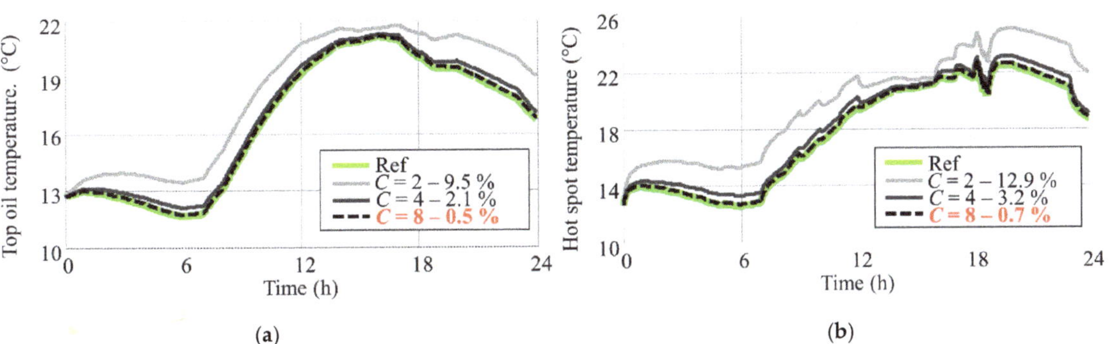

Figure 7. PWL performances for different numbers of blocks—(**a**) top oil temperature; (**b**) hot spot temperature.

The approximation error is computed through the normalized **root-mean-square error** (in %) for different numbers of PWL blocks. This error decreases significantly with more blocks considered. A value of $C = 6$ is ultimately chosen as it allows us to reach a deviation of less than 1% with the reference.

3.4. Approximate Aging Model

As introduced in Section 3.1, the normalized aging is computed with the hot-spot temperature through the function denoted $f^{AEQ}(\theta^h) = 2^{(\theta^h - 98)/6}$. For this nonlinear equation, we apply the same linearization process as for the temperature functions. The f^{AEQ} is linearized by the introduction of additional variables $\delta\theta^{hAEQ}_{t,c}$ to represent the contribution of each block that should ultimately be equal to the estimated hot-spot temperature at every time step (θ^h_t). Finally, the degradation over the simulated period is computed with the PWL approximation summed over the time horizon and should be below 1 to avoid an accelerated aging (12).

$$\begin{cases} \sum_{c \in C} \delta\theta^{hAEQ}_{t,c} = \theta^h_t \\ 0 \leq \delta\theta^{hAEQ}_{t,c} \leq \theta^{hAEQ}_c - \theta^{hAEQ}_{c-1} \\ AEQ \leftarrow \sum_{t \in T} \left(f^{AEQ}\left(\theta^h_0\right) + \sum_{c \in C} a^{fAEQ}_c \times \delta\theta^{hAEQ}_{t,c} \right) \leq 1 \end{cases} \tag{12}$$

Figure 8a shows a three-block PWL for the nonlinear function f^{AEQ} as we previously did for temperature-related nonlinear functions f^1 and f^2. However, the aging function grows exponentially with hot-spot temperatures over 80 °C Consequently, an error in the temperature estimation at $\theta^h_t > 80$ °C would incur larger deviations in the AEQ estimation. For instance, Figure 8b shows that only 5% error in the temperature at 120 °C leads to 100% error in AEQ.

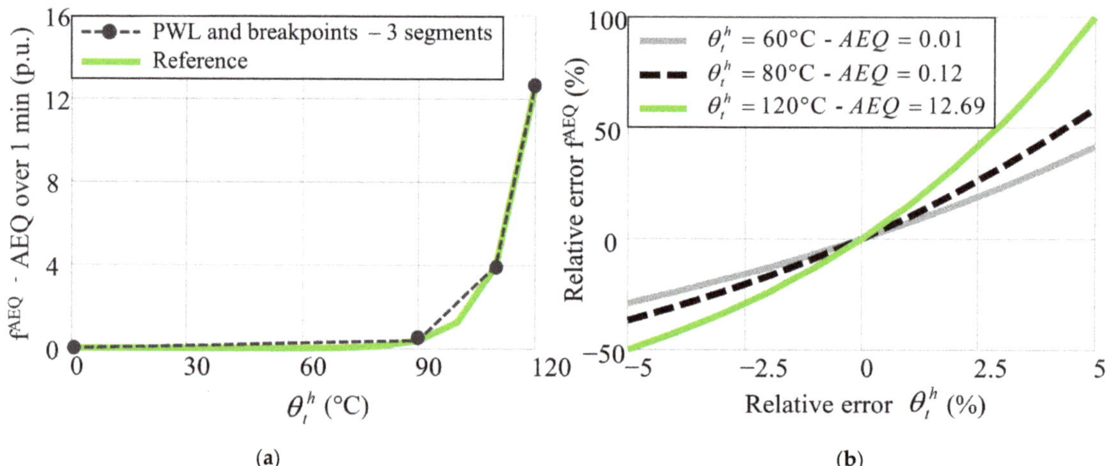

Figure 8. PWL process: (**a**) functions fitting and breakpoints for f^{AEQ}; (**b**) aging function—f^{AEQ} error versus θ^h error.

Since the PWL formulation of the degradation is expressed with the linearized hot-spot temperature, the approach could ultimately lead to significant errors in the AEQ in case of lower numbers of PWL blocks. Thus, a set of preliminary test runs over a single day is performed while varying the numbers of PWL segments for the approximation of the temperature on one side and for the estimation of the AEQ on the other. Note that the AEQ computed with the PWL in the DR design is always equal to 1 as the aging is a binding constraint for the considered test day. For every simulation, this value is compared to the AEQ computed with the exact profiles for the transformer temperatures (i.e., via IEC standard [42]) and the temperature profiles approximated in the PWL process. The results displayed in Table 4 show around 33% difference with three PWL blocks for the AEQ, while six segments are still considered for the temperature (1.00 versus 0.67 in terms of AEQ values). Increasing the number of blocks for the AEQ estimation allows us to reduce the error compared to the reference. However, it is not worth increasing the number of PWL segments for the AEQ without increasing the number of segments for the temperature (i.e., when increasing from 12 to 18 blocks for the AEQ computation). Thus, the test with 12 blocks for both AEQ and temperature shows the best results. Note that due to the overestimation of the PWL for convex function, the AEQ computed with the PWL will be always higher than the one computed with the PWL estimation for θ_t^o and θ_t^h, which is itself always higher than the exact value. It allows us to give an additional safety margin despite the estimation error.

Table 4. Aging EQuivalent (AEQ) computation for different numbers of PWL segments.

Number of PWL blocks for θ^o and θ^h	6	6	6	6	9	12
Number of PWL blocks for AEQ	3	6	12	18	12	12
AEQ computed with exact θ^o and θ^h	0.67	0.88	0.94	0.94	0.95	0.96
AEQ computed with PWL θ^o and θ^h	0.70	0.91	0.97	0.97	0.97	0.97
AEQ computed with PWL	1.00	1.00	1.00	1.00	1.00	1.00

3.5. Multiple Time Sets

Additional preliminary tests, performed with a given DR design (in terms of power and energy), show numerical oscillations especially for the DR contribution P_t^{DR} (Figure 9a) and the hot-spot temperature (Figure 9b). This is due to the load profile, which is originally at 1 h resolution and resampled at 1 min to fit in the time resolution for the transformer thermal model. Then, the load (i.e., the transformer loading without DR) is assumed constant between two successive 60-min intervals. Then, there is a wide range of DR power profiles that would lead to the same energy balance (from the transformer perspective) and similar temperature values at the end of the intervals. One way to overcome the observed oscillations would be to assume ramping constrains or ramping penalties for the DR flexibility power to supply/absorb a given amount of energy over a single 60-min period. The choice is made here to decouple the time sets (or "time grids") with a 1-min resolution for the transformer model and 1-h resolution ($t^d \in T^D$) for the equations related to the DR operation. This approach is often used in generation expansion planning similar to [58] where investments on the installed capacity are done periodically over decades and the operation/dispatch of the assets is computed on a finer resolution between two investment periods.

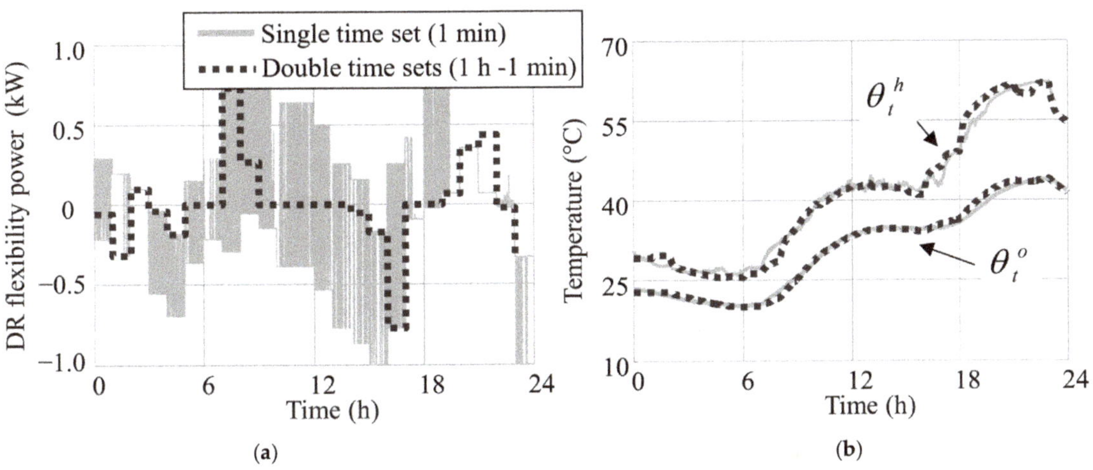

Figure 9. Simulation with single and multiple time sets: (**a**) DR flexibility power; (**b**) transformer temperatures.

Then, the constraints for the power dispatch (i.e., power balance (7)) and DR operations (6)–(8) are rewritten following the newly introduced time set. Similarly to the DR power, the transformer loading will be defined along the time set T^D. Finally, the link between the two time sets is ensured by modifying the constraints for the PWL with the contributions in the block (along the set T) that should match the operating point of the transformer (along the set T^D) in (13). Figure 9 displays the smoothed results obtained when moving from a single time set to the optimization of two time sets. Additionally, the computational time is also significantly shortened from 40 s down to only 2 s for simulation over a single day interval. The complexity of the problem is reduced despite similar numbers of variables and constraints: 50,000 variables/23,000 constraints instead of 59,000 variables/26,000 constraints.

$$\begin{cases} \sum_{c\in C} \delta p_{t,c}^{f2} = P_{t^d}^{tr} \\ \sum_{c\in C} \delta p_{t,c}^{f1} = P_{t^d}^{tr} \end{cases} \forall t, t^d \text{ with } t \in \left[t^d, t^{d+1} \right] \tag{13}$$

4. Results and Discussion

The reader could refer to Section 4.1 to see the validation runs for the integrated management and design of DR.

Otherwise, the reader could pass directly to the obtained results presented in Section 4.2.

4.1. Validation Runs for the Integrated Management and Design of DR

Before investigating different reserve margins with the methodology introduced in Section 2, different validation runs are performed for a better understanding of the DR optimization problem and results of its solution. A first simulation is run over a single day interval. The objective is to validate the DR design and management block, which were introduced in Section 3. At first, the DR flexibility is operated under "energy shifting" conditions. Note that the aging constraint (i.e., AEQ < 1) is not considered here. Figure 10 shows the results without DR and with DR for a given value of reserve margin (i.e., with a given amount of surplus load).

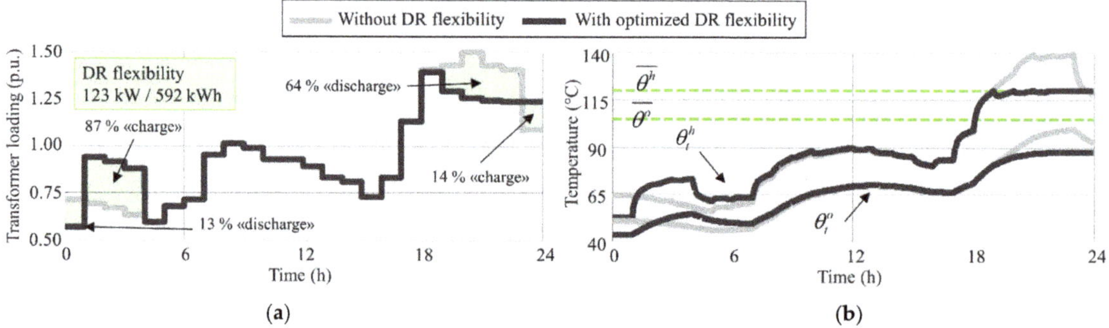

Figure 10. Validation run with "energy shifting": (**a**) transformer loadings; (**b**) temperatures.

The reader can see in Figure 10 that adding a DR flexibility allows us to keep the hot-spot temperature below its limit. At the same time, the oil temperature always remains below the limit with or without the use of flexibility. To avoid the winding overheating during the evening, the load shifting is applied by lowering the peak load after 18:00 and transferring some of the load to the morning (Figure 10a). Note that at the beginning of the simulation, the loading is decreased to reduce the temperature profiles before the DR flexibility is fully charged (to ensure the "energy conservation" constraints). This ultimately leads to higher temperatures at night, but it is still far from the overheating limits. Finally, the DR flexibility follows a "charge/discharge/charge" pattern, and 64% of the estimated capacity (592 kWh here) is necessary to shave the peak loading. Note that DR flexibility cannot be fully "discharged" as the virtual state of charge should return back to 50% at the end of the simulated period.

The second simulation is performed given the same optimization inputs but with a DR flexibility operating as a typical "energy shedding" and no AEQ constraints. Figure 11 displays the results with the DR activated only in the evening to shave the peak transformer loading similar to the previous simulation. Then, the loading and temperature profiles (Figure 11b) remain unchanged for the rest of the considered day, i.e., prior to the peak shaving. This load shaving is equal to 377 kWh. This shed energy corresponds to the optimized capacity of the installed DR flexibility in the "energy shedding" mode. This expected capacity is obviously much lower than the one computed in the case of "energy shifting", since there is no need to shift/recover the shed load at other time during the day.

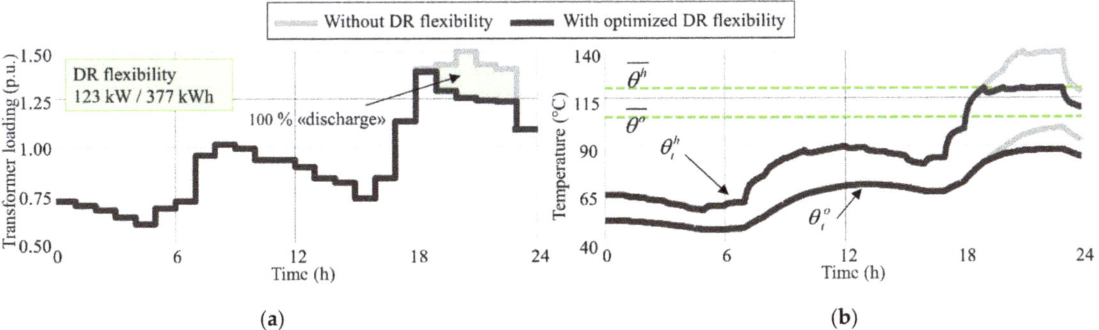

Figure 11. Validation run with "energy shedding": (**a**) transformer loadings; (**b**) temperatures.

Final validation runs consist of introducing the aging constraint for the same simulated day in case of DR flexibility operated under "energy shedding". The obtained results show that the amount of curtailed energy is greater than the one observed in the previous runs (Figure 12a) so that the hot-spot temperature remains far below the limit which would otherwise incur an over degradation of the winding insulation (Figure 12b). The oil temperature is reduced as expected. Note that the DR capacity estimated at 637 kWh is almost twice as much as in the case with no aging constraint.

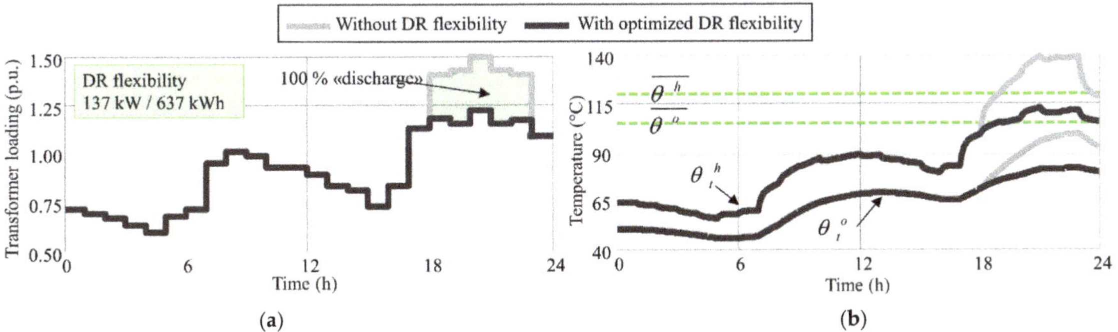

Figure 12. Validation run with "energy shedding" and aging: (**a**) transformer loadings; (**b**) temperatures.

4.2. Results for Different Reserve Margins

Having performed the validation runs in Section 4.1, this subsection addresses results considering a full representative year that were obtained with the methodology introduced in Section 2. Figure 13 illustrates typical results while comparing three scenarios: the base case scenario (i.e., base load), the base case after adding a given reserve (75% here), and the last case considering the application of DTR/DR ("energy-shedding mode").

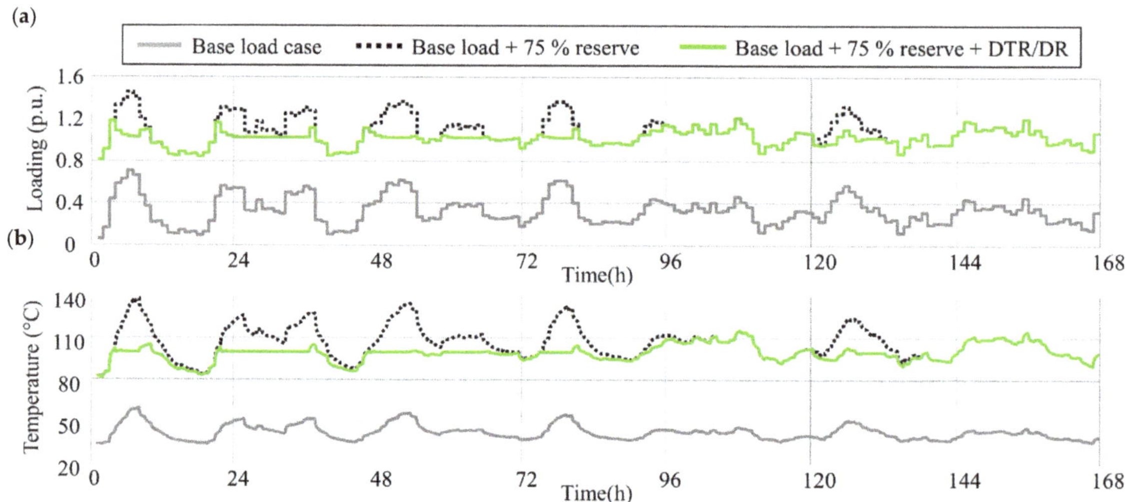

Figure 13. One-week profiles in January, comparison of the base case and Dynamic Thermal Rating (DTR)/DR in "energy-shedding mode": (**a**) transformer loading; (**b**) hot-spot temperature.

Figure 13 shows that the total load has increased significantly after connecting a constant load (corresponding to reserve 75%) over the whole year. Although the current limit (1.5 p.u.) is not violated, adding that load leads to heavy violations of the hot-spot temperature up to 140 °C (Figure 13b). Note that the curves are given for a week in January corresponding to the peak load period. As previously mentioned and observed, the appropriate DR design and management allows adjusting the transformer loading, and the hot-spot temperature remains well below the limit at 120 °C to fulfill the aging constraints. Thus, the DR is activated almost every day here, which is not the case for the rest of the year, as it is discussed further.

Then, different reserve margins can be investigated and the yearly profile reconstructed with optimized DTR/DR in every case following the methodology of Section 2. Figure 14 shows the main results obtained with a DR in "energy-shedding" mode. As expected, DR volumes in kW (Figure 14a) and kWh (Figure 14b) tend to increase with greater reserve margins. Note that the optimization problem is not tractable for reserve above 75% due to the length of the studied intervals (and consequent size of matrix constraints with over 3.10^6 variables). Specifically, the green curve represents a full formulation of optimization problem i.e., with aging, power and temperature constraints. The formulations with thermal and power constraints (i.e., without aging constraint) is shown by the black curve. Then, the green curve is always equal or higher than the black one due to the higher DR required to mitigate the aging constraints. Specific attention should be given to the gray curves in Figure 14a,c which represent DR volumes calculated with a conventional DTR considering a design winding temperature (98 °C) as a temperature limit. As it was discussed earlier, this assumption is often taken in the papers dealing with DTR; however, the design temperature is not a temperature limit and therefore should not be considered as such. The obtained results prove that if fulfilling the design temperature as a constraint, more DR in terms of kW and kWh is required, and, as mentioned earlier, fewer reserve margins could be managed without DR at all (i.e., reserves 30% at 98 °C limit versus 50% with a 120 °C limit). The reader can note that there is no need of DR for reserve margins below 45% (the green and black curves remain flat in Figure 14a,b). This means that the thermal capacity of one distribution transformer alone is sufficient to withstand the connected load (i.e., as we mentioned earlier without need to apply DR). In other words, if any load, corresponding to reserve margins below 45%, would be connected to transformers, the transformer total load will not violate any temperature or aging limits (see Figure 3 for specific values of temperature and aging).

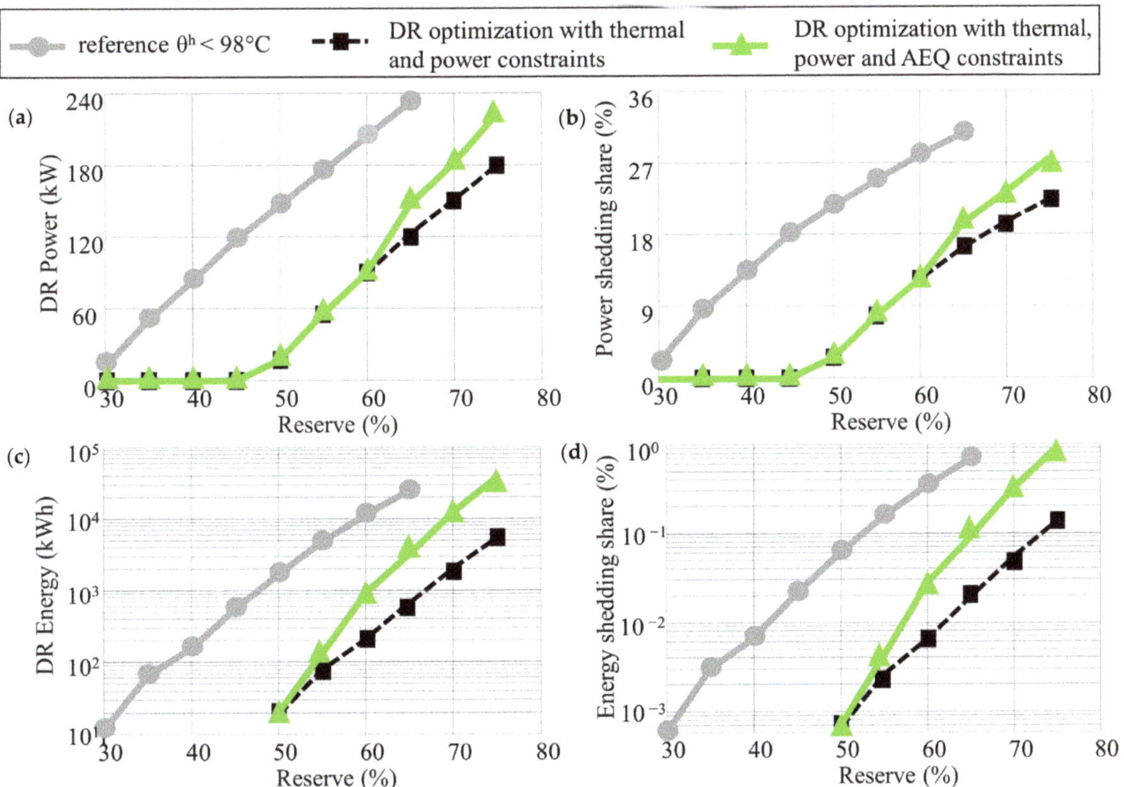

Figure 14. Obtained results for different reserve margins and DR in "energy-shedding" mode: (**a**) DR rated power; (**b**) DR power share compared to the added load; (**c**) DR rated energy; (**d**) DR energy share compared to the total energy of load.

One significant result of this paper can be observed in Figure 14d that displays the total curtailed energy compared to the total consumption over the simulated year. Results show that only 1% of total consumption needs to be curtailed to connect up to 75% of the additional load (for the transformer already loaded on 86% in N-1 mode). We remind that results are obtained for a very strict hypothesis: the constant load profile of new consumers, the maximum ambient temperature, and N-1 condition during the whole year. Even if it is necessary to shed almost 50% of nominal power of the transformer (Figure 14a or around 30% of the peak load in Figure 14b), the curtailed energy remains marginal, and the full DR capacity is activated only for a few hours of the year. DR operation is further depicted in the histograms of Figure 15.

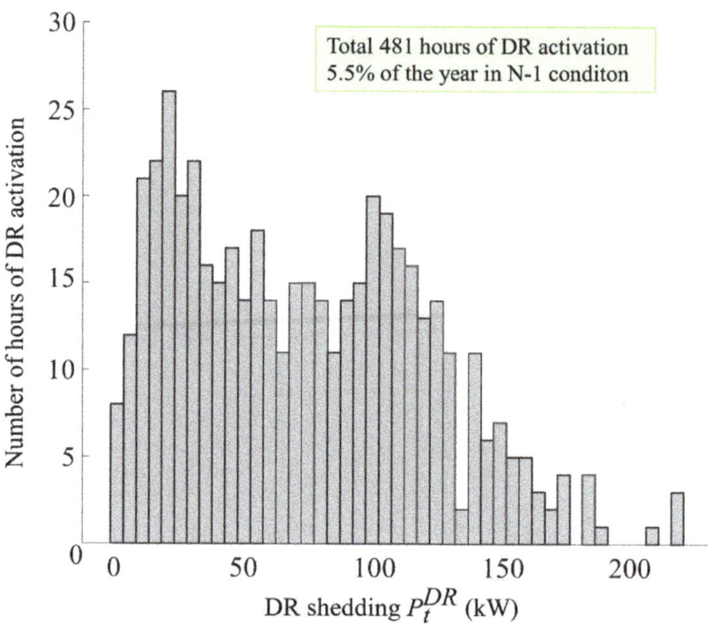

Figure 15. Power shedding over the year—75% reserve in "energy-shedding" mode.

The maximum DR shedding (the last bar at the right side) is only activated for 2–3 h per year. In total, the DR is required around 6% of the year if aggregating all the hours of activation. Once again, it is necessary to remind that such time of DR application would be required only in the N-1 condition, which is unlikely to happen all year long. Figure 16 displays the duration curves for the hot-spot temperature over the year and for different simulations with 75% of reserve (i.e., added 375 kVA to the 500-kVA transformer already loaded for 215 kVA in N mode and 430 kVA in N-1). In normal operation, the temperature remains well below the limit, and no power shedding is actually required. However, under the N-1 condition, significant overheating above 150 °C is observed and can be avoided with appropriate DR design and operation with regard to thermal constraints. If the aging is considered in the DR optimization, the DSO should ensure the transformer operation even at lower temperatures (see the green curve in Figure 16).

Another set of simulations is performed with the DR operating in "energy-shifting" mode. The results displayed in Figure 17 show slightly higher DR capacities if compared to the case with the "energy-shedding" mode. However, it is impossible to consider more than 60% reserve. Indeed, due to the energy conservation constraint, any load shedding during the peak shall be shifted and compensated at other time steps. At the first order, the thermal model of the transformer can be considered as dependent with the integral of the loading. Thus, even for different power profiles, if important amounts of loaded energy are considered, overheating can no longer be avoided.

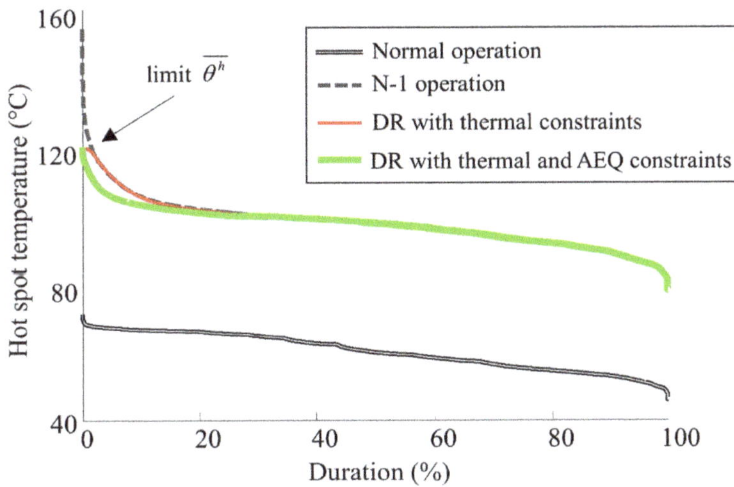

Figure 16. Yearly temperature duration curves: 75% reserve in "energy-shedding" mode leading to less than 1% energy curtailment, as seen in Figure 14d.

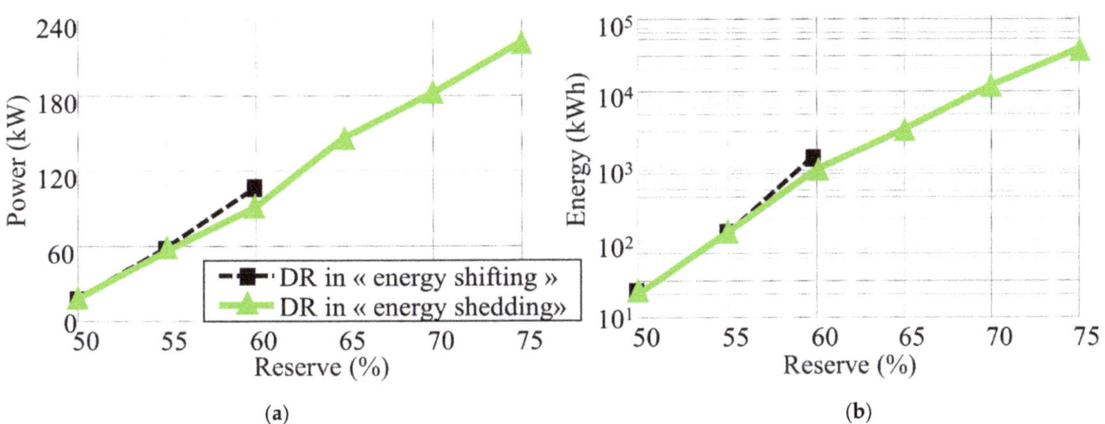

Figure 17. Results: DR modeled with or without payback effect/energy conservation: **(a)** DR power in kW; **(b)** DR energy in kWh.

The last but not least result: Figure 18 shows that the suggested PWL formulation of the optimization problem allows reducing computation times in comparison to the nonlinear formulation. As it was mentioned earlier, the nonlinearity is caused by f^1 and f^2 formulas used for temperature calculations and by f^{AEQ} used for the aging calculation. Moreover, the high computational times and non-systematic convergences of nonlinear formulation happen due to the large size of the optimization problem, which is caused by the high (1-min) time resolution (required by IEC standard [42] for the numerical stability). Thanks to the PWL of f^1, f^2 and f^{AEQ}, it is possible to reduce computation times while still keeping the required high (1-min) time resolution for load and ambient temperature profiles.

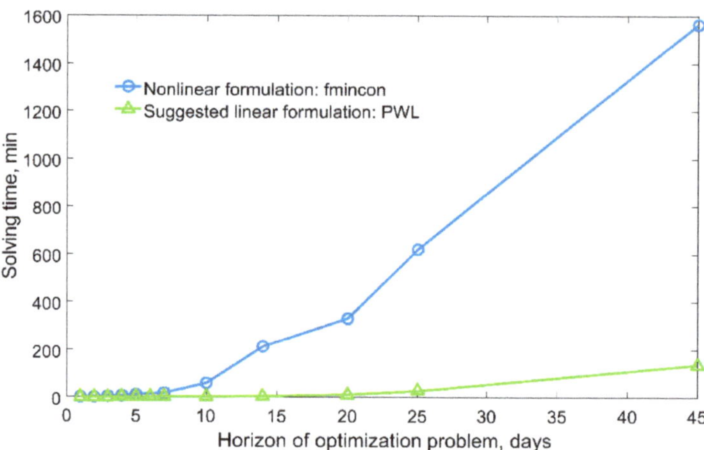

Figure 18. Computation times for the nonlinear formulation (the blue line) and for the suggested PWL formulation (the green line). Both formulations were tested for the "energy-shedding" case ($SOC_0^{DR} = 100\%$ and $SOC_{t=t}^{DR} = 0\%$).

The reader can see that for horizons up to 1 week, both formulations have approximately the same computation time. Thus, it does not matter which formulation is used if the connected load (a reserve margin) leads to transformer overheating less than 1 week per year. However, when a horizon of the optimization problem overpasses 1 week (i.e., for particular reserve margins), the difference between the two formulations becomes even more evident. For instance, the PWL optimization problem at the horizon of 45 days is solved for about 2 h, whereas the same problem in the nonlinear formulation would be solved for more than 1 day. Despite the benefits of PWL over the nonlinear formulation, it is still impossible to solve the optimization problem at the horizon of one year. The optimization problem at the year-wise horizon, even in PWL formulation, becomes intractable. It seems that the high resolution of data required by IEC standard [42] is a main barrier for solving the optimization at such long horizons with nonlinearities f^1, f^2, and f^{AEQ} of the transformer thermal model. Therefore, other approaches e.g., [59] can be envisaged together with PWL to enlarge the studied horizons (and thus reserve margins), but such study is outside the scope of this paper.

5. Conclusions

This paper presents the methodology to increase the available reserve of a transformer using Demand Response and Dynamic Thermal Rating. The maximum reserve estimation relies on a linear programming that simultaneously optimizes the DR volume and operation over a given time interval. The mathematical formulation accounts for the thermal limits of the transformer, the maximum power/current, and the aging effects. The most noticeable result shows that relatively small DR volumes ($\leq 1\%$ of total energy consumption) could ensure high reserve margins of transformers. Although DR volumes in kW could reach 30% of peak loads, such high DR volumes will be needed only if the transformer operates in N-1 conditions and for only a few hours per year. In the N mode, no DR is required at all; no thermal stress of the transformer is observed even if high reserve margins are studied. Additionally, those results are obtained despite very strict hypotheses: the constant load profile of a new consumer, historical maximum ambient temperature over the whole month, and normal cyclic limits. Thus, it is very likely that if DSO adopts the methodology to assess the reserve with consideration of exact load profile, then even a large increase of consumption (reserves) can be approved compared with the results obtained in this paper.

The paper also shows that the optimization problem formulated in this paper becomes intractable for the long horizons (i.e., for high reserve margins). This happens due to the inherit feature of the transformer thermal models, which require a high (few minutes) time resolution of data to keep the numerical stability of temperature and aging calculations. Due to using high resolution over the long horizons, the number of variables and constraints of the optimization problem increase substantially. PWL can reduce the computation times drastically in comparison to the nonlinear formulation but still cannot cope with year-wise horizons. Thus, more research is required to allow solving the integrated design and management problem on the long horizons.

As a conclusion, the observed results are very valuable for DSO and consumers since they could be used to establish a variable network access also known as "flexible network connection agreements" [60]. The general idea of such agreements is that the DSO does not provide a firm capacity all the time for certain consumers (or generators). Depending on different incentives (e.g., lower connections costs), the consumer agrees to have a limited access to the distribution network during certain time/events. Such agreements are already used in the United Kingdom for generators, and they are tested in France [60]. For the considered test case, all consumers have access to the distribution network in the N mode, and no transformer overheating occurs. However, if in case of the N-1 mode, consumers could have a limited access during 5–6% of time, as it was earlier illustrated in Figure 14. At the same moment, we remind that apart from Demand Response, the DSO could also use automation [41], load transfer and reconfiguration [61], volt-Var control [62], electric vehicles management [31,63,64], and standby transformers [65] to quickly mitigate the lack of the transformer capacity in the N-1 mode. Thus, the actual time of limited access for consumers could be further reduced. Another legal possibility for implementing those DR operations is the introduction of interruptible contracts [16,25,41]. The interruptible contracts allow DSO to shed some consumer load in exchange for financial payment to consumers. It is believed that interruptible contracts and flexible network connection agreements could be a legal foundation to connect more load to existing transformers while deferring large investments for reinforcements. Moreover, the recent study [33] shows that the operation of existing transformers (with electric vehicles) ensures less CO_2 emission against reinforced transformers. This additionally justifies the utilization of the existing transformers instead of their reinforcement.

The results also showed that for DR application, it is more beneficial to apply a DTR based on Hot Spot Temperature (HST) limit (120 °C) rather than DTR based on design HST (98 °C), which is widely used in other papers on DTR [66–69]. Specifically, the reader can see in Figure 14 that if DTR based on HST limit is used, then DSO needs to apply less DR volumes both in power and energy terms for studied reserve margins. The authors would like to point out that transformers, thanks to using the HST limit (120 °C) instead of the design HST (98 °C), could ensure a better utilization of capacity rather than other network elements. Transformers, even in normal mode, can exceed a design HST (98 °C) for short time (without exceeding the aging), whereas lines are not supposed to exceed their designed operating temperatures during normal operation [70]. From this point of view, DSO can better utilize a transformer capacity in normal mode and therefore have an additional degree of freedom. However, it is also true that the line's DTR could be twice as great as the line's static thermal rating in MegaVolt Ampere (MVA) [70], whereas a maximal MVA rating of transformers would be limited by a current limit of 1.5 p.u. from IEC standard and even lower current limits [71]. The reader could refer to [72] for details on the difference between HST limit and the design HST as well as their impact on transformer capacity. Permission for lines with higher maximum temperatures is under discussion [73,74], but to the author's knowledge, exceeding the designed operating temperature of lines is not yet approved for normal operation in the standards [75,76] (in contrast to transformers standards [42]).

Author Contributions: I.D.: Conceptualization, Methodology, Software, Writing—Original Draft, Writing—Review and Editing, Visualization, Investigation, Validation, Formal analysis, Funding acquisition. R.R.-M.: Conceptualization, Methodology, Software, Writing—Original Draft, Writing—Review and Editing, Visualization, Investigation, Validation, Formal analysis. M.-C.A.-H.: Supervision, Funding acquisition. R.C.: Writing—Review and Editing, Validation, Supervision, Funding acquisition. A.P.: Writing—Review and Editing, Validation, Supervision, Funding acquisition. All authors have read and agreed to the published version of the manuscript.

Funding: This research was funded by Conference des Grandes Ecoles, Tomsk Polytechnic University Competitiveness Enhancement Program, IDEX Université Grenoble Alpes, ATER Grenoble INP.

Institutional Review Board Statement: Not applicable.

Informed Consent Statement: Not applicable.

Data Availability Statement: The data presented in this study are available on request from the authors.

Acknowledgments: We deeply appreciate Conference des Grandes Ecoles, who first supported our research and laid the ground for its further development. The research is also funded from Tomsk Polytechnic University Competitiveness Enhancement Program grant. The authors thank the IDEX for funding Ildar DAMINOV travel grant. We gratefully appreciate the funding received from Grenoble INP in the framework of ATER position of Ildar DAMINOV. We thank Egor GLADKIKH (a founder of TOPPLAN start-up) for his help with financing the Ph.D. thesis. Authors thank MeteoBlue for provided data on ambient temperature in Grenoble, France.

Conflicts of Interest: The authors declare no conflict of interest. The funders had no role in the design of the study; in the collection, analyses, or interpretation of data; in the writing of the manuscript, or in the decision to publish the results.

References

1. International Renewable Energy Agency. *Global Energy Transformation: A Roadmap to 2050*; International Renewable Energy Agency: Abu Dhabi, United Arab Emirates, 2018; ISBN 978-92-9260-059-4.
2. International Renewable Energy Agency. *Electrification with Renewables. Driving the Transformation of Energy Services*; International Renewable Energy Agency: Abu Dhabi, United Arab Emirates, 2019; ISBN 9789292601089.
3. van Ruijven, B.J.; De Cian, E.; Sue Wing, I. Amplification of future energy demand growth due to climate change. *Nat. Commun.* **2019**, *10*, 2762. [CrossRef] [PubMed]
4. Cavlovic, M. Universal Procedure for Determining the Optimal Connection to the Distribution Network. In Proceedings of the CIRED 2019, Madrid, Spain, 3–6 June 2019.
5. Mason, T.; Curry, T.; Wilson, D. *Capital Costs for Transmission and Substations. Recommendations for WECC Transmission Expansion Planning Western Electricity Coordinating Council*; Black & Veatch Corporation: Overland Park, KS, USA, 2012.
6. DNV GL Energy Transition Outlook 2018—DNV GL. Available online: https://eto.dnv.com/2018/ (accessed on 1 March 2021).
7. Willis, H.L.; Schrieber, R. *Aging Power Delivery Infrastructures*, 2nd ed.; Power Engineering (Willis); CRC Press: Boca Raton, FL, USA, 2013; Volume 20132771, ISBN 978-1-4398-0403-2.
8. Esmat, A.; Usaola, J.; Moreno, M.A. Congestion Management in Smart Grids with Flexible Demand Considering the Payback Effect. In Proceedings of the 2016 IEEE PES Innovative Smart Grid Technologies Conference Europe (ISGT-Europe), Slovenia Ljubljana, Slovenia, 9–12 October 2016; IEEE: Scottsdale, AZ, USA, 2016; pp. 1–6.
9. Li, R.; Wang, W.; Chen, Z.; Jiang, J.; Zhang, W. A Review of Optimal Planning Active Distribution System: Models, Methods, and Future Researches. *Energies* **2017**, *10*, 1715. [CrossRef]
10. Chiodo, E.; Lauria, D.; Mottola, F.; Pisani, C. Lifetime characterization via lognormal distribution of transformers in smart grids: Design optimization. *Appl. Energy* **2016**, *177*, 127–135. [CrossRef]
11. Flexibility: The Path to Low-Carbon, Low-Cost Electricity Grids—CPI. Available online: https://www.climatepolicyinitiative.org/publication/flexibility-path-low-carbon-low-cost-electricity-grids/ (accessed on 1 March 2021).
12. Martínez Ceseña, E.A.; Good, N.; Mancarella, P. Electrical network capacity support from demand side response: Techno-economic assessment of potential business cases for small commercial and residential end-users. *Energy Policy* **2015**, *82*, 222–232. [CrossRef]
13. Martinez Cesena, E.A.; Mancarella, P. Distribution Network Capacity Support from Flexible Smart Multi-Energy Districts. In Proceedings of the 2016 IEEE PES Transmission & Distribution Conference and Exposition-Latin America (PES T&D-LA), Morelia, Mexico, 21–24 September 2016; IEEE: Scottsdale, AZ, USA, 2016; pp. 1–6.
14. Flexibility Model of Integrated Energy Sources for the Strategic Planning of Distribution Networks. Available online: http://www.cired.net/publications/workshop2018/pdfs/Submission%200525%20-%20Paper%20(ID-21419).pdf (accessed on 1 March 2021).
15. Esmat, A.; Usaola, J. DSO congestion management using demand side flexibility. In Proceedings of the CIRED Workshop 2016, Helsinki, Finland, 14–15 June 2016; Institution of Engineering and Technology, IET: Stevenage, UK, 2016; Volume 2016, p. 197.

16. Jiang, S.; Zeng, Y.; Cheng, L.; Tian, L.; Zhao, J.; Yan, B.; Wang, X.; Li, C. Substation Capacity Planning Method Considering Interruptible Load. In Proceedings of the 2018 International Conference on Power System Technology (POWERCON), Guangzhou, China, 6–8 November 2018; IEEE: Scottsdale, AZ, USA, 2018; pp. 496–500.
17. Mullen, C. Interactions between Demand Side Response, Demand Recovery, Peak Pricing and Electricity Distribution Network Capacity Margins. Ph.D. Thesis, School of Electrical and Electronic Engineering, Newcastle University, Newcastle upon Tyne, UK, 2018.
18. Elmakis, D.; Braunstein, A.; Naot, Y. A probabilistic method for establishing the transformer capacity in substations based on the loss of transformer expected life. *IEEE Trans. Power Syst.* **1988**, *3*, 353–360. [CrossRef]
19. Elmakis, D.; Braunstein, A.; Naot, Y. A Probabilistic Method for Establishing the Transformer Capacity Reserve Criteria for HV/MV Substations. *IEEE Trans. Power Syst.* **1986**, *1*, 158–164. [CrossRef]
20. Sen, P.K.; Pansuwan, S.; Malmedal, K.; Martinoo, O.; Simoes, M.G. Transformer Overloading and Assessment of Loss-of-Life for Liquid-Filled Transformers. *Power Syst. Eng. Res. Cent.* **2011**, 1–106. Available online: https://pserc.wisc.edu/publications/reports/2011_reports/T-25_Final-Report_Feb-2011.pdf (accessed on 1 March 2021).
21. Bunn, M.; Das, B.P.; Seet, B.C.; Baguley, C. Empirical Design Method for Distribution Transformer Utilization Optimization. *IEEE Trans. Power Deliv.* **2019**, *34*, 1803–1813. [CrossRef]
22. Kostin, V.N.; Minakova, T.E.; Kopteva, A.V. Urban Substations Transformers allowed Loading. In Proceedings of the 2018 IEEE Conference of Russian Young Researchers in Electrical and Electronic Engineering (EIConRus), St. Petersburg, Russia, 29 January–1 February 2018; IEEE: Scottsdale, AZ, USA, 2018; Volume 2018, pp. 692–695.
23. Daminov, I.; Prokhorov, A.; Moiseeva, T.; Caire, R.; Alvarez-Herault, M.-C. Application of Dynamic Transformer Ratings to Increase the Reserve of Primary Substations for New Load Interconnection. Available online: https://www.cired-repository.org/handle/20.500.12455/708 (accessed on 1 March 2021).
24. Gourisetti, S.N.G.; Kirkham, H.; Sivaraman, D. A review of transformer aging and control strategies. In Proceedings of the 2017 North American Power Symposium (NAPS), Morgantown, WV, USA, 17–19 September 2017; IEEE: Scottsdale, AZ, USA, 2017; pp. 1–6.
25. Sousa, J.C.S.; Saavedra, O.R.; Lima, S.L. Decision Making in Emergency Operation for Power Transformers With Regard to Risks and Interruptible Load Contracts. *IEEE Trans. Power Deliv.* **2018**, *33*, 1556–1564. [CrossRef]
26. Teja S, C.; Yemula, P.K. Architecture for demand responsive HVAC in a commercial building for transformer lifetime improvement. *Electr. Power Syst. Res.* **2020**, *189*, 106599. [CrossRef]
27. Davison, P.J.; Lyons, P.F.; Taylor, P.C. Temperature sensitive load modelling for dynamic thermal ratings in distribution network overhead lines. *Int. J. Electr. Power Energy Syst.* **2019**, *112*, 1–11. [CrossRef]
28. Zhou, B.; Cao, Y.; Li, C.; Wu, Q.; Liu, N.; Huang, S.; Wang, H. Many-criteria optimality of coordinated demand response with heterogeneous households. *Energy* **2020**, *207*, 118267. [CrossRef]
29. Van Der Klauw, T.; Gerards, M.E.T.; Hurink, J.L. Using demand-side management to decrease transformer ageing. In Proceedings of the IEEE PES Innovative Smart Grid Technologies Conference Europe, Ljubljana, Slovenia, 9–12 October 2016.
30. Liu, J.; Xiao, J.; Zhou, B.; Wang, Z.; Zhang, H.; Zeng, Y. A two-stage residential demand response framework for smart community with transformer aging. In Proceedings of the 2017 IEEE PES Asia-Pacific Power and Energy Engineering Conference (APPEEC), Bangalore, India, 8–10 November 2017; IEEE: Scottsdale, AZ, USA, 2017; pp. 1–6.
31. Soleimani, M.; Kezunovic, M. Mitigating Transformer Loss of Life and Reducing the Hazard of Failure by the Smart EV Charging. *IEEE Trans. Ind. Appl.* **2020**, *56*, 5974–5983. [CrossRef]
32. Mohsenzadeh, A.; Kapourchali, M.H.; Chengzong, P.; Aravinthan, V. Impact of smart home management strategies on expected lifetime of transformer. In Proceedings of the 2016 IEEE/PES Transmission and Distribution Conference and Exposition (T&D), Dallas, TX, USA, 3–5 May 2016; IEEE: Scottsdale, AZ, USA, 2016; pp. 1–5.
33. Brinkel, N.B.G.; Schram, W.L.; AlSkaif, T.A.; Lampropoulos, I.; van Sark, W.G.J.H.M. Should we reinforce the grid? Cost and emission optimization of electric vehicle charging under different transformer limits. *Appl. Energy* **2020**, *276*, 115285. [CrossRef]
34. Humayun, M.; Ali, M.; Safdarian, A.; Degefa, M.Z.; Lehtonen, M. Optimal use of demand response for lifesaving and efficient capacity utilization of power transformers during contingencies. In Proceedings of the IEEE Power and Energy Society General Meeting 2015, Denver, CO, USA, 26–30 July 2015.
35. Humayun, M.; Degefa, M.Z.; Safdarian, A.; Lehtonen, M. Utilization improvement of transformers using demand response. *IEEE Trans. Power Deliv.* **2015**, *30*, 202–210. [CrossRef]
36. Humayun, M.; Safdarian, A.; Ali, M.; Degefa, M.Z.; Lehtonen, M. Optimal capacity planning of substation transformers by demand response combined with network automation. *Electr. Power Syst. Res.* **2016**, *134*, 176–185. [CrossRef]
37. Humayun, M.; Safdarian, A.; Degefa, M.Z.; Lehtonen, M. Demand Response for Operational Life Extension and Efficient Capacity Utilization of Power Transformers During Contingencies. *IEEE Trans. Power Syst.* **2015**, *30*, 2160–2169. [CrossRef]
38. Salehi, J.; Haghifam, M.-R. Determining the optimal reserve capacity margin of Sub-Transmission (ST) substations using Genetic Algorithm. *Int. Trans. Electr. Energy Syst.* **2014**, *24*, 492–503. [CrossRef]
39. Kannan, S.; Au, M.T. Probabilistic approach in sizing distribution transformers. In Proceedings of the 2010 IEEE 11th International Conference on Probabilistic Methods Applied to Power Systems, Singapore, 14–17 June 2010; IEEE: Scottsdale, AZ, USA, 2010; pp. 599–603.

40. Helmi, B.A.; D'Souza, M.; Bolz, B.A. The application of power factor correction capacitors to reserve spare capacity of existing main transformers. In Proceedings of the Industry Applications Society 60th Annual Petroleum and Chemical Industry Conference, Chicago, IL, USA, 23–25 September 2013; IEEE: Scottsdale, AZ, USA, 2013; pp. 1–6.

41. Turnham, V.; Blair, S.M.; Booth, C.D.; Turner, P. Increasing Distribution Network Capacity using Automation to Reduce Carbon Impact. In Proceedings of the 12th IET International Conference on Developments in Power System Protection (DPSP 2014), Copenhagen, Denmark, 31 March–3 April 2014; Volume 2014, pp. 1–6.

42. IEC. *IEC 60076-7 Power Transformer—Part 7: Loading Guide for Oil-Immersed Power Transformers*; IEC: Geneva, Switzerland, 2018.

43. Norris, E.T. The thermal rating of transformers. *J. Inst. Electr. Eng.* **1928**, *66*, 841–854. [CrossRef]

44. Sealey, W.C.; Hodtum, J.B. Overloading of Transformers—Cases Not Covered by the General Rules. *Trans. Am. Inst. Electr. Eng.* **1944**, *63*, 149–153. [CrossRef]

45. Norris, E.T. Loading of power transformers. *Proc. Inst. Electr. Eng.* **1967**, *114*, 228. [CrossRef]

46. Sandels, C.; Widén, J.; Nordström, L. Forecasting household consumer electricity load profiles with a combined physical and behavioral approach. *Appl. Energy* **2014**, *131*, 267–278. [CrossRef]

47. House Load Electricity—File Exchange—MATLAB Central. Available online: https://fr.mathworks.com/matlabcentral/fileexchange/63375-house-load-electricity (accessed on 1 March 2021).

48. MeteoBlue Ambient Temperature History over 30 Years for Grenoble Region. Available online: https://www.meteoblue.com/en/historyplus (accessed on 1 March 2021).

49. Viafora, N.; Holboll, J.; Kazmi, S.H.H.; Olesen, T.H.; Sorensen, T.S. Load dispatch optimization using dynamic rating and optimal lifetime utilization of transformers. In Proceedings of the 2019 IEEE Milan PowerTech, PowerTech 2019, Milan, Italy, 23–27 June 2019.

50. Degefa, M.Z.; Humayun, M.; Safdarian, A.; Koivisto, M.; Millar, R.J.; Lehtonen, M. Unlocking distribution network capacity through real-time thermal rating for high penetration of DGs. *Electr. Power Syst. Res.* **2014**, *117*, 36–46. [CrossRef]

51. Watson, A.; Weller, R. Temperature correction factor applied to substation capacity analysis. In Proceedings of the 18th International Conference and Exhibition on Electricity Distribution (CIRED 2005), Turin, Italy, 6–9 June 2005; IET: Stevenage, UK, 2005; Volume 2005, pp. 1–4.

52. Susa, D.; Nordman, H. IEC 60076-7 loading guide thermal model constants estimation. *Int. Trans. Electr. Energy Syst.* **2013**, *23*, 946–960. [CrossRef]

53. Das, B.; Jalal, T.S.; McFadden, F.J.S. Comparison and extension of IEC thermal models for dynamic rating of distribution transformers. In Proceedings of the 2016 IEEE International Conference on Power System Technology (POWERCON), Wollongong, NSW, Australia, 28 September–1 October 2016; IEEE: Scottsdale, AZ, USA, 2016; pp. 1–8.

54. Wang, L.; Zhang, X.; Villarroel, R.; Liu, Q.; Wang, Z.; Zhou, L. Top-oil temperature modelling by calibrating oil time constant for an oil natural air natural distribution transformer. *IET Gener. Transm. Distrib.* **2020**, *14*, 4452–4458. [CrossRef]

55. Rigo-Mariani, R.; Chea Wae, S.O.; Mazzoni, S. Impact of the Economic Environment Modelling for the Optimal Design of a Multi-Energy Microgrid. In Proceedings of the 46th Annual Conference of the IEEE Industrial Electronics Society IECON, Singapore, 18–21 October 2020.

56. Rigo-Mariani, R.; Chea Wae, S.O.; Mazzoni, S.; Romagnoli, A. Comparison of optimization frameworks for the design of a multi-energy microgrid. *Appl. Energy* **2020**, *257*, 113982. [CrossRef]

57. Lofberg, J. YALMIP: A toolbox for modeling and optimization in MATLAB. In Proceedings of the 2004 IEEE International Conference on Robotics and Automation (IEEE Cat. No.04CH37508), New Orleans, LA, USA, 26 April–1 May 2004; IEEE: Scottsdale, AZ, USA, 2004; pp. 284–289.

58. Liu, Y.; Sioshansi, R.; Conejo, A.J. Multistage Stochastic Investment Planning With Multiscale Representation of Uncertainties and Decisions. *IEEE Trans. Power Syst.* **2017**, *33*, 781–791. [CrossRef]

59. Mégel, O.; Mathieu, J.L.; Andersson, G. Scheduling distributed energy storage units to provide multiple services under forecast error. *Int. J. Electr. Power Energy Syst.* **2015**, *72*, 48–57. [CrossRef]

60. Flexibility in the Energy Transition a Toolbox for Electricity Dsos Flexibility in the Energy Transition i a Toolbox for Electricity DSOs. Available online: https://www.edsoforsmartgrids.eu/flexibility-in-the-energy-transition-a-toolbox-for-electricity-dsos/ (accessed on 1 March 2021).

61. Dorostkar-Ghamsari, M.R.; Fotuhi-Firuzabad, M.; Safdarian, A.; Hoshyarzade, A.S.; Lehtonen, M. Improve capacity utilization of substation transformers via distribution network reconfiguraion and load transfer. In Proceedings of the EEEIC 2016—International Conference on Environment and Electrical Engineering, Florence, Italy, 7–10 June 2016.

62. Kumar, K.; Satsangi, S.; Kumbhar, G.B. Extension of life of distribution transformer using Volt-VAr optimisation in a distribution system. *IET Gener. Transm. Distrib.* **2019**, *13*, 1777–1785. [CrossRef]

63. Godina, R.; Rodrigues, E.M.G.; Matias, J.C.O.; Catalão, J.P.S. Smart electric vehicle charging scheduler for overloading prevention of an industry client power distribution transformer. *Appl. Energy* **2016**, *178*, 29–42. [CrossRef]

64. Powell, S.; Kara, E.C.; Sevlian, R.; Cezar, G.V.; Kiliccote, S.; Rajagopal, R. Controlled workplace charging of electric vehicles: The impact of rate schedules on transformer aging. *Appl. Energy* **2020**, *276*, 115352. [CrossRef]

65. Xueqiang, Z.; Pengfei, Z.; Haoyi, T.; Sanshan, Z. Research and Practical Application of Transformer Emergency Standby Calculation and Assessment Method. In Proceedings of the 2018 International Conference on Power System Technology (POWERCON), Guangzhou, China, 6–8 November 2018; IEEE: Scottsdale, AZ, USA, 2018; pp. 3659–3666.

66. Biçen, Y.; Aras, F. Loadability of power transformer under regional climate conditions: The case of Turkey. *Electr. Eng.* **2014**, *96*, 347–358. [CrossRef]
67. Yang, J.; Chittock, L.; Strickland, D.; Harrap, C. Predicting practical benefits of dynamic asset ratings of 33KV distribution transformers. *IET Int. Conf. Resil. Transm. Distrib. Netw.* **2015**, *2015*, 60354.
68. Yang, J.; Bai, X.; Strickland, D.; Jenkins, L.; Cross, A.M. Dynamic Network Rating for Low Carbon Distribution Network Operation—A UK Application. *IEEE Trans. Smart Grid* **2015**, *6*, 988–998. [CrossRef]
69. Salama, M.M.M.; Mansour, D.E.A.; Abdelmaksoud, S.M.; Abbas, A.A. Impact of Long-Term Climatic Conditions on the Ageing and Cost Effectiveness of the Oil-Filled Transformer. In Proceedings of the 2018 20th International Middle East Power Systems Conference, MEPCON 2018—Proceedings, Cairo, Egypt, 18–20 December 2018.
70. Metwaly, M.K.; Teh, J. Probabilistic Peak Demand Matching by Battery Energy Storage Alongside Dynamic Thermal Ratings and Demand Response for Enhanced Network Reliability. *IEEE Access* **2020**, *8*, 181547–181559. [CrossRef]
71. Pasricha, A.; Crow, M.L. A method of improving transformer overloading beyond nameplate rating. In Proceedings of the 2015 North American Power Symposium (NAPS), Charlotte, NC, USA, 4–6 October 2015; IEEE: Scottsdale, AZ, USA, 2015; pp. 1–6.
72. Daminov, I.; Prokhorov, A.; Caire, R.; Alvarez-Herault, M.C. Assessment of dynamic transformer rating, considering current and temperature limitations. *Int. J. Electr. Power Energy Syst.* **2021**, *129*, 106886. [CrossRef]
73. Olsen, R. Dynamic Loadability of Cable Based Transmission Grids. Ph.D. Thesis, Department of Electrical Engineering, Technical University of Denmark, Copenhagen, Denmark, 2013.
74. Alghamdi, A.S.; Desuqi, R.K. A study of expected lifetime of XLPE insulation cables working at elevated temperatures by applying accelerated thermal ageing. *Heliyon* **2020**, *6*, e031120. [CrossRef] [PubMed]
75. IEC. *IEC Standard 60287-2-1*; IEC: Geneva, Switzerland, 2015; ISBN 2831886376.
76. IEEE Standard for Calculating the Current-Temperature Relationship of Bare Overhead Conductors-738-2012. Available online: https://ieeexplore.ieee.org/document/6692858 (accessed on 1 March 2021).

MDPI
St. Alban-Anlage 66
4052 Basel
Switzerland
Tel. +41 61 683 77 34
Fax +41 61 302 89 18
www.mdpi.com

Energies Editorial Office
E-mail: energies@mdpi.com
www.mdpi.com/journal/energies